AQA
GCSE physics

Authors

Graham Bone

Jim Newall

Contents

How to use this book

Welcome to your AQA GCSE Physics course. This book has been specially written by experienced teachers and examiners to match the 2011 specification.

On these two pages you can see the types of pages you will find in this book, and the features on them. Everything in the book is designed to provide you with the support you need to help you prepare for your examinations and achieve your best.

Unit openers

Specification matching grid: This shows you how the pages in the unit match to the exam specification for GCSE Physics, so you can track your progress through the unit as you learn.

Why study this unit: Here you can read about the reasons why the science you're about to learn is relevant to your everyday life.

You should remember: This list is a summary of the things you've already learnt that will come up again in this unit. Check through them in advance and see if there is anything that you need to recap on before you get started.

Opener image: Every unit starts with a picture and information on a new or interesting piece of science that relates to what you're about to learn.

Main pages

Learning objectives: You can use these objectives to understand what you need to learn to prepare for your exams. Higher Tier only objectives appear in pink text.

Key words: These are the terms you need to understand for your exams. You can look for these words in the text in bold or check the glossary to see what they mean.

Questions: Use the questions on each spread to test yourself on what you've just read.

Higher Tier content: Anything marked in pink is for students taking the Higher Tier paper only. As you go through you can look at this material and attempt it to help you understand what is expected for the Higher Tier.

Worked examples: These help you understand how to use an equation or to work through a calculation. You can check back whenever you use the calculation in your work.

Summary and exam-style questions

Every summary question at the end of a spread includes an indication of how hard it is. These indicators show which grade you are working towards. You can track your own progress by seeing which of the questions you can answer easily, and which you have difficulty with.

When you reach the end of a unit you can use the exam-style questions to test how well you know what you've just learnt. Each question has a grade band next to it.

→E Working towards Grade E

→C Working towards Grade C

→A* Working towards Grade A*

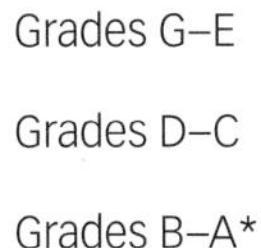

G–E Grades G–E

D–C Grades D–C

B–A* Grades B–A*

Course catch-ups

Revision checklist: This is a summary of the main ideas in the unit. You can use it as a starting point for revision, to check that you know about the big ideas covered.

Visual summary: Another way to start revision is to use a visual summary, linking ideas together in groups so you can see how one topic relates to another. You can use this page as a start for your own summary.

Upgrade: Upgrade takes you through an exam question in a step-by-step way, showing you why different answers get different grades. Using the tips on the page you can make sure you achieve your best by understanding what each question needs.

Exam-style questions: Using these questions you can practice your exam skills, and make sure you're ready for the real thing. Each question has a grade band next to it, so you can understand what level you are working at and focus on where you need to improve to get your target grade.

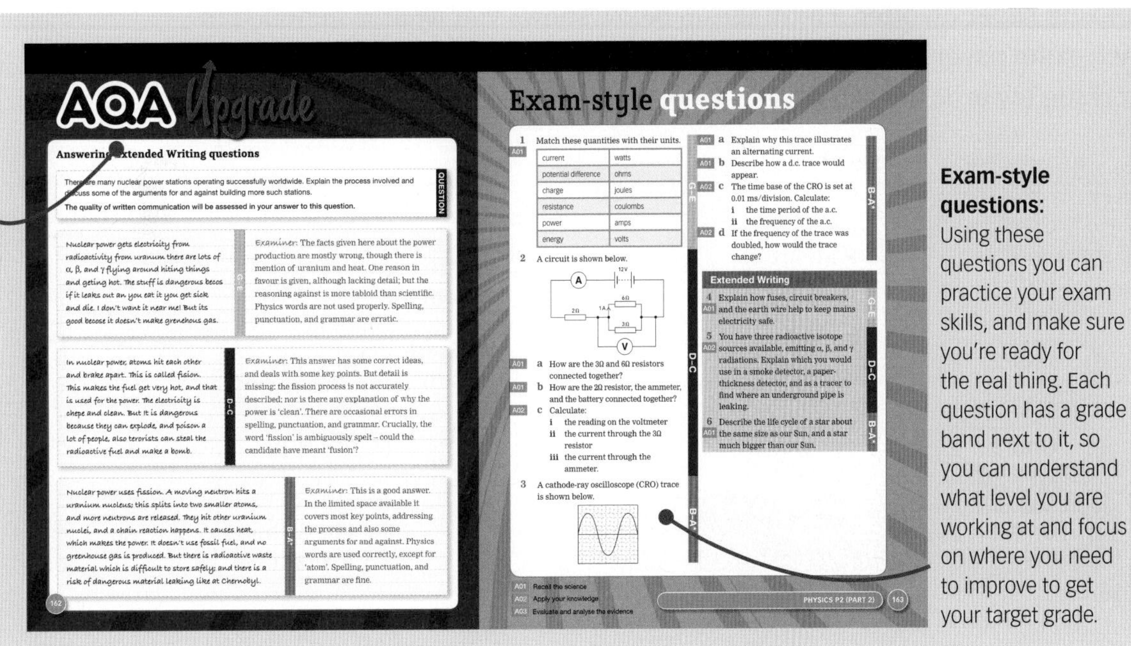

AQA Upgrade

Answering Extended Writing questions

QUESTION: There are many nuclear power stations operating successfully worldwide. Explain the process involved and discuss some of the arguments for and against building more such stations.

The quality of written communication will be assessed in your answer to this question.

G–E: Nuclear power gets electricity from radioactivity from uranum there are lots of α, β, and γ flying around hiting things and geting hot. The stuff is dangerous becos if it leaks out an you eat it you get sick and die. I don't want it near me! But its good becose it doesn't make grenehous gas.

Examiner: The facts given here about the power production are mostly wrong, though there is mention of uranium and heat. One reason in favour is given, although lacking detail; but the reasoning against is more tabloid than scientific. Physics words are not used properly. Spelling, punctuation, and grammar are erratic.

D–C: In nuclear power, atoms hit each other and brake apart. This is called fision. This makes the fuel get very hot, and that is used for the power. The electricity is chepe and clean. But It is dangerous because they can explode, and poison a lot of people, also terorists can steal the radioactive fuel and make a bomb.

Examiner: This answer has some correct ideas, and deals with some key points. But detail is missing: the fission process is not accurately described; nor is there any explanation of why the power is 'clean'. There are occasional errors in spelling, punctuation, and grammar. Crucially, the word 'fission' is ambiguously spelt – could the candidate have meant 'fusion'?

B–A*: Nuclear power uses fission. A moving neutron hits a uranium nucleus; this splits into two smaller atoms, and more neutrons are released. They hit other uranium nuclei, and a chain reaction happens. It causes heat, which makes the power. It doesn't use fossil fuel, and no greenhouse gas is produced. But there is radioactive waste material which is difficult to store safely; and there is a risk of dangerous material leaking like at Chernobyl.

Examiner: This is a good answer. In the limited space available it covers most key points, addressing the process and also some arguments for and against. Physics words are used correctly, except for 'atom'. Spelling, punctuation, and grammar are fine.

162

Exam-style questions

1 (A01) Match these quantities with their units.

current	watts
potential difference	ohms
charge	joules
resistance	coulombs
power	amps
energy	volts

2 A circuit is shown below.

a (A01) How are the 3Ω and 6Ω resistors connected together?

b (A01) How are the 2Ω resistor, the ammeter, and the battery connected together?

c (A02) Calculate:

i the reading on the voltmeter

ii the current through the 3Ω resistor

iii the current through the ammeter.

3 A cathode-ray oscilloscope (CRO) trace is shown below.

a (A01) Explain why this trace illustrates an alternating current.

b (A01) Describe how a d.c. trace would appear.

c (A02) The time base of the CRO is set at 0.01 ms/division. Calculate:

i the time period of the a.c.

ii the frequency of the a.c.

d (A02) If the frequency of the trace was doubled, how would the trace change?

Extended Writing

4 (A01) Explain how fuses, circuit breakers, and the earth wire help to keep mains electricity safe.

5 (A02) You have three radioactive isotope sources available, emitting α, β, and γ radiations. Explain which you would use in a smoke detector, a paper-thickness detector, and as a tracer to find where an underground pipe is leaking.

6 (A01) Describe the life cycle of a star about the same size as our Sun, and a star much bigger than our Sun.

A01 Recall the science

A02 Apply your knowledge

A03 Evaluate and analyse the evidence

PHYSICS P2 (PART 2) 163

Routes and assessment

Matching your course

The units in this book have been written to match the specification for **AQA GCSE Physics**.

In the diagram below you can see that the units and part units can be used to study either for **GCSE Physics** or as part of **GCSE Science** and **GCSE AQA Additional Science** courses.

	GCSE Biology	GCSE Chemistry	GCSE Physics
GCSE Science	B1 (Part 1)	C1 (Part 1)	P1 (Part 1)
	B1 (Part 2)	C1 (Part 2)	P1 (Part 2)
GCSE Additional Science	B2 (Part 1)	C2 (Part 1)	P2 (Part 1)
	B2 (Part 2)	C2 (Part 2)	P2 (Part 2)
	B3 (Part 1)	C3 (Part 1)	P3 (Part 1)
	B3 (Part 2)	C3 (Part 2)	P3 (Part 2)

GCSE Physics assessment

The units in this book are broken into two parts to match the different types of exam paper on offer. The diagram below shows you what is included in each exam paper. It also shows you how much of your final mark you will be working towards in each paper.

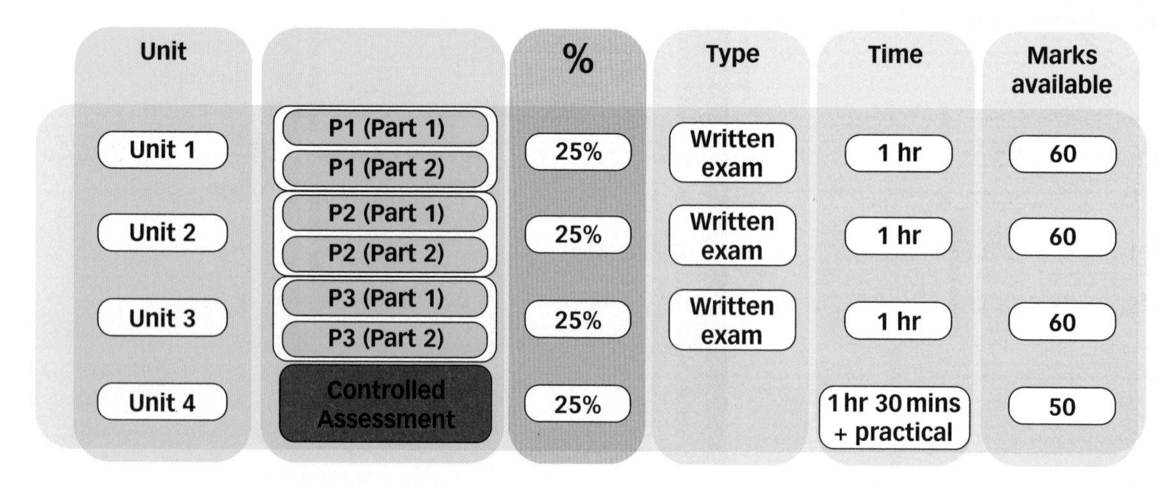

Unit		%	Type	Time	Marks available
Unit 1	P1 (Part 1) P1 (Part 2)	25%	Written exam	1 hr	60
Unit 2	P2 (Part 1) P2 (Part 2)	25%	Written exam	1 hr	60
Unit 3	P3 (Part 1) P3 (Part 2)	25%	Written exam	1 hr	60
Unit 4	Controlled Assessment	25%		1 hr 30 mins + practical	50

Understanding exam questions

When you read the questions in your exam papers you should make sure you know what kind of answer you are being asked for. The list below explains some of the common words you will see used in exam questions. Make sure you know what each word means. Always read the question thoroughly, even if you recognise the word used.

Calculate
Work out your answer by using a calculation. You can use your calculator to help you. You may need to use an equation; check whether one has been provided for you in the paper. The question will say if your working must be shown.

Describe
Write a detailed answer that covers what happens, when it happens, and where it happens. The question will let you know how much of the topic to cover. Talk about facts and characteristics. (Hint: don't confuse with 'Explain')

Explain
You will be asked how or why something happens. Write a detailed answer that covers how and why a thing happens. Talk about mechanisms and reasons. (Hint: don't confuse with 'Describe')

Evaluate
You will be given some facts, data or other information. Write about the data or facts and provide your own conclusion or opinion on them.

Outline
Give only the key facts of the topic. You may need to set out the steps of a procedure or process – make sure you write down the steps in the correct order.

Show
Write down the details, steps or calculations needed to prove an answer that you have been given.

Suggest
Think about what you've learnt in your science lessons and apply it to a new situation or a context. You may not know the answer. Use what you have learnt to suggest sensible answers to the question.

Write down
Give a short answer, without a supporting argument.

Top tips

Always read exam questions carefully, even if you recognise the word used. Look at the information in the question and the number of answer lines to see how much detail the examiner is looking for.

You can use bullet points or a diagram if it helps your answer.

If a number needs units you should include them, unless the units are already given on the answer line.

Controlled Assessment in GCSE Physics

As part of the assessment for your GCSE Physics course, you will undertake a Controlled Assessment task.

What is Controlled Assessment?

Controlled Assessment has taken the place of coursework for the new 2011 GCSE Science specifications. The main difference between coursework and Controlled Assessment is that you will be supervised by your teacher when you carry out your Controlled Assessment task.

What will my Controlled Assessment task look like?

Your Controlled Assessment task will be made up of four sections. These four sections make up an investigation, with each section looking at a different part of the scientific process.

	What will I need to do?	How many marks are available?
Research	• Independently develop your own hypothesis. • Research two methods for carrying out an experiment to test your hypothesis. • Prepare a table to record your results. • Carry out a risk assessment.	
Section 1	• Answer questions relating to your own research.	20 marks
Practical investigation	• Carry out your own experiment and record and analyse your results.	
Section 2	• Answer questions relating to the experiment you carried out. • Select appropriate data from data supplied by AQA and use it to analyse and compare with your hypothesis. • Suggest how ideas from your investigation could be used in a new context.	30 marks
	Total	**50 marks**

How do I prepare for my Controlled Assessment?

Throughout your course you will learn how to carry out investigations in a scientific way, and how to analyse and compare data properly.

On the next three pages there are Controlled Assessment-style questions matched to the content in P1, P2, and P3. You can use them to test yourself, and to find out which areas you want to practise more before you take the Controlled Assessment task itself.

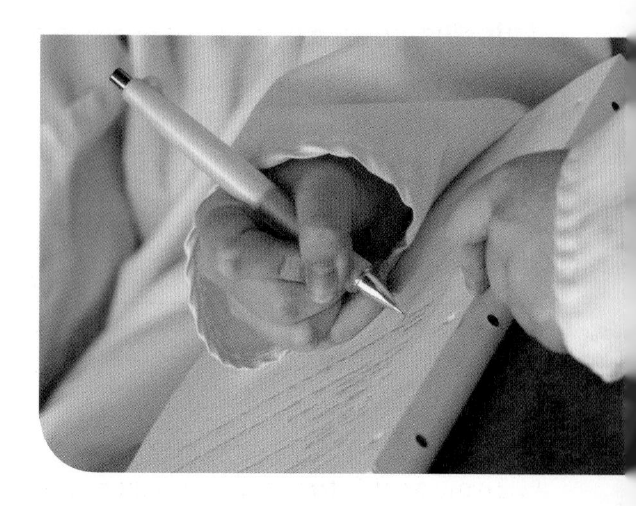

P1 Controlled Assessment-style questions

Hypothesis: It is suggested that the colour of a beaker affects the amount of infrared radiation (IR) it emits. You can investigate this using two copper beakers, one painted matt black, the other painted white. The more IR the beaker emits, the faster it cools.

Download the Research Notes and Data Sheet for P1 from **www.oxfordsecondary.co.uk/aqacasestudies**.

Research

*Record your findings in the **Research Notes table**.*

1. Research two different methods that could be used to test the hypothesis.
2. Find out how the results of the investigation might be useful in designing clothing for expeditions to the North Pole.

Section 1 — Total 20 marks

Use your research findings to answer these questions.

1. **(a)** Name two sources that you used for your research.
   (b) Which of these sources did you find more useful, and why? [3]
2. **(a)** Identify one control variable.
   (b) Briefly describe a preliminary investigation to find a suitable value for this variable, and explain how the results of this work will help you decide on the best value for it. [3]
3. *In this question you will be assessed on using good English, organising information clearly, and using specialist terms where appropriate.* Describe how to carry out an investigation to test the hypothesis. Include the equipment needed and how to use it, the measurements to make, how to make it a fair test, and a risk assessment. [9]
4. Use your research to outline another possible method, and explain why you did not choose it. [3]
5. Draw a table to record data from the investigation. You may use ICT if you wish. [2]

Section 2 — Total 30 marks

*Use the **Data Sheet** to answer these questions.*

1. **(a)** Name the independent and dependent variables, and give one control variable. [3]
   (b) The smallest scale division on a measuring instrument is called its resolution. What was the resolution of the instrument you used, and was this resolution suitable for your experiment? [3]
   (c) Display the ***Group A data*** in a line graph of temperature decrease against the colour of the beaker. *This data has been provided for you to use instead of data that you would gather yourself.* [4]
   (d) Does the ***Group A data*** support the hypothesis? Explain how. [3]
   (e) Describe the similarities and differences between the ***Group A data*** and the ***Group B data***. Suggest one reason why the results of the two groups may be different. *The **Group B data** has been provided for you to use instead of data that would be gathered by others in your class.* [3]
   (f) Suggest how your method might have helped Group A identify a clear pattern in their results. [3]
2. **(a)** Sketch a line graph of the results in ***Case study 1***. [2]
   (b) Explain to what extent the data from ***Case studies 1–3*** support the hypothesis. [3]
   (c) Use ***Case study 4*** to describe the relationship between the surface area of an object and the amount of IR it emits. Explain how well the data supports your answer. [3]
3. The context of this investigation (the topic it relates to) is designing clothing for expeditions to the North Pole. Describe how your results may be useful in this context. [3]

P2 Controlled Assessment-style questions

Overview: A force acting on an object may cause a change in shape of the object. When a force is applied to a spring it changes its length. **You must develop your own hypothesis to test.** You will be provided with a retort stand, clamp, boss, a small spring, 100 g masses, eye protection, and a metre rule.

Download the Research Notes and Data Sheet for P2 from **www.oxfordsecondary.co.uk/aqacasestudies**.

Research

*Record your findings in the **Research Notes table**.*

1. Research two different methods to find out how the force applied to a spring changes its length.
2. Find out how the results of the investigation might be useful in designing a simple force meter to measure the weight of rock samples.

Section 1 — Total 20 marks

Use your research findings to answer these questions.

1. **(a)** Name the two most useful sources that you used for your research.
   (b) Explain why these sources were the most useful. [3]
2. Write a hypothesis about how the force applied to a spring might affect its extension. Use your research findings to explain why you made this hypothesis. [3]
3. Describe how to carry out an investigation to test your hypothesis. Include the equipment needed and how to use it, the measurements to make, how to make it a fair test, and a risk assessment. [9]
4. Use your research to outline another possible method, and explain why you did not choose it. [3]
5. Draw a table to record data from the investigation. You may use ICT if you wish. [2]

Section 2 — Total 30 marks

*Use the **Data Sheet** to answer these questions.*

1. Display the ***Group A data*** on a graph. *This data has been provided for you to use instead of data that you would gather yourself.* [4]
2. **(a)** What conclusion can you draw from the ***Group A data*** about a link between force and extension? Use any pattern you can see in the ***Group A data*** and quote figures from it. [3]
   (b) (i) Compare the ***Group A*** and ***Group B data***. Do you think the ***Group A data*** is reproducible? Explain why. *The **Group B data** has been provided for you to use instead of data that would be gathered by others in your class.* [3]
   (ii) Explain how you could use the repeated results from Group B to obtain a more accurate answer. [3]
   (c) Look at the ***Group A data***. Are there any anomalous results? Quote from the data. [3]
3. **(a)** Sketch a graph of the results in ***Case study 1***. [2]
   (b) Explain to what extent the data from ***Case studies 1–3*** support or contradict your hypothesis. [3]
   (c) Compare the ***Group A data*** to the ***Case study 4 data***. Explain how far the ***Case study 4 data*** supports or contradicts your hypothesis. [3]
4. A spring manufacturing company claims that force applied to a spring is directly proportional to its extension up to a certain point.
   (a) Does the ***Group A data*** support or contradict this hypothesis? Quote figures from the data to explain your answer. [3]
   (b) Suggest how the results of your investigation and the ***Case studies*** might be useful in designing a simple force meter to measure the weight of rock samples. [3]

P3 Controlled Assessment-style questions

Overview: The length of a pendulum affects the frequency of the swing. **You must develop your own hypothesis to test**. You will be provided with a length of string, a pendulum bob, a stopwatch, retort stand, clamp, boss, and a metre rule.

Download the Research Notes and Data Sheet for P3 from **www.oxfordsecondary.co.uk/aqacasestudies**.

Research

*Record your findings in the **Research Notes table**.*

1. Research two methods to investigate how the length of a pendulum affects the frequency of the swing.
2. Find out how the results of the investigation might be useful in designing a simple swing for a children's playground.

Section 1 — Total 20 marks

Use your research findings to answer these questions.

1. **(a)** Name the two most useful sources that you used for your research.
   (b) Explain why these sources were the most useful. [3]
2. Write a hypothesis about how the length of a pendulum affects its frequency. Use your research findings to explain why you've made this hypothesis. [3]
3. Describe how to carry out an investigation to test your hypothesis. Include the equipment needed and how to use it, the measurements to make, how to make it a far test, and a risk assessment. [9]
4. Use your research to outline another possible method, and explain why you did not choose it. [3]
5. Draw a table to record data from the investigation. You may use ICT if you wish. [2]

Section 2 — Total 30 marks

1. **(a)** Display the ***Group A data*** on a graph. *This data has been provided for you to use instead of data that you would gather yourself.* [4]
   (b) What conclusion can you draw from the ***Group A data***? Use any pattern you can see in the ***Group A data*** and quote figures from it to support your answer. [3]
2. **(a) (i)** Compare the ***Group A*** and ***Group B data***. Do you think your results are reproducible? Explain why. *The **Group B data** has been provided for you to use instead of data that would be gathered by others in your class.* [3]
   (ii) Explain in detail what might account for the slight differences between the readings from each group. [3]
   (c) Look at the ***Group A data***. Are there any anomalous results? Quote from the data to explain your answer. [3]
3. **(a)** Sketch a graph of the results in ***Case study 1***. [2]
   (b) Explain to what extent the data from ***Case studies 1–3*** support or contradict your hypothesis. Use the data to support your answer. [3]
   (c) Explain whether the ***Case study 4 data*** supports or contradicts your hypothesis. State any other conclusion that can be drawn from the data. Use examples of data from ***Case study 4*** and the ***Group A data***. [3]
4. **(a) (i)** Explain how the length of a pendulum with a time period of exactly 1.00 s might be found from the data in the case studies. [2]
   (ii) In order to find a more precise value for this length, suggest three more lengths to be tested. [1]
   (b) Suggest how the results of your investigation and the case studies might be useful in designing a simple swing for a children's playground. [3]

P1 Part 1

Energy and efficiency

AQA specification match		
P1.1	1.1	Infrared radiation and surfaces
P1.1	1.2	Solids, liquids, and gases
P1.1	1.3	Conduction
P1.1	1.4	Convection
P1.1	1.5	Evaporation and condensation
P1.1	1.6	Energy transfer by heating
P1.1	1.7	Comparing energy transfers
P1.1	1.8	Heating and insulating buildings
P1.1	1.9	Solar panels and payback time
P1.1	1.10	Specific heat capacity
P1.1	1.11	Uses of specific heat capacity
P1.2	1.12	Understanding energy
P1.2	1.13	Useful energy and energy efficiency
P1.3	1.14	Using electricity
P1.3	1.15	Paying for electricity
Course catch-up		
AQA Upgrade		

Why study this unit?

Energy can be transferred in different ways by heating processes. The clothes that you wear on a cold day keep you warm by reducing the transfer of heat energy from your body. Houses are insulated to reduce the amount of heat transferred to the outside. But some appliances need to transfer lots of energy – central heating radiators transfer energy into a room.

In this unit you will look at the different ways that energy can be transferred by conduction, convection, radiation, evaporation, and condensation. You will learn how these transfers are minimised in homes and some of the things you use, but maximised in other appliances. You will also learn about the useful and wasted energy transferred by appliances and how to calculate the efficiency of an appliance.

Finally, you will learn about how different appliances in your home transfer electrical energy into a range of different forms of energy, and how much these appliances cost to run.

You should remember

1. Energy cannot be created or destroyed.
2. Energy can be transferred by a number of different methods.
3. The hotter an object, the higher its temperature.
4. Heating or cooling involves a transfer of energy.
5. There are three states of matter: solids, liquids, and gases.

This house in Kent is so well insulated that it does not need any central heating. The house uses energy from the Sun and stores it for when the weather is cold. The house uses less than 10% of the energy that a conventional three-bedroom house uses for heating rooms and water. Most of the energy used by this house is for heating hot water. The house has a ventilation system which takes in fresh air from outside and transfers heat energy to it from the stale air that is about to be pumped outside.

1: Infrared radiation and surfaces

Learning objectives

After studying this topic, you should be able to:

- ✔ understand that all objects both emit and absorb infrared radiation
- ✔ investigate which surfaces are good or bad at absorbing infrared radiation

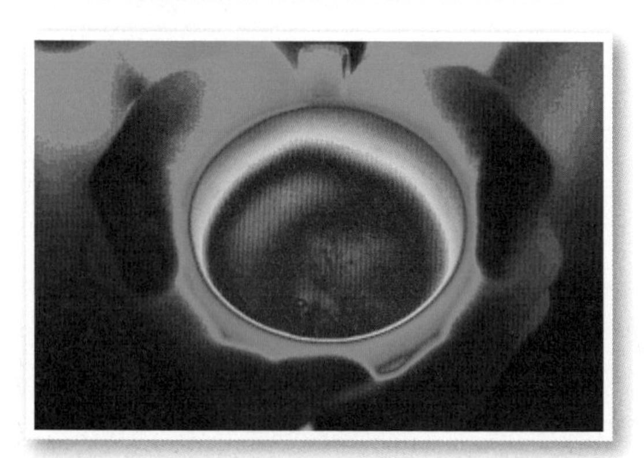

▲ A thermogram of a hot drink

Key words

infrared radiation, thermogram, emit, medium

▲ This cooker ring is emitting infrared radiation and light

Infrared radiation

When you put your hands on a cup that contains a hot drink, heat energy is transferred to your hands by conduction. If you take your hands off the cup, but keep them close to it, you can still feel heat energy being transferred from the cup.

This energy is being transferred by **infrared radiation** (IR). Infrared radiation is a type of electromagnetic wave.

A **thermogram** shows how much infrared radiation something is emitting. You take a picture using a special camera which is sensitive to infrared radiation but not visible light. The photo on the left shows a thermogram of a hot drink. The hotter areas are red. The cooler areas are blue.

A What is infrared radiation?

All objects **emit** and absorb infrared radiation.

When something is glowing then it is also emitting light radiation. The electric cooker ring in the photo is glowing red – it is emitting light as well as infrared radiation.

Sometimes you feel infrared radiation but no light radiation is emitted. For example, you can feel the heat being radiated by a dish that has just come out of the oven, but the dish does not glow.

B Can you tell if an object is emitting infrared radiation just by looking at it? Explain your answer.

If an object is warmer than its surroundings, it will radiate infrared radiation. If an object is cooler than its surroundings, it will absorb infrared radiation. The hotter something is, the more infrared radiation it will radiate in a given time. If two objects are the same size and one of them is hotter, the hotter one will emit more infrared radiation.

Objects can absorb radiation from the Sun. For example, if you put an object in sunlight, it will absorb infrared radiation from the Sun and its temperature will increase.

The surface of the object affects how much energy it absorbs. A dark matt surface is good at absorbing infrared radiation. A light-coloured or shiny surface is bad at absorbing infrared radiation. This is why the inside of a vacuum flask is shiny.

A dark matt surface is also a good emitter of radiation and a light shiny surface is a poor emitter of radiation.

▲ These buildings are designed to stay cool in the summer as they reflect heat

Energy transfer by infrared radiation

Infrared radiation is a type of electromagnetic wave, like light. It can travel through a vacuum, like light does. It does not need a **medium** to travel through.

Questions

1 Make a table summarising which surfaces are good and which are poor at absorbing and emitting infrared radiation. (E)

2 Why does an electric fire usually have a shiny panel behind the heating element?

3 Look at the picture of the houses. What features help to keep the houses cool in summer?

4 Why is the inside of a vacuum flask shiny? (C)

5 What does a motion detector need to do to set off an alarm? (A*)

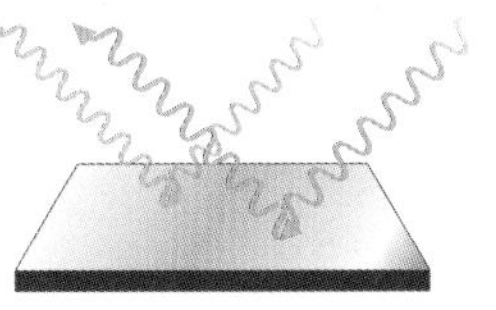

light shiny surface

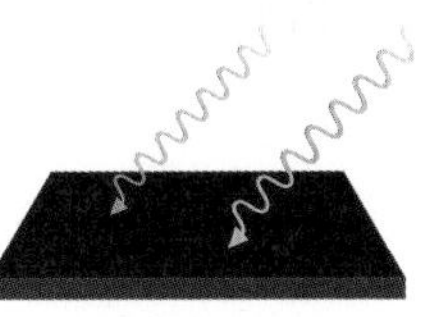

matt black surface

▲ How infrared radiation is absorbed by different surfaces

Did you know...?

A patio heater is cleverly designed to warm the area around it, whilst minimising energy transfer from the top surface. The top surface is shiny. This reduces the infrared radiation emitted to the surroundings.

▲ This patio heater uses infrared radiation

Exam tip

✔ Remember that the hotter an object is, the more infrared radiation it emits.

2: Solids, liquids, and gases

Learning objectives

After studying this topic, you should be able to:

- ✔ use kinetic theory to explain the three states of matter

Did you know...?

All big structures have expansion joints. Structures expand and contract as the air temperature changes. For example, if a bridge did not have expansion joints, it could buckle and break in very hot weather.

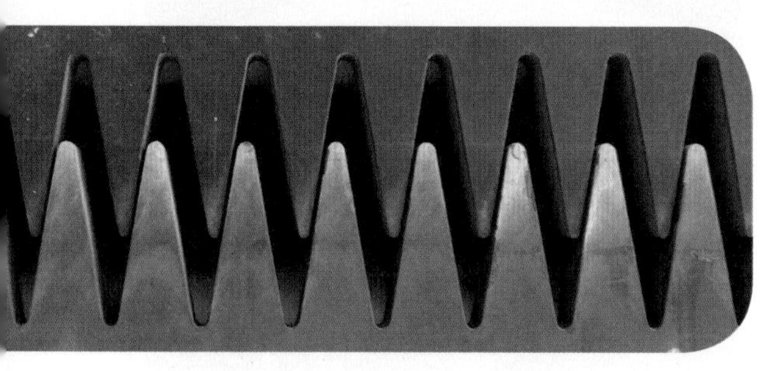

▲ An expansion joint in a bridge

A What are the differences between a solid and a liquid?

B What are the similarities between a liquid and a gas?

States of matter

The three **states of matter** are solid, liquid, and gas. Solids, liquids, and gases are all made up of tiny particles.

In a **solid**, the particles are packed close together in a regular pattern. They are held by attractive forces between them. Each particle vibrates about its fixed position. The shape and the volume of the solid is fixed.

In a **liquid**, the particles still attract one another and are still packed close together, but not in a regular way. They vibrate more and are free to move around. This is why liquids can flow.

In a **gas**, the particles are too spread out to attract one another much. The particles move around freely at high speed and collide with one another and with the walls of the container.

This **kinetic theory** explains the properties of the states of matter. The properties are given in the table.

State	Properties
solid	fixed shape, fixed volume
liquid	no fixed shape, fixed volume
gas	no fixed shape (takes the shape of its container), no fixed volume

States and energy

The particles in solids, liquids, and gases have different amounts of energy.

In gases, the particles have the most energy and are moving around at high speed. Particles in solids are fixed and can only vibrate from side to side. They have less energy than the particles in liquids, which are free to move.

solid

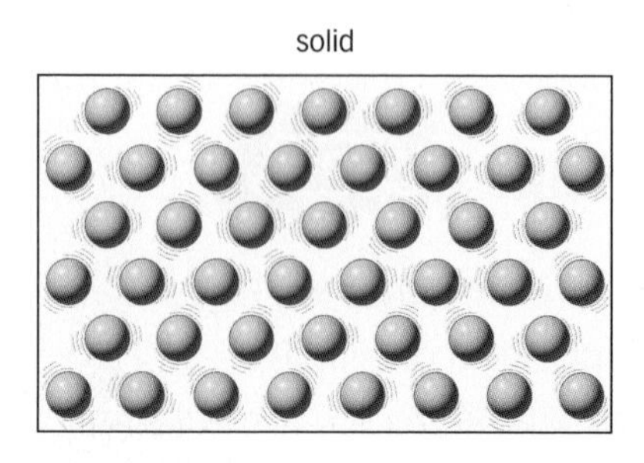

liquid

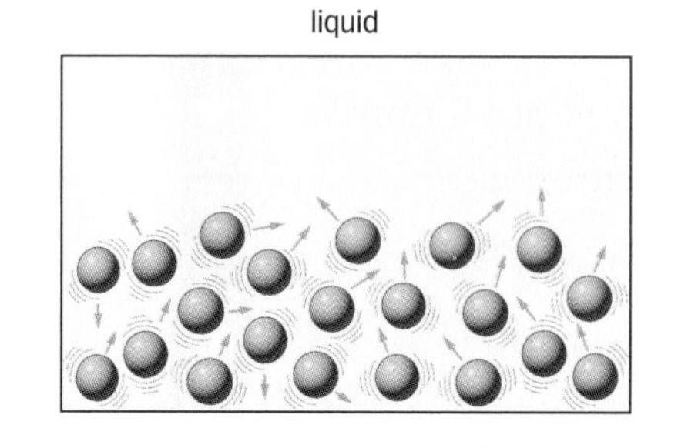

gas

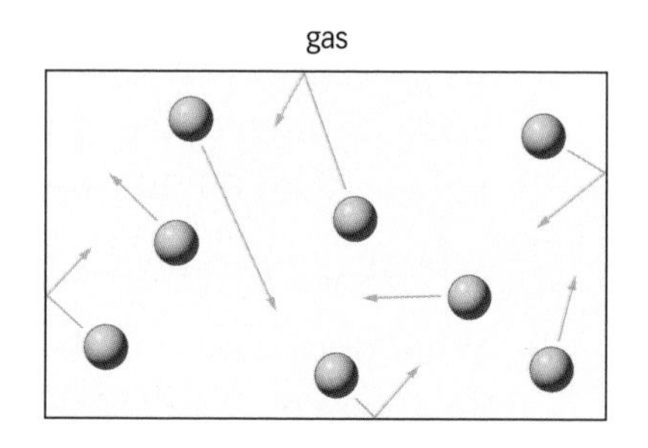

▲ Particles in a solid, a liquid, and a gas. In each case, the amount of energy is related to the temperature of the substance.

As the temperature increases, the size of the vibrations, or the speed of the particles, increases. The particles are taking up more space and the substance expands.

When the temperature decreases, the particles have less energy. In a solid, they vibrate less and so take up less space. The solid contracts.

In liquids and gases, the particles also have less energy and so move more slowly. The gas or liquid becomes more dense as the temperature decreases.

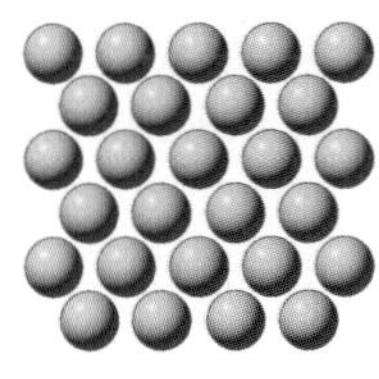

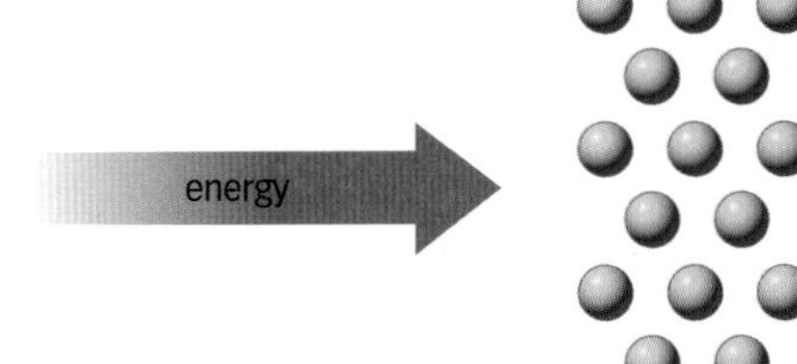

▲ Particles take up more space when the temperature increases

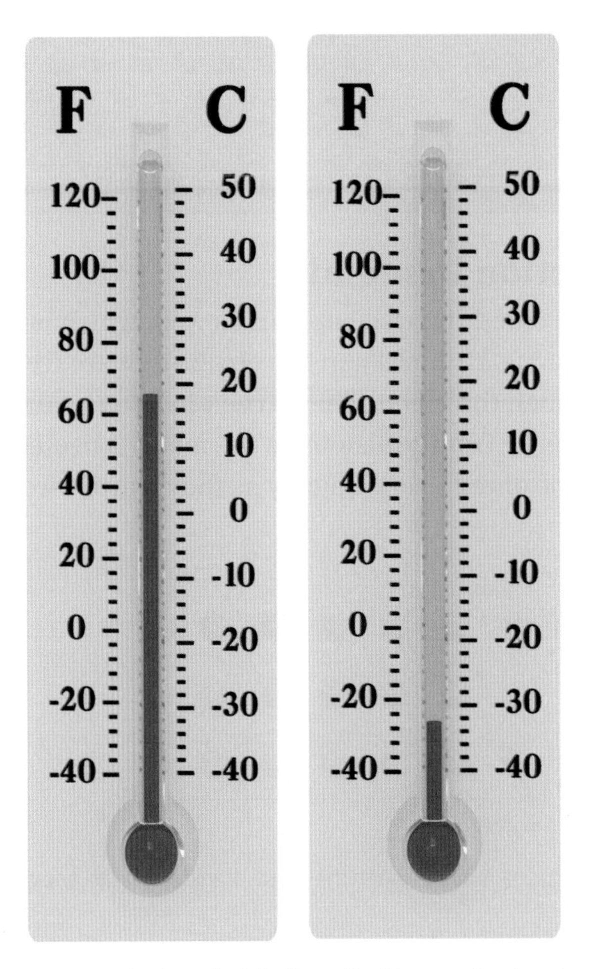

▲ A thermometer uses the principle that a liquid expands when it gets hotter

Key words

states of matter, solid, liquid, gas, kinetic theory

C What happens to the particles in a liquid when the temperature increases?

Questions

1. Make a table summarising how the particles behave in solids, liquids, and gases. (E)
2. Describe what happens to the particles in a gas as the temperature increases. (C)
3. Use the kinetic theory to explain why liquids can flow.
4. Explain how a thermometer works. (A*)

3: Conduction

Learning objectives

After studying this topic, you should be able to:

- ✔ understand how energy is transferred by conduction

Key words

free electrons, thermal conductor, thermal insulator, conduction

Thermal conduction

Heat energy is transferred through solid materials by conduction. The particles in a solid are always vibrating. When the particles are hotter, they have more energy and so vibrate more.

When you heat one end of an object, the particles start vibrating more. The particles collide with neighbouring particles. Energy is transferred from one particle to another, in the same way that energy is transferred from the cue ball when it collides with another ball on a pool table. Energy can be conducted through the solid by the particles colliding with each other.

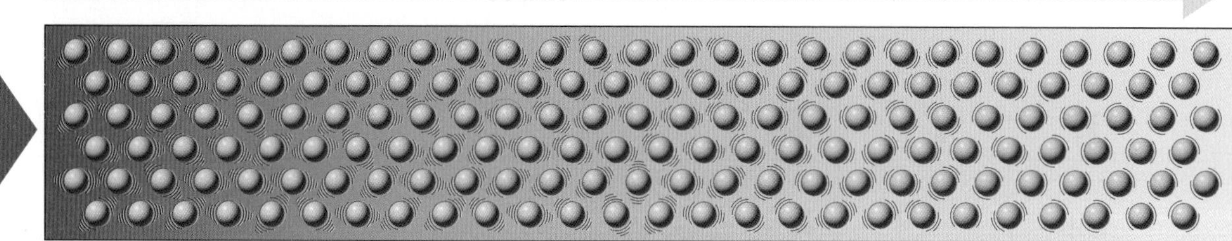

▲ Heat is conducted through a solid

Conduction in metals

In metals something else happens as well, which is more important. Metals have many electrons that are free to move through the metal. These **free electrons** gain more kinetic energy from collisions as the metal is heated. They transfer the energy very quickly as they travel through the metal.

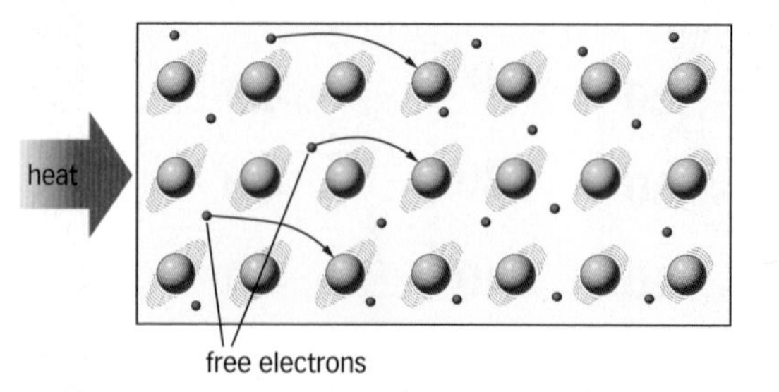

▲ Free electrons move through the metal, transferring the energy more quickly

Did you know...?

Diamond is an even better thermal conductor than metals, even though it does not have free electrons. This is because its particles are arranged in a very regular and stable pattern. This also makes it one of the hardest known substances.

Conductors and insulators

Metals are good **thermal conductors**. Energy is conducted through them easily.

Some materials do not conduct heat very well – they are called **thermal insulators**. Materials such as wood and plastics are insulators and they transfer energy slowly. The particles are close together, but the pattern is not as regular as in a metal. This means that energy only passes slowly from one particle to another.

You can tell whether a material is a conductor or an insulator by touching it. A conductor feels cold when you touch it because energy is conducted away from your skin quickly. Insulators feel warmer because the energy is not conducted away from your skin.

Liquids and gases are poor thermal conductors. The particles in a liquid do not have a regular arrangement, so it is much more difficult for energy to be passed on by **conduction**. In a gas, the particles are far apart so it is a very poor conductor.

A Imagine a metal spoon and a wooden spoon in a metal saucepan over a gas flame. Both spoons are touching the pan. Which spoon will heat up more quickly?

B Describe how the energy gets from the gas flame to the handle of the metal spoon in Question A.

Questions

1. Give an example of a good thermal conductor. (E)
2. An object feels cool when you touch it. Is it likely to be a thermal insulator or a thermal conductor? Explain your answer.
3. Why is the handle of the kettle made of plastic?

The kettle has a plastic handle

4. Why are liquids and gases poor thermal conductors? (C)
5. Explain how free electrons transfer heat in a metal. (A*)

Exam tip

✔ Remember:
a good thermal insulator is a poor heat conductor
a poor thermal insulator is a good heat conductor.

4: Convection

Learning objectives

After studying this topic, you should be able to:

- ✔ understand how energy is transferred by convection

Key words

fluid, convection, convection current

Did you know...?

Fur works by trapping pockets of air, so that energy cannot be transferred away from the surface of the skin by convection.

▲ This polar bear keeps warm because its fur minimises convection near its skin

Transferring energy in liquids and gases

You have already learned that **fluids** (liquids and gases) are poor thermal conductors. But they can still transfer energy, because the particles are free to move. When moving particles carry energy from one place to another this is called **convection**.

When the particles in one part of a fluid gain more energy, they move faster and that part expands because the particles are taking up more space. There is still the same mass of fluid, but it is taking up a larger volume. The density of that part of the fluid decreases because there is the same number of particles but in a larger space. Its lower density causes it to rise.

The cooler fluid nearby is denser than the heated fluid. The denser cooler fluid falls to the bottom, and the less dense hotter fluid rises. This movement is called a **convection current**.

The diagram shows how a radiator can heat all of the air in a room.

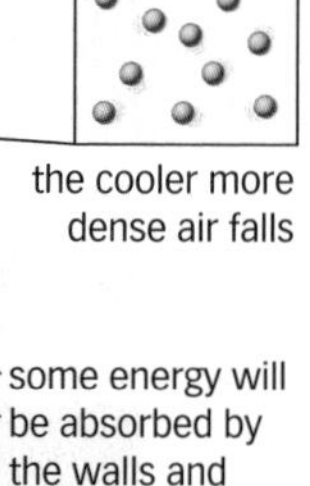

▲ Convection currents are used to heat a room

A How is energy transferred from the radiator to the particles in the air?

B What happens to air when it gains energy?

Convection currents

Sometimes you can see the effects of a convection current. For example, when a plant is on a windowsill above a radiator, the convection current set up by the radiator can be strong enough to move the leaves of the plant.

Convection currents can also be deflected. If there is an object in the way, the convection current will flow around it.

Some central heating systems use a convection current to transfer energy from the boiler to the radiators. Convection currents are also used in some domestic hot water systems.

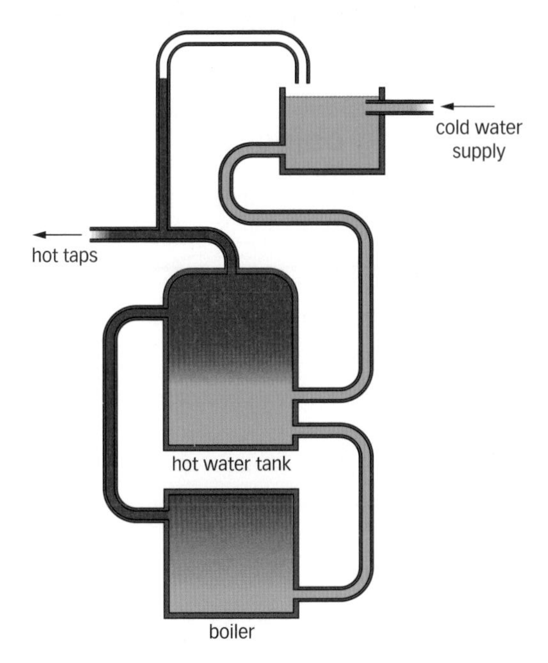

▲ Domestic hot water system

C Why do convection currents occur in liquids?

Questions

1 Draw a diagram and label it to show how convection currents transfer energy through the water in the saucepan shown in the picture.

◀ Saucepan of water being heated on stove

E

2 How is convection used in the hot water system shown in the diagram?

3 Harry is sitting in the chair in a room like the one in the picture on the opposite page, when the heating is switched off on a cold day. Why does he feel cold?

C

4 An ice cube is placed in a glass of water. In terms of energy and particles, explain why the water beneath the ice cube begins to sink.

A*

5: Evaporation and condensation

Learning objectives

After studying this topic, you should be able to:

- understand how energy can be transferred by evaporation and condensation
- understand how to change rates of evaporation and condensation

Key words

evaporation, **condensation**

▲ The puddles will dry up because of evaporation

Exam tip AQA

- Remember that evaporation has a cooling effect.

Evaporation

The ground doesn't stay wet for very long after rain – it dries by **evaporation**. The kinetic theory helps to explain how this happens.

In a liquid, the particles do not all have the same energy. They have different energies and so some of them will be moving faster than others. Particles with a higher energy than the average are constantly escaping from the surface of the liquid. Some particles fall straight back into the liquid, but others with a higher energy escape into the air. So the total amount of energy in the liquid decreases. As the escaping particles have a higher energy than the average energy of a particle left in the liquid, the average energy of the particles left behind decreases. The temperature of the liquid goes down.

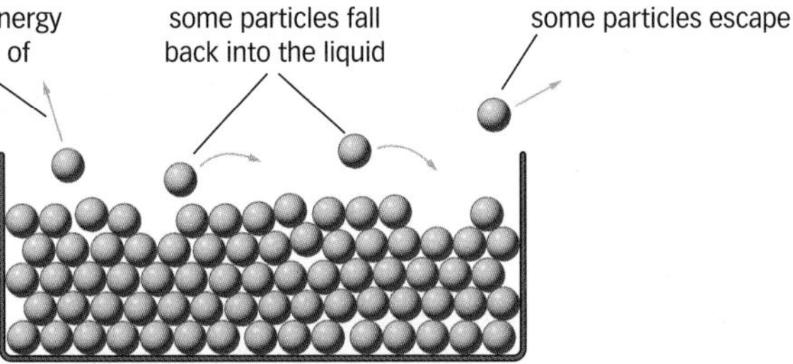

▲ Evaporation from the surface of a liquid

A How does a liquid cool down by evaporation from its surface?

Condensation

Condensation can be explained in a similar way. Particles in a gas have more energy than particles in a liquid. If the particles lose enough energy they will condense to form a liquid.

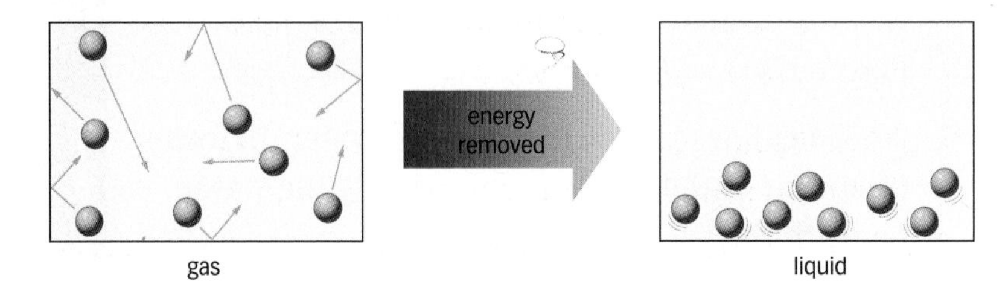

▲ When gas or vapour particles lose energy, they condense into a liquid

Changing the rate of evaporation

You can change the rate of evaporation in these ways:

- Increase the flow of air, for example by blowing a fan over the surface.
- Add energy, by heating the liquid.
- Increase the surface area of the liquid.

How does that work?

Increasing the flow of air means that the air particles that have escaped from the surface of the liquid are carried away before they have a chance to fall back into the liquid.

When you give the particles in the liquid more energy, they have a higher speed and so are more likely to escape from the surface of the liquid.

Increasing the surface area means that more particles are near the surface.

An electric hand dryer dries your hands by blowing air over them. The moving air is also heated, so energy can be transferred to the water on your skin and increase the rate of evaporation.

B How could you change the rate of evaporation?

C What are the best conditions for drying washing in the fresh air?

Moisture evaporating from a damp tree in a forest

Questions

E
1. Why does blowing over a hot drink help to cool it down?

C
2. Explain why you can feel cold when you get out of a swimming pool. Use the kinetic theory in your explanation.
3. Explain how a tumble dryer dries clothes.
4. Draw and label a diagram to show how the floating water distiller works.

A*
5. Describe how evaporation can be used to keep things cool, and give an example.

Exam tip

✔ Remember that condensation is the reverse process of evaporation.

Learning objectives

After studying this topic, you should be able to:

- ✔ understand that the shape and size of an object affects how quickly energy is transferred to and from it
- ✔ understand that temperature difference affects how quickly energy is transferred

Exam tip

- ✔ Remember that energy always moves from a warmer place to a cooler place, never the other way round.

▲ What will happen to this cold drink?

Heating and cooling

The hot drink shown in the photo contains energy. The hot drink will cool down because energy is transferred from it to its surroundings. Energy is transferred until the temperature of the drink is the same as its surroundings.

▲ This hot drink will cool down as energy is transferred to its surroundings

Changing the rate

The difference in temperature between an object and its surroundings affects the rate at which energy is transferred. When the temperature difference is greater, the rate of energy transfer is higher.

Size and shape also affect the rate at which energy is transferred. If two objects have the same volume but different shapes, the one with a larger surface area will cool down more quickly.

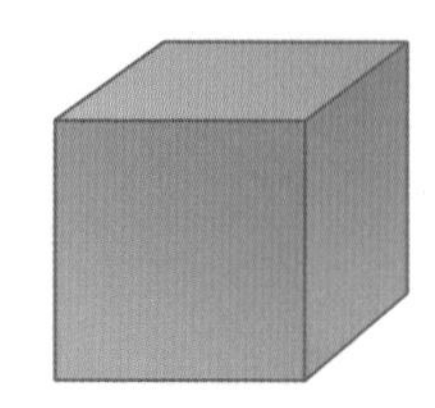

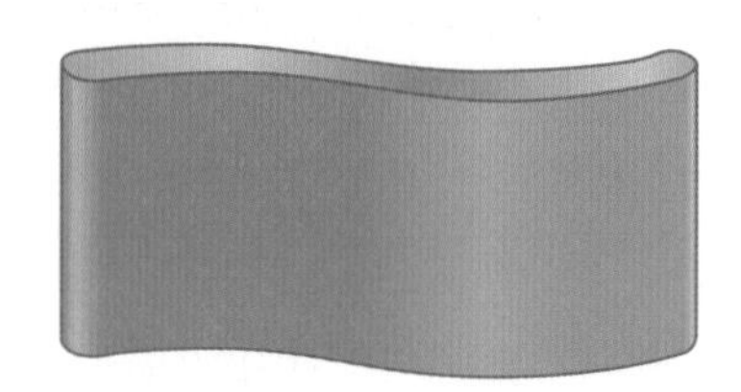

▲ These two objects have the same volume but different shapes

The type of material that something is made of also affects how quickly it transfers heat energy. For example, an object that is made from a good thermal conductor will transfer heat energy more quickly than one that is similar but is made of a thermal insulator.

Transfer of heat energy also depends on what an object is in contact with. For example, you have already learnt that a good conductor feels cool when you touch it because heat energy is quickly conducted away from your hand. Heat energy is transferred much more quickly along a metal spoon than along a wooden one.

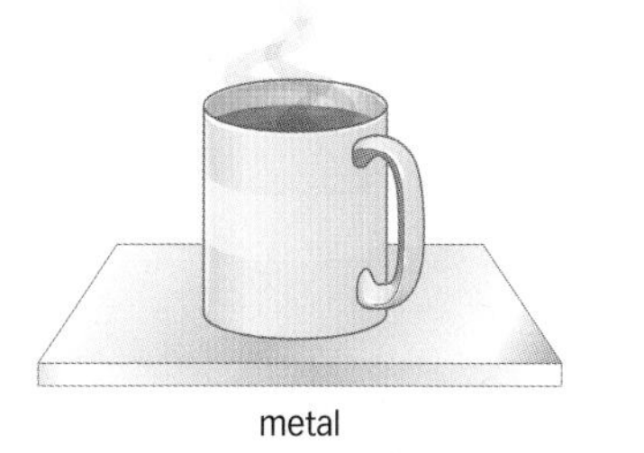

▲ Two identical objects – one touching a thermal conductor, the other a thermal insulator

Questions

1. Look at the cold drink shown in the picture. Describe what will happen in terms of the temperature of the drink and the flow of energy. ↓E
2. Draw a table to summarise all the things that can change the rate at which heat energy is transferred by heating.
3. How could you reduce the rate of heat energy transfer from a hot drink? ↓C
4. What can you say about the rate of heat energy transfer in an object that is at the same temperature as its surroundings? ↓A*

A Which will cool down to 50 °C more quickly – a kettle of water at 100 °C or a hot drink at 70 °C? Explain your answer.

B Which contains more heat energy – a kettle full of water at 60 °C or a tank full of water at 60 °C?

C Which object in the diagram on the left will cool down more quickly? Explain your answer.

Did you know...?

African elephants live in a hot climate and need to cool down. An elephant has large ears that help to control its body temperature. The ears have a large surface area and so more energy can be transferred to the elephant's surroundings.

▲ African elephants use their ears to control their body temperature

7: Comparing energy transfers

Learning objectives

After studying this topic, you should be able to:

- ✔ compare the ways in which energy is transferred between objects and their surroundings by heating
- ✔ understand how to vary the rate of energy transfer by heating
- ✔ evaluate the design of everyday appliances that transfer energy by heating

Key words

dissipate, evaluate

▲ The cooling fins on a motorbike engine transfer energy to the surrounding air

Sometimes we want to transfer as much energy by heating as possible, for instance with radiators. In other cases we want to minimise energy transfers, for example with double-glazed windows.

Animal adaptations for energy transfer

Some animals in hot countries need to be able to **dissipate** excess energy from their bodies. The cape fox below, for example, has large ears. Blood is pumped through the ears and energy is transferred to them. As the ears have a large surface area, more energy can be transferred to the air. In contrast, the arctic fox has small ears to prevent it losing precious body heat.

▲ A cape fox

▲ An arctic fox

Design for transfer of energy

Central heating radiators are designed to transfer as much energy by heating as possible. Radiators are long and thin so that the surface area is maximised, and some have fins which are bent at the top. This means that the convection current that is caused by the radiator is directed into the room rather than up the wall.

A What feature of the cape fox's ears increases the rate of energy transfer by heating?

B What features of the cooling fins in the picture are designed to transfer energy quickly?

A vacuum flask is designed to keep things hot or cold. The transfer of energy by heating is minimised.

The double walls of the glass bottle inside the flask are silvered, which reduces the amount of energy transferred by radiation. There is a vacuum between the two silvered surfaces, so there are no particles to transfer energy by conduction or convection. The glass bottle is held in place inside the outer casing by pads that are made of a poor conductor. The flask also has a tight-fitting screw cap made of a poor conductor (like plastic). This reduces losses by conduction, convection, and radiation.

However, some energy can still be transferred through the vacuum by radiation. Some energy will still be lost through the pads and the cap.

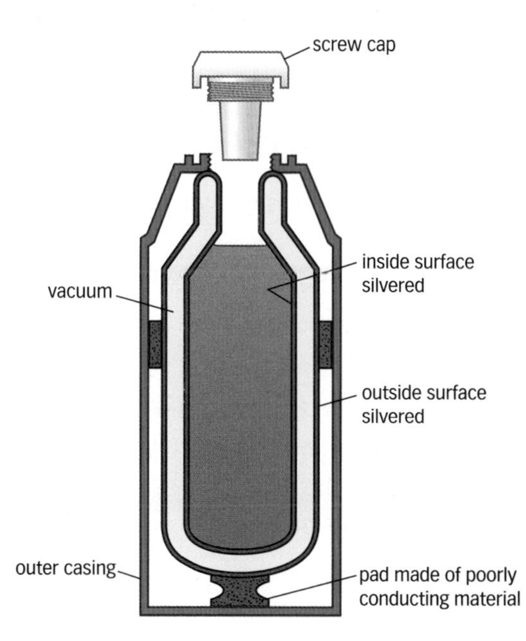

▲ Cross-section of a vacuum flask

Questions

1 How could the engine in the picture be changed to increase the rate of energy transfer?

E

▲ Video card used in personal computers

C

2 Look at the picture of the video card. The video card produces energy which needs to be taken away. How do the cooling fins do this?

3 Describe in detail how a vacuum flask keeps a drink cool.

4 What kind of material would you put at the top of the vacuum flask?

5 Why should you fill up a vacuum flask rather than half-fill it?

A*

Did you know...?

Many appliances that dissipate energy will also have a fan to increase the flow of air over them. For example, car engines have a fan that starts automatically in hot weather to increase the flow of air over the radiator when the engine gets too hot.

Exam tip

✔ When you are asked to **evaluate** the energy transfers in an appliance, first look at what the appliance is designed for. Is it to increase or to decrease the rate of energy transfer by heatng? What types of heat transfer might be happening? Then look at the characteristics of the appliance. Does it have a large surface area?

P1

8: Heating and insulating buildings

Learning objectives

After studying this topic, you should be able to:

- describe how to reduce heat transfer in buildings
- understand what a U-value is
- understand that better thermal insulators have a lower U-value

Key words

U-value

Did you know...?

Closing curtains when it gets dark in winter can help keep a room much warmer by reducing the energy transferred through the windows.

outer wall of bricks

lightweight concrete block

plaster

expanded polystyrene

Cross-section of a cavity wall

Insulating houses

Houses transfer energy to their surroundings. It costs money to heat them and so home owners try to reduce the cost of heating by insulating them better. The thermogram below shows that the middle house in the terrace is much better insulated than its neighbours.

The areas transferring the most energy are red (warmest), then yellow, then green, with blue being the coolest

When you insulate a house you reduce energy transfers through conduction, convection, and radiation.

A How do you reduce the energy transfers from a house to its surroundings?

The diagram shows a cross-section of a cavity wall. There are two parts to the wall with a cavity between them. The inner wall is made from a material that does not conduct energy well. If there is nothing in the cavity between the walls, a convection current can be set up which transfers energy from the inner wall to the outer wall. So the cavity is filled with an insulating material. This is usually something with small pockets of trapped air. The trapped air prevents convection currents forming.

U-values

Different parts of a building will conduct energy out of a building at different rates. For example, the windows in your house will usually transfer energy more quickly than the walls. We can measure the rate of losing energy for each part of a building. This is called its **U-value**. The higher the U-value, the more quickly energy is transferred. A part or a material with a low U-value is a good insulator.

House builders try to use materials with the lowest U-values. You can also reduce U-values by adding more insulation in the roof or by using glass that transfers less energy.

Part of building	U-value
270 mm cavity wall, no insulation	1.0
270 mm cavity wall with insulation	0.6
single-glazed window	5.0
double-glazed window	2.9
roof material, 50 mm insulation	0.6
roof material, 100 mm insulation	0.3

▲ Some typical U-values found in a building

B Look at the table above. Which part of the building is insulated best?

Exam tip 

✔ You might be asked to evaluate features of a house by looking at the U-values. Remember that a better insulator has a lower U-value.

Questions

1 Describe how the double-glazed window reduces the transfer of energy.

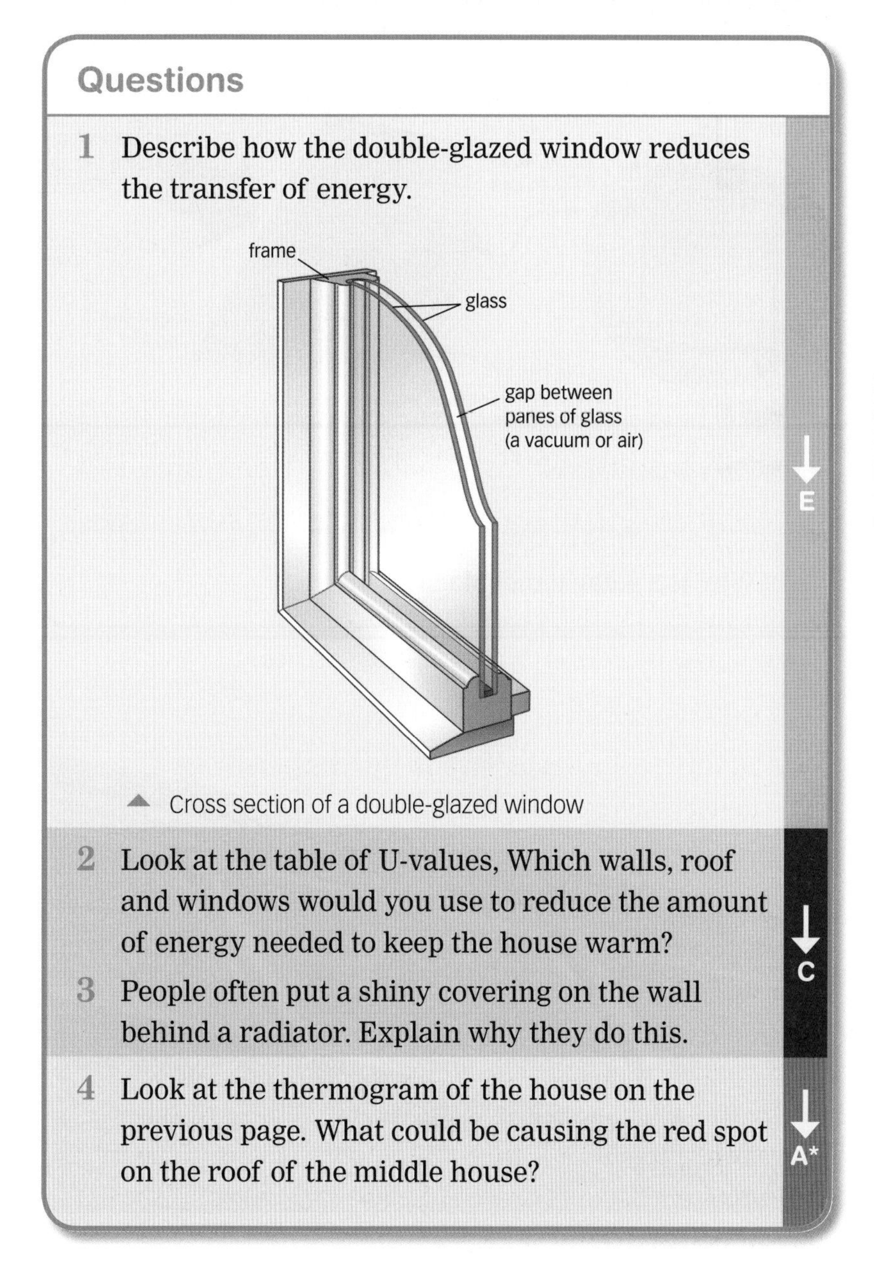

▲ Cross section of a double-glazed window

2 Look at the table of U-values, Which walls, roof and windows would you use to reduce the amount of energy needed to keep the house warm?

3 People often put a shiny covering on the wall behind a radiator. Explain why they do this.

4 Look at the thermogram of the house on the previous page. What could be causing the red spot on the roof of the middle house?

9: Solar panels and payback time

Learning objectives

After studying this topic, you should be able to:

- ✔ understand how solar panels work
- ✔ calculate payback time

Key words

solar panel, payback time, cost-effective

Solar panels

Solar panels absorb infrared radiation from the Sun. The energy is transferred from the panel to the metal pipes in the panel, and from there to the water in the pipes.

Usually, a pump is not needed to move the water through the system. A natural convection current occurs in the system which transfers the energy to the hot water tank.

When the Sun is shining on a cold day, there is still infrared radiation from the Sun. This means that solar panels can still heat water when the outside temperature is below freezing on a sunny day.

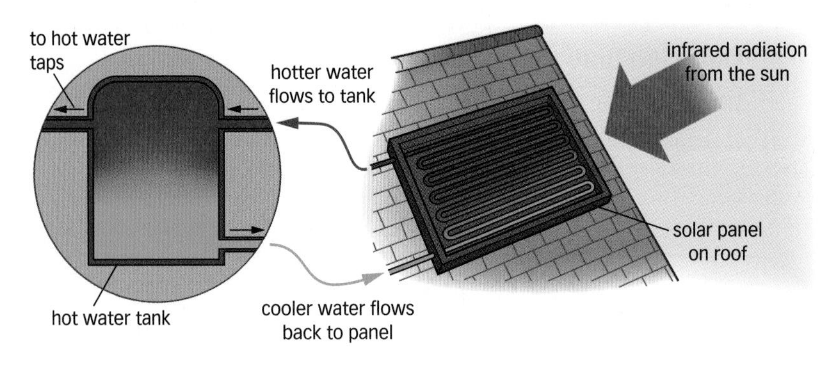

How a solar panel works

A Why is a solar panel black?

Solar panels on a roof

Did you know...?

Solar panels can still heat water on a cloudy day. But they don't transfer as much energy to the water as they do on a sunny day.

Payback time

You can cut down your energy bills by reducing the amount of energy that you use. Some ways of doing this do not cost anything. For example, you could turn the thermostat on your central heating down by 1 °C or draw the curtains when it gets dark.

Other methods do cost money. To decide whether they are worth doing, you need to calculate how long it will take to recover the amount of money you spend. This is the **payback time**. The method with the shortest payback time is the most **cost-effective**.

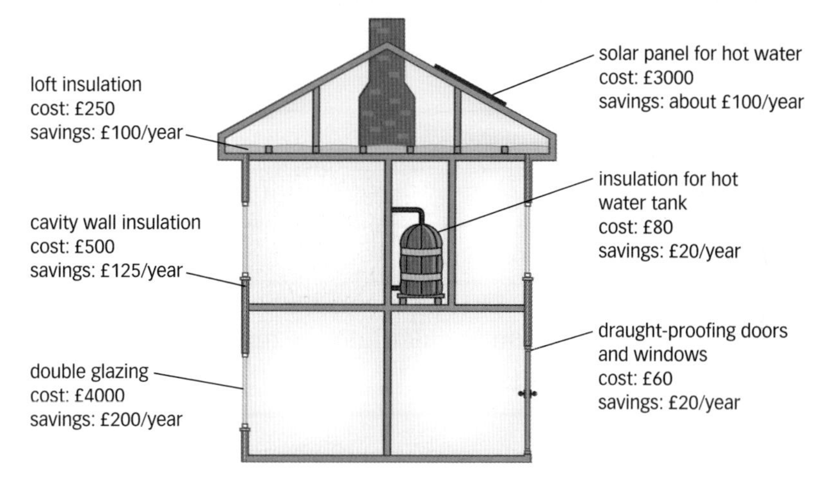

▲ Different ways of reducing your energy consumption

Worked example

Jack currently has 50 mm insulation in his loft. It will cost £250 to increase the thickness of insulation to the recommended 270 mm. His heating bill would decrease by £100 per year. What is the payback time for adding the extra insulation?

$$\text{payback time} = \frac{\text{cost}}{\text{savings per year}} = \frac{£250}{£100 \text{ per year}} = 2.5 \text{ years}$$

Exam tip 

✔ You do not need to remember any particular values for payback calculations. In exam questions, you will be given the data and asked to calculate the payback time to evaluate which methods are the most cost-effective.

B What is payback time?

Questions

1. Describe all the energy transfers that take place in a solar panel system. (E)
2. Look at the diagram of the house. Which method of reducing energy consumption is:
   (a) most cost-effective?
   (b) least cost-effective?
3. Sometimes you can get grants to help with the cost of some kinds of insulation. Work out the payback time for:
   (a) solar panels, when there is a grant of £500
   (b) loft insulation, when there is a grant of £200. (C)
4. What is the ideal direction for a solar panel to be facing in the UK? (A*)

10: Specific heat capacity

Learning objectives

After studying this topic, you should be able to:

- understand the idea of specific heat capacity of a material
- use specific heat capacity to work out how much energy is needed

A What is meant by the specific heat capacity of a material?

How much energy is required to heat the oil to fry food?

B What energy transfer is needed to heat a 500 g aluminium pan containing 1 kg of cooking oil from 20 °C to 170 °C?

C The amount of energy you have calculated is less than what would actually be needed. Explain why.

The amount of energy an object stores is related to its mass and temperature. It also relates to the material the object is made of. For example, it takes more energy to raise the temperature of 1 kg of water than it does to raise the temperature of 1 kg of aluminium.

The **specific heat capacity** of a material is the amount of energy needed to raise the temperature of 1 kg of the material by 1 °C. Its units are J/kg °C.

Some specific heat capacities are given in the table below.

Material	Specific heat capacity (J/kg°C)
water	4200
aluminium	880
copper	380
cooking oil	1200 (about)

You can use this equation to calculate the amount of energy needed to heat an object:

$$E = m \times c \times \theta$$

energy transferred (joules, J) = mass (kilograms, kg) × specific heat capacity (J/kg °C) × temperature change (degrees Celsius, °C)

Calculating the amount of heat transferred

Worked example

A kettle contains 1.5 kg of water at a temperature of 18 °C. How much energy is needed to bring the water to the boil?

energy needed = mass × specific heat capacity of water × temperature change

= 1.5 kg × 4200 J/kg °C × (100 − 18) °C

= 1.5 × 4200 × 82 J

= 516 600 J = 516.6 kJ

Calculating specific heat capacity

You can work out the specific heat capacity of a material if you know the amount of energy transferred, the mass of the object and the change in temperature.

$$\text{specific heat capacity (J/kg°C)} = \frac{\text{mass (kg)}}{\text{temperature difference (°C)}}$$

Key words

specific heat capacity

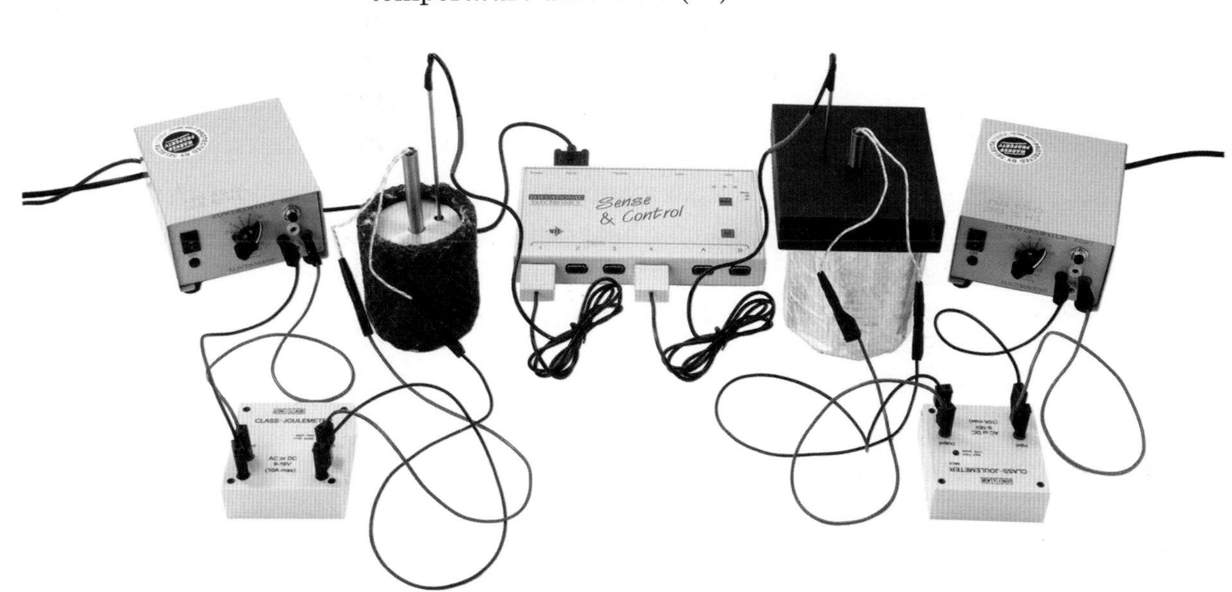

This apparatus is being used to find the specific heat capacity of the material in a metal block

Questions

1. What are the units of specific heat capacity? (E)
2. Look at the apparatus in the picture above. Why is the metal block wrapped in an insulating cover? Explain your answer.
3. Which needs more energy: to heat 5 kg of copper from to 20 °C to 50 °C or to heat 0.5 kg of water from 20 °C to 50 °C? (C)
4. 50 kJ of energy was transferred to a material with a mass of 5 kg. The temperature increased from 20 °C to 60 °C.
What is the specific heat capacity of the material? (A*)

Exam tip

✔ Don't forget to work out the temperature difference when you are calculating how much energy is needed.

11: Uses of specific heat capacity

Learning objectives

After studying this topic, you should be able to:

- ✔ understand how materials with different specific heat capacities are used

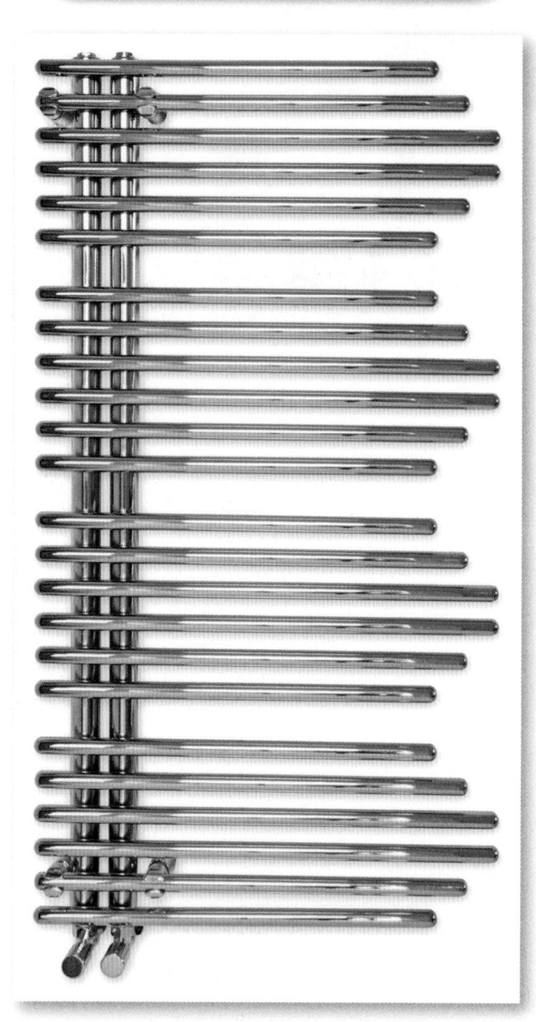

▲ Water is used to transfer energy from the boiler to radiators like this one

Using different materials

You know that the specific heat capacity of a material tells us how much energy it can store. Different materials are used for particular purposes because of their specific heat capacities.

Water is used in central heating systems to transfer energy from the boiler to the radiators. Water is used because it has a very high specific heat capacity. It can transfer much larger amounts of energy than a liquid with a lower specific heat capacity. This also means that the water does not need to be pumped very quickly around the central heating circuit.

A Why is water used in central heating systems?

Electric storage heaters contain blocks of concrete or bricks that store energy. The heater uses cheap electricity during the night to heat the blocks. The energy is released slowly during the day.

Concrete has a lower specific heat capacity than water. It is used because it is simpler to make the heaters than if water is used and is easier to maintain. The storage heater can store a lot of energy because it contains large blocks of concrete. This also means that the temperature of the blocks can be lower than if less material was used.

▲ This heater contains blocks of concrete or bricks that store energy

Oil-filled radiators can also store energy. They are designed to provide a gentle steady heat. They do this by using oil which has a high specific heat capacity. Lots of energy is stored in the oil and then radiated and carried by convection from the surface of the heater.

Material	Specific heat capacity (J/kg°C)
water	4200
concrete	880
oil	1500
glass	500–840
wood	1700

▲ Specific heat capacities of some materials

▲ This heater is filled with oil which has a high specific heat capacity

B Why does a storage heater need to store lots of energy?

Questions

1 Water is used to cool car engines. Explain why water is used and not some other substance.

E

2 In hot countries, some buildings have very thick walls. How does this help to keep the inside of the building cool?

3 Wood has a high specific heat capacity. Why would you not use it in a heater?

C

4 Alex suggests replacing the oil in the heater with a material that has a much lower specific heat capacity. What changes would you have to make to the heater?

A*

Did you know...?

Some office buildings use blocks of concrete to help keep them cool in hot weather. The ceiling of each storey is a large block of concrete. It absorbs excess heat during the day. It then cools down by transferring the energy back into the air at night, when there is no-one in the building.

Exam tip

✔ You need to be able to apply the principles of specific heat capacity. In the exam, you may be given one of these examples or a completely new example.

12: Understanding energy

Learning objectives

After studying this topic, you should be able to:

- describe energy transfers
- understand that energy is not created or destroyed, only transferred

Key words

kinetic energy, chemical energy, nuclear energy, elastic potential energy, gravitational potential energy, transfer, useful energy, wasted energy, law of conservation of energy

As well as heat, energy takes many forms such as light, sound, **kinetic energy**, and electrical energy.

Energy can be stored in different ways:

- **chemical energy** is stored inside food, fuels and batteries
- **nuclear energy** is stored inside the nucleus of atoms
- **elastic potential energy** is stored in anything that is stretched or squashed like a stretched rubber band or a coiled spring
- **gravitational potential energy** is stored in any object that is higher than its surroundings.

All of these store chemical energy

You have already seen how energy can be transferred by conduction, convection, and radiation. Many appliances or machines **transfer** energy from one form to another. For example:

- a kettle transfers electrical energy into heat energy and sound energy
- a petrol mower transfers chemical energy stored in petrol or diesel into kinetic energy, heat energy, and sound energy.

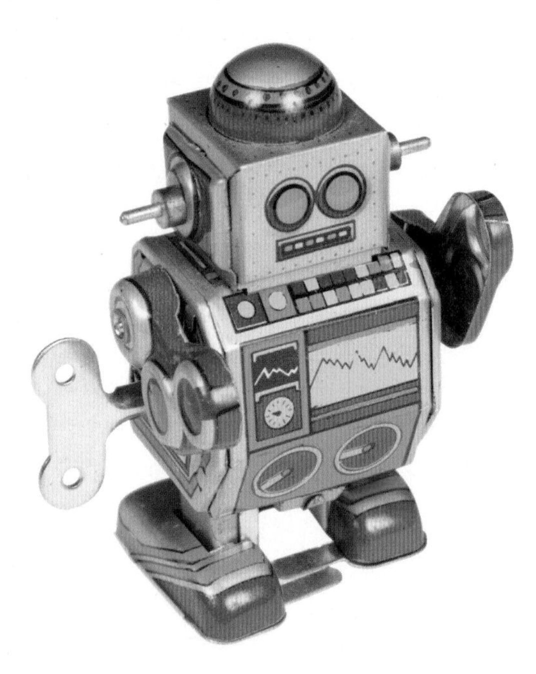

This wind-up toy stores elastic potential energy

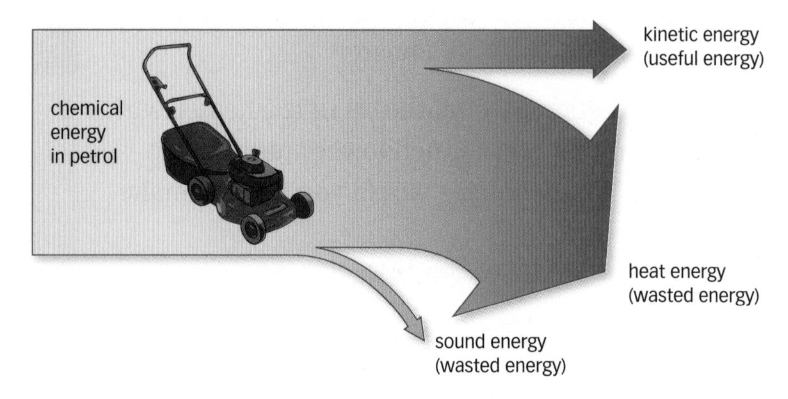

Energy transfers in a petrol mower

A What are the energy transfers when you release a stretched spring?

B What energy transfers happen when you light a barbecue?

Useful and wasted energy

Machines usually transfer energy into more than one form. Often only one of these transfers is useful. For example, when a kettle transfers electrical energy into heat energy, this is a **useful energy** transfer. The sound energy is not useful – this is **wasted energy.**

C In Questions A and B, which energy transfers are useful and which ones produce wasted energy?

You can reduce your energy consumption by using appliances that transfer less energy into wasted forms. For example, you could replace old light bulbs with low-energy bulbs which transfer much less energy into wasted heat energy.

Modern cars also transfer less energy than older ones into wasted heat energy.

Conservation of energy

The total amount of energy before and after these energy transfers is always the same. All the different forms of energy output from the mower shown in the diagram on the left add up to the amount of chemical energy that was supplied to the mower.

Energy cannot be created or destroyed. It can only be transferred from one form to another. This is the **law of conservation of energy**.

LED light bulbs

Questions

1. What energy transfers take place in the wind-up toy pictured?

E

2. What energy transfers take place in:
   (a) gas-powered hair curling tongs?
   (b) a wind-up radio?

   In each case state which of the transfers are wanted and which are unwanted.

C

3. A computer transfers 120 J of energy every second.
   (a) Write down the energy transfers that are taking place.
   (b) What is the total energy output per second in all these forms of energy?

4. An energy-saving light bulb uses 11 W and produces the same amount of light as a 75 W filament light bulb. Why should you use the energy-saving bulb instead of the filament bulb?

A*

Did you know...?

You can now buy LED lights that produce the same amount of light energy as energy-saving light bulbs, but use even less electrical energy.

13: Useful energy and energy efficiency

Learning objectives

After studying this topic, you should be able to:

- ✔ understand more about useful and wasted energy
- ✔ understand that energy becomes increasingly spread out
- ✔ understand what the efficiency of an appliance is
- ✔ calculate the efficiency of an appliance

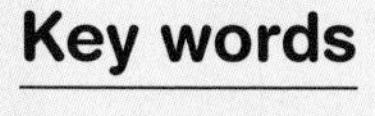

Key words

efficiency

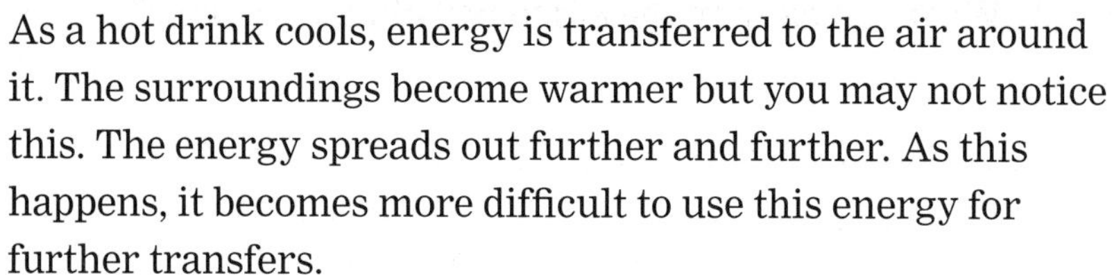

As a hot drink cools, energy is transferred to the air around it. The surroundings become warmer but you may not notice this. The energy spreads out further and further. As this happens, it becomes more difficult to use this energy for further transfers.

This hot drink transfers energy to its surroundings

An energy-efficient light bulb (centre left), two conventional filament bulbs, and a halogen bulb (centre right)

Efficiency

You have already seen that when a machine transfers energy, only some of the transferred energy is useful and the rest is wasted.

For example, a light bulb transfers electrical energy into heat as well as light. Light is useful energy, but the heat is wasted energy. The Sankey diagram shows these transfers in a light bulb.

Sankey diagram for a light bulb

A What is meant by the efficiency of an appliance?

We can work out how efficient the light bulb is, using

$$\textbf{efficiency} = \frac{\text{useful energy transferred}}{\text{total energy supplied}}$$

Remember that the units for energy transferred and energy supplied should be the same – they should both be joules or both in kilojoules.

You can also give the efficiency as a percentage by multiplying the answer by 100%.

Worked example

What is the efficiency of the light bulb shown in the diagram?

useful energy transferred = 300 J

total energy supplied = 1500 J

$$\text{efficiency} = \frac{300}{1500} = 0.2 \text{ or } 20\%$$

Some of these light bulbs transfer electrical energy into light energy much more efficiently than the others

Questions

1 Which label in the picture on the right shows the most efficient light bulb? **E**

2 A kettle is supplied with 500 kJ of electrical energy. It transfers 400 kJ of heat energy to the water in it, 99 kJ of heat energy to the kettle itself and its surroundings, and 1 kJ into sound energy.

(a) What is the efficiency of the kettle?

(b) How could you make the kettle more efficient?

(c) Draw a Sankey diagram to show the energy transfers that are happening in the kettle. **C**

3 A halogen light bulb transfers 250 J of electrical energy into 220 J of heat energy and the rest into light energy. What is its efficiency? **A***

Did you know...?

All electrical appliances must now have an energy efficiency label. The label shows how much energy you would expect to use in a year. It also grades the appliance in one of seven categories, from A to G.

Exam tip

- Remember that efficiency can never be greater than 100%. If your calculations produce an efficiency of greater than 100%, go back and check them – you will have done something wrong!
- If you write efficiency as a ratio, it can never be greater than 1.

P1

14: Using electricity

Learning objectives

After studying this topic, you should be able to:

- give examples of energy transfers in electrical appliances
- understand that the amount of energy used depends on the appliance's power and how long it is switched on

Key words

joule, power, watt, kilowatt, power rating

Energy transformations

We use electrical energy a great deal in our everyday lives, because it can readily be transferred into other types of energy.

Each electrical appliance is designed to bring about energy transfers. Some of these will be unwanted. For example, a toaster produces heat energy, but it will also produce some light energy (from the glowing elements).

Some appliances may be designed to produce more than one energy transfer. For example, an MP3 player transfers electrical energy into sound energy. It also transfers electrical energy into light energy, so that you can see which track is being played.

▲ These appliances all transfer electrical energy into other forms of energy

Different types of electrical appliances have advantages and disadvantages. For example, a battery-powered radio will run for a long time, but eventually you need to replace the batteries. A clockwork radio does not need batteries, but you need to wind it up regularly.

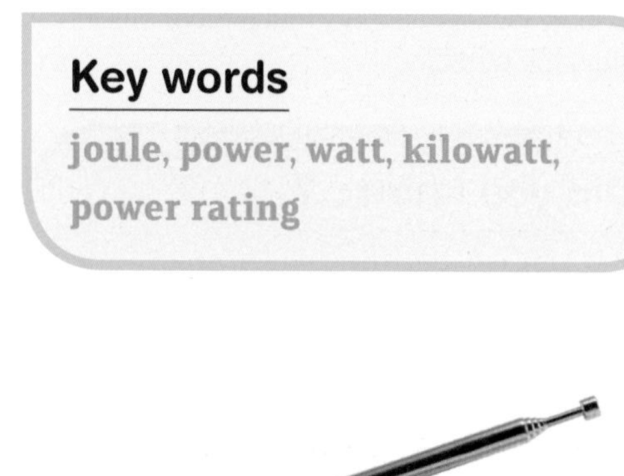

▲ A traditional (top) and a clockwork (bottom) radio. A clockwork radio offers a number of advantages over a traditional battery-powered one, but is not without its own problems.

Power

You have already learned that energy can be measured in **joules** (J). The amount of energy an appliance transfers per second is called its **power**. An electric blanket and an electric heater both transfer electrical energy to heat energy, but the electric heater transfers many more joules of energy per second than the electric blanket. The power of the heater is higher than the power of the blanket.

Power can be measured in joules per second or in **watts** (symbol W). 1 joule per second is the same as 1 watt. 1000 W is 1 **kilowatt** (symbol kW).

Many electrical appliances have labels showing how much power is needed to run them. This is called the **power rating** of the appliance.

The total amount of energy transferred by an appliance depends on how long it is switched on for, as well as its power. You can work out how much electrical energy an appliance transfers:

$$\text{energy transferred} = \text{power} \times \text{how long the appliance is switched on for}$$

$$E = P \times t$$

Worked example

A kettle takes 3 minutes to boil some water. The power of the kettle is 3 kW. How much energy is transferred by the kettle?

3 minutes = 3 × 60 = 180 seconds

3 kilowatts (kW) = 3000 watts (W)

energy transferred = power × time

= 3000 W × 180 seconds

= 540 000 J (or 540 kJ)

A The rating of the heater in the photo is 2000 W. What is this in kilowatts?

B The power rating of a computer is 125 W, and it is switched on for eight hours. The power rating of a toaster is 1.2 kW, and it is switched on for ten minutes. Which appliance transfers more energy?

Power rating label on a heater

Exam tip

- In an exam, when you are asked to compare different electrical appliances, the question will give you all the data you need.

Questions

1. What is the power of an electrical appliance?
2. Draw up a table and list the appliances shown in the photo on the previous page. List all the useful energy transfers and wasted energy transfers that each appliance makes.
3. What are the advantages and disadvantages of mains-operated fans and battery-operated fans?
4. List all of the things you could not do in your everyday life:
   (a) if you could only use appliances with batteries
   (b) if you could not use any electrical appliances.

P1

15: Paying for electricity

Learning objectives

After studying this topic, you should be able to:

- work out the amount of energy transferred from the mains supply by an appliance
- work out the cost of energy transferred from the mains supply

Key words

kilowatt-hour, unit

Energy transferred

You have to pay for the amount of electricity that you use, but how can you work this out? You have already learned that the total amount of energy used depends on the power of the appliance and how long it is switched on for. Or we can say:

$$E = P \times t$$

$$\text{energy transferred} = \text{power} \times \text{time}$$

If the power is given in kilowatts (kW), and the time is in hours, then the amount of energy used is measured in **kilowatt-hours (kWh)**:

$$\text{energy transferred (kWh)} = \text{power (kW)} \times \text{time (hours)}$$

On your electricity bill, kilowatt-hours are called **units** of electricity.

You can also calculate the amount of energy used in joules. The power is given in watts, and the time is given in seconds:

$$\text{energy transferred (J)} = \text{power (W)} \times \text{time (s)}$$

Exam tip AQA

- If power is given in watts, don't forget to convert it to kilowatts by dividing by 1000 when you substitute it into the equation.

Worked example 1

A computer uses 250 watts and is switched on for 5 hours. How much energy does it use?

energy transferred = power × time

= (250 ÷ 1000) kW × 5 hours

= 1.25 kWh

A A light bulb uses 20 W and is switched on for ten hours. How much energy does it use? Give your answer in kWh

Light bulb switched on

Cost of energy transferred

The amount of electricity you use at home is recorded by an electricity meter. Electricity is charged by the unit, or kilowatt-hour. You can work out the cost of the electricity that an appliance uses if you know how much energy it transfers.

The cost of electricity used is given by the equation:

cost = energy transferred × cost per unit

Worked example 2

A kettle transfers 3.5 kWh of energy. A unit of electricity costs 10.2p. What is the cost of the energy transferred by the kettle?

cost of electricity = energy transferred × cost per unit

= 3.5 kWh × 10.2p/kWh

= 35.7p

B What is the cost of the energy transferred by the light bulb in Question A? Assume that a unit of electricity costs 10.2p.

Questions

1. What is a kilowatt-hour?

E

2. An electric heater has two settings. The first setting uses 800 W and the second 1.5 kW. Work out the cost of using the fire:
   (a) for five hours at the 800 W setting
   (b) for two hours at the 1.5 kW setting.
   Assume that a unit of electricity costs 10.2p.
3. The meter in the photo shows the reading on 31 July. The reading on 31 October is 09231. The first 250 kWh are charged at 16.5p and the remainder at 10.2p.
   Work out the cost of the electricity bill for this period.

C

4. A kettle has a power of 3 kW. It takes three minutes to boil when it is full. A unit of electricity costs 10.5p. What is the cost of boiling a full kettle?

A*

An electricity meter

This meter shows how much the electricity you are using costs

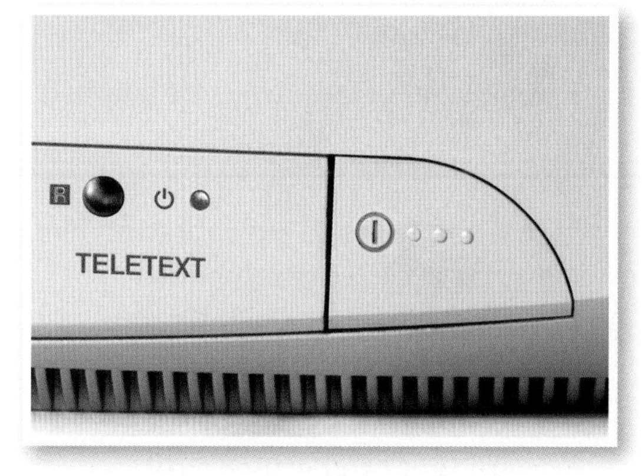

This TV is in standby mode

Did you know...?

When an electrical device is in standby mode, it still uses power. The only way to stop it using any power is to switch it off at the wall socket.

Course catch-up

Revision checklist

- The transfer of energy from a hot object to a cool object by electromagnetic waves is infrared radiation. Dark matt surfaces emit and absorb infrared radiation better than light shiny surfaces.
- The three states of matter, solids, liquids, and gases, are explained by the kinetic theory of particles.
- Metals conduct energy well due to free electrons. Some solids do not conduct energy very well and are thermal insulators.
- Liquids conduct energy, their particles are free to move. This is convection. Convection currents develop when denser cool fluid falls and less dense warmer fluid rises.
- Particles in liquids have different amounts of energy. Energetic particles escape from the surface. When these particles lose energy condensation occurs. Evaporation is increased by giving the liquid more energy.
- Energy transfers from a hot object to its surroundings at a greater rate if the temperature difference is greater. A larger surface area and volume increase the rate of transfer.
- Rate of energy transfer also depends on the material from which an object is made, and the nature of the surface with which it is in contact. A vacuum flask is designed to minimise energy transfer.
- Animals in hot countries are adapted to dissipate energy quickly, unlike those in cold countries.
- Buildings lose energy by conduction, convection and radiation. Insulation helps to prevent this. The rate of energy loss of a material is called its U-value.
- Solar panels provide warm water when infrared radiation from the sun is transferred to water in pipes. Solar panels and insulation keep heating costs down.
- The amount of energy required to change the temperature of one kilogram of the substance by one degree Celsius is its specific heat capacity (SHC). It is different for all substances.
- Water has a high SHC and is useful in central heating systems. Concrete blocks are also used, but have a lower SHC.
- Energy has many forms: kinetic, chemical, light, sound, elastic and gravitational potential, nuclear, and electrical. Energy is neither created nor destroyed.
- Calculating the efficiency of energy transfers within appliances can help to prevent energy being wasted as heat and sound.

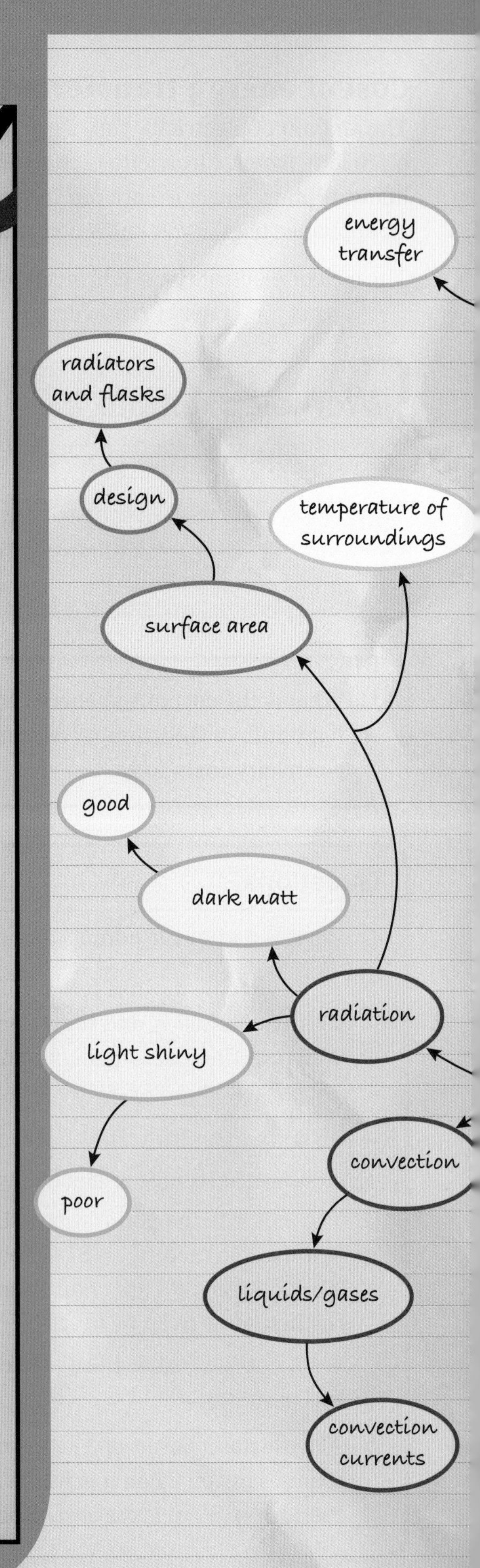

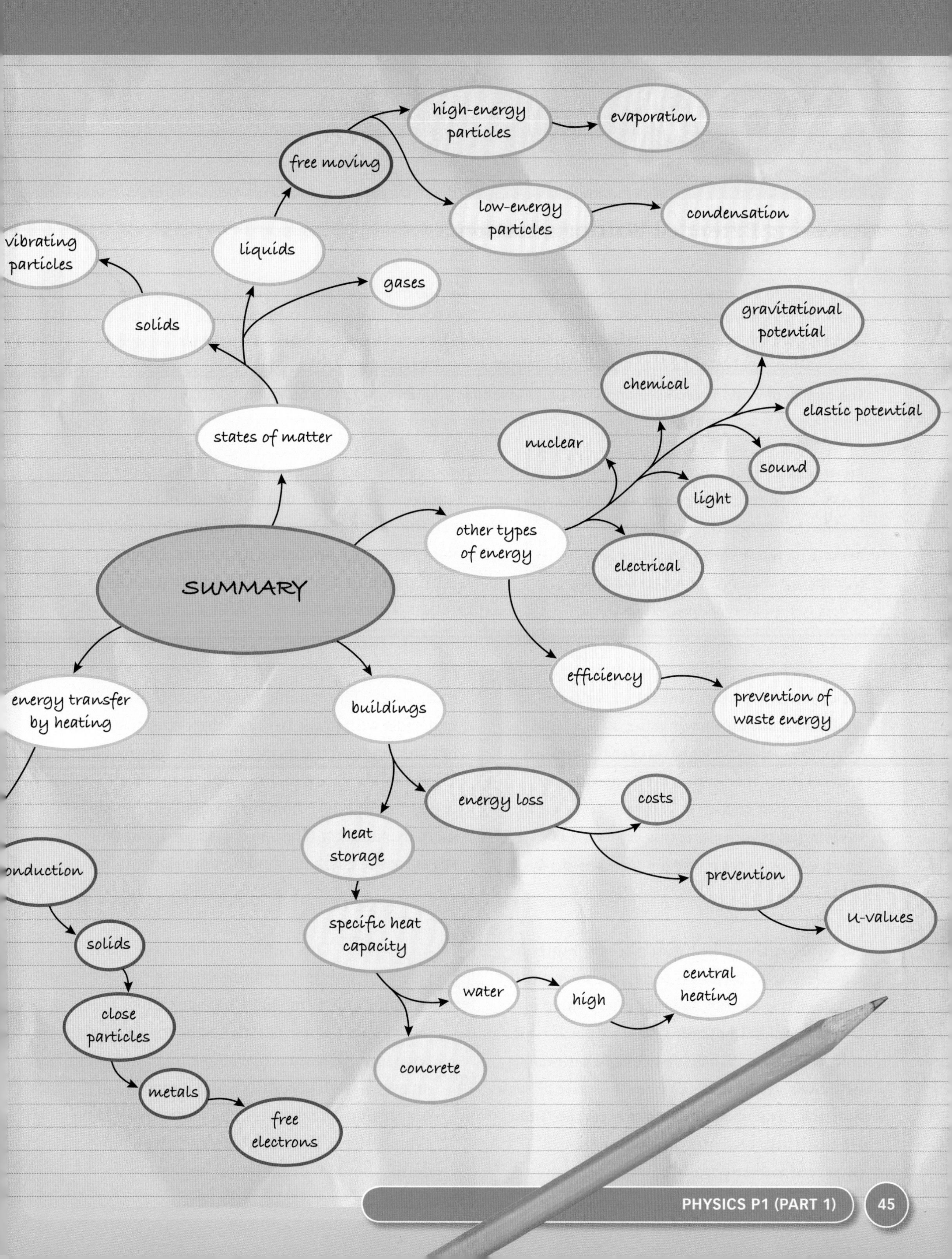
SUMMARY
states of matter
solids
vibrating particles
liquids
free moving
high-energy particles
evaporation
low-energy particles
condensation
gases
other types of energy
gravitational potential
chemical
elastic potential
nuclear
sound
light
electrical
efficiency
prevention of waste energy
energy transfer by heating
buildings
energy loss
costs
prevention
U-values
heat storage
specific heat capacity
water
high
central heating
concrete
onduction
solids
close particles
metals
free electrons

AQA Upgrade

Answering Extended Writing questions

QUESTION

Dave wants to insulate his loft. He looks up the U-values for 100 mm thickness of three possible insulation materials: sheep wool – 0.39 W/m²K; recycled plastic – 0.42 W/m²K; fibreglass – 0.43 W/m²K.
Outline how Dave can use the U-values to help him choose between the three insulation materials. As well as the U-values, what other factors might Dave consider in making his choice?

The quality of written communication will be assessed in your answer to this question.

G–E

Fibreglass is best because it has a hi U-value. It is cruel to use sheep wool for insulation. Recycled plastic is best becuase no much carbon diokside goes into the air when it is made, so making it does not make globel warming worse.

Examiner: The candidate makes a good point about the environmental benefits of using recycled plastic. However, the candidate is wrong to think that materials with higher U-values are better insulators. No credit is given for the comment about wool because it needs more explanation. There are several spelling errors.

D–C

The U-values are nearly the same, but sheep wool have a slightly lower value, so is a little better. Producing wool causes global warming, because sheep breathe out carbon dioxide, the plastic is good because it is recycled. You need to wear gloves with fibreglass because it has sharp bits in it, the others are safer to put in.

Examiner: The candidate understands that materials with lower U-values are better insulators, but has not used the terms 'insulator' or 'insulation'. The point about producing sheep wool is well explained, but the comment about plastic needs more explanation. The spelling is good, but there are errors in grammar and punctuation.

B–A*

The U-value shows how quickly each material transfers heat. Based on U-values, Dave should choose the sheep wool, because it's U-value is lowest. He should also consider payback time. This is the time it takes to get back the money he spends. So he needs to know the price of each material, and what savings to expect per year. He could also consider the environmental impact of making the insulation.

Examiner: This answer includes the main scientific points. It includes a clear explanation of the term 'payback time', and shows that the candidate knows that materials with lower U-values are better insulators. The answer is well organised, and includes only one grammatical error. The answer would be even better if the candidate had given more detail about environmental impacts.

Exam-style questions

1 A01 Match each description with the correct energy type below, and write each pair out in a list.

Energy type	Description
heat	stored inside fuel, food, and batteries
light	stored inside atoms
sound	energy stored inside a stretched rubber band
electrical	stored energy because of its height above the ground
nuclear	produced by visible electromagnetic radiation
chemical	energy due to movement of an object
kinetic	produced from a drum, guitar, speaker, etc
gravitational potential	energy transferred by electrons in a conductive wire
elastic potential	energy produced when particles vibrate rapidly

G–E

2 A03 Match up the sentences showing which method of energy transfer is prevented by which means in a vacuum flask.

Method of energy transfer	Means of preventing energy transfer
conduction	silvered glass walls
convection	tight fitting plastic screw cap
radiation	vacuum within the glass layers
	inner casing separated by air gap to outer casing

D–C

3 A01 Draw Sankey diagrams on graph paper using the data for the following machines which are doing work:

Machine	% efficiency	% wasted energy
car petrol engine	15	60 – heat 10 – sound 15 – moving parts
fossil fuel electrical power station	35	45 – heat 10 – moving parts 10 - sound
train diesel engine	36	45 – heat 10 – moving parts 9 – sound

B–A*

Extended Writing

4 A02 Isabelle and Reegan both noticed that after a hot summer's day the bricks of their houses gave off a lot of energy during the evening.

Why does this happen?

G–E

5 A03 Tom is trapped on a small desert island with no fresh water to drink. He remembers from his science lessons that he may be able to evaporate sea water to get pure drinking water. What could Tom do to ensure that he has a fresh supply of water?

D–C

6 A02 The diagram below shows a small beaker of water placed inside a larger beaker of water. The temperature of the water in beaker A is initially at 90 °C and the temperature of water in beaker B is at 19 °C. Explain what will happen in time.

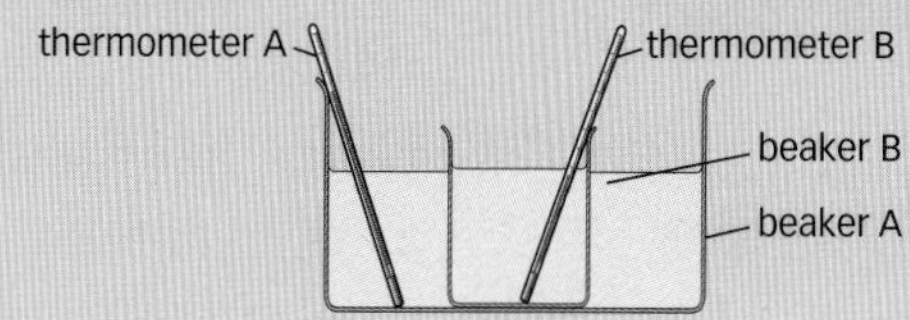

B–A*

A01 Recall the science

A02 Apply your knowledge

A03 Evaluate and analyse the evidence

P1 Part 2

Electrical energy and waves

AQA specification match		
P1.4	1.16	Generating electricity
P1.4	1.17	Fossil fuels and carbon capture
P1.4	1.18	Nuclear power
P1.4	1.19	Biomass and biofuels
P1.4	1.20	Solar and wind power
P1.4	1.21	Energy from water
P1.4	1.22	The National Grid
P1.4	1.23	Matching supply and demand
P1.5	1.24	Waves all around us
P1.5	1.25	Transverse and longitudinal waves
P1.5	1.26	Experiments with waves
P1.5	1.27	The electromagnetic spectrum
P1.5	1.28	Communications and optics
P1.5	1.29	Sound waves
P1.5	1.30	The Doppler effect
P1.5	1.31	Red-shift
P1.5	1.32	The Big Bang theory
Course catch-up		
AQA Upgrade		

Why study this unit?

Human activity is leading to changes in climate, and we are rapidly using up our natural resources. Scientists are working on solutions to these problems. We consume vast amounts of electricity every day, powering a range of appliances including our TVs and computers. In the future, how will we generate enough to meet our needs?

In this unit you will learn about how electricity is generated, and the advantages and disadvantages of the different technologies, from large coal-fired power stations to small solar cells on calculators.

You will also learn more about waves. How, when we talk, our mobile phones convert sound waves into microwaves before beaming them at high speed to the nearest mobile phone mast. Finally, you will learn about how electromagnetic waves provide the evidence for the Big Bang theory. This is one of the most important of scientific ideas; it describes how everything around us, the entire Universe, was formed.

You should remember

1. All human activity has an impact on the environment.
2. The conservation of energy states that energy cannot be created or destroyed.
3. Electricity is generated in different types of power station.
4. Waves, like light and sound, transfer energy from one place to another.
5. Light and sound can be reflected off objects, and refracted when they travel from one material to another.

The world's largest power station is the Three Gorges Dam on the Yangtze River in China. The project has been in development since 1994. When it reaches full capacity it is expected to be able to produce as much as 22 500 000 000 watts of power (22.5 gigawatts, GW). That's enough electricity for every person in the UK and Australia to watch their own large plasma TV at the same time.

The world's smallest 'power station' is a phytoplankton – a single-celled aquatic organism that converts sunlight into chemical energy to create living biomass.

16: Generating electricity

Learning objectives

After studying this topic, you should be able to:

- explain that, in some power stations, fuel is used to heat water to produce steam
- describe how steam drives a turbine connected to a generator
- describe how naturally occurring steam can be used to drive turbines

Generating electricity

Electricity is generated using sources of energy. A great deal of the electricity in the UK is generated by using an energy source to heat water. The energy source can be **fossil fuels,** biomass or the Sun. This type of power station is called a **thermal power station**.

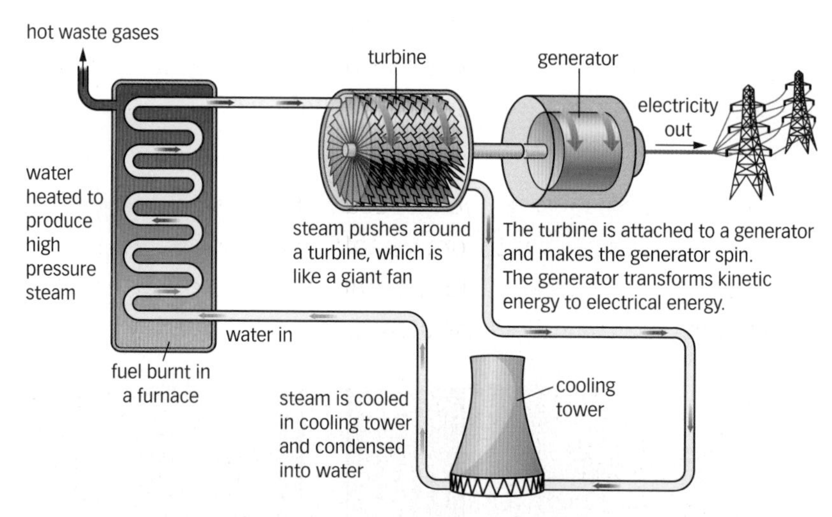

A conventional power station where an energy source is used to heat water

A cooling tower at a power station

A turbine in a power station being repaired. The turbine is about 5 m across.

Did you know...?

The shape of cooling towers helps to set up a natural convection current. Cool air is drawn in at the base and flows over radiators. Heat from the steam is transferred to the air.

A Why is electricity not a source of energy?

Geothermal power

In some parts of the world, such as Iceland, where there are volcanic areas, **geothermal** energy can be used to **generate** electricity.

Steam rises to the surface of the Earth, or is not far below it. The steam can be collected and piped to a power station to drive the **turbines** directly.

Geothermal energy does not have any fuel costs, but money is spent on building the power station and maintaining it. These power stations can also be started up and stopped relatively easily.

Geothermal energy can also be used to heat houses. In Iceland, the waste steam from a geothermal power station is piped to houses to heat them. Boreholes can also be drilled into the ground to collect the steam.

▲ In New Zealand, boreholes are drilled into the ground to collect the steam

B What is geothermal energy?

Key words

fossil fuel, thermal power station, geothermal power, generator, turbine

Exam tip 

- ✔ Remember that electricity is not a source of energy, but is a form of energy.

Questions

1. What are the differences between a conventional power station and a geothermal power station? (E)
2. What does the turbine do?
3. What does the generator do?
4. What is the function of the cooling towers? (C)
5. Draw a flow diagram to show all of the energy transfers taking place in a conventional power station. (A*)

17: Fossil fuels and carbon capture

Learning objectives

After studying this topic, you should be able to:

- ✔ list some of the advantages and disadvantages of using different fossil fuels to generate electricity
- ✔ describe the potential benefits of carbon capture technology

▲ Coal is a fossil fuel

Did you know...?

Fossil fuels are a concentrated store of chemical energy. One kilogram of coal stores a lot more energy than one kilogram of wood. So when the fuels are burnt in a power plant they can generate very large amounts of electricity. In Yorkshire, the Drax coal-fired power station produces 7% of the UK's electricity needs, supplying over 2 million homes. You would need around 4000 of the largest wind turbines to generate the same amount of electricity. (There are only around 3000 wind turbines currently in use in the UK.)

Different fossil fuels

Coal, crude oil and natural gas are examples of **non-renewable** energy resources. All three are formed in similar ways. They are the remains of living organisms which died millions of years ago; this is why they are called fossil fuels. Their remains have been squashed and heated in layers of the Earth's crust. This takes a long time, and we are using up fossil fuels far more quickly than they are being formed. They will eventually run out.

All three fuels must be mined and transported, and this can be bad for the environment. However, using fossil fuels for generating electricity has some advantages. The power plants use well established and reliable technology and produce large amounts of electricity. In 2009, around 77% of the UK's electricity came from fossil fuels (mainly coal and natural gas).

A Give two advantages of using fossil fuels to generate electricity.

In every fossil fuel power station, fuel is burnt and the heat produced turns water to steam. This steam turns a turbine. (Sometimes air is heated directly in gas-fired stations.)

Burning the fuel releases pollutants into the atmosphere.

▲ Burning any fossil fuel releases carbon dioxide into the atmosphere

One of the gases released is carbon dioxide. This contributes to global warming. The different fuels produce different amounts of carbon dioxide, as shown in the table below.

The table shows other differences too. For example, gas power plants have a short start up time. They can very quickly increase the amount of electricity they generate, to cope with sudden increases in demand.

Fossil fuel	Substance heated	Carbon dioxide production	Start up time	% UK electricity supply
coal	water	very high	long	31
oil	water	high	long	1
natural gas	water or air	medium	short	46

Carbon capture

One idea for reducing the amount of carbon dioxide that burning fossil fuels releases into the atmosphere is **carbon capture**. The carbon dioxide is trapped and stored before it enters the atmosphere. One suggestion is that carbon dioxide should be stored in old oil and gas fields such as those found under the North Sea.

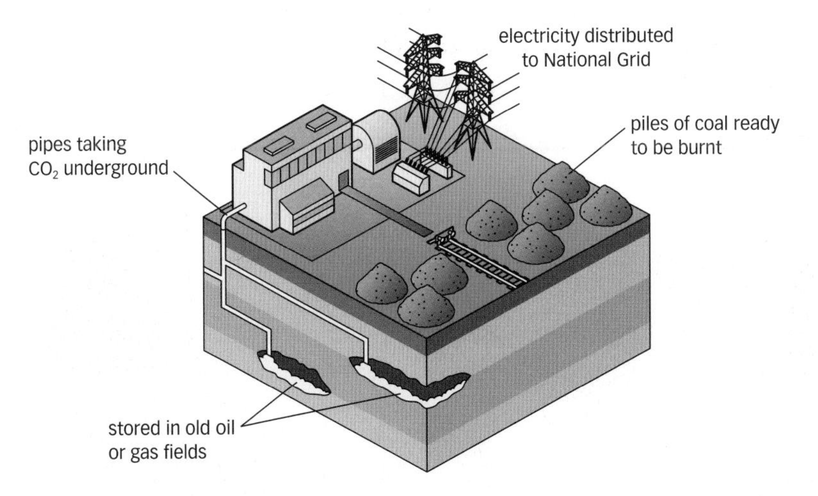

▲ Carbon capture will reduce the amount of carbon dioxide that enters the atmosphere

Some scientists have calculated that carbon capture could reduce the amount of carbon dioxide entering the atmosphere from a coal power plant by over 80%. However, this is a rapidly evolving technology and there are currently no carbon capture power stations in the UK.

Key words

non-renewable, carbon capture

B Which fossil fuel produces the least amount of carbon dioxide when burnt in a power plant?

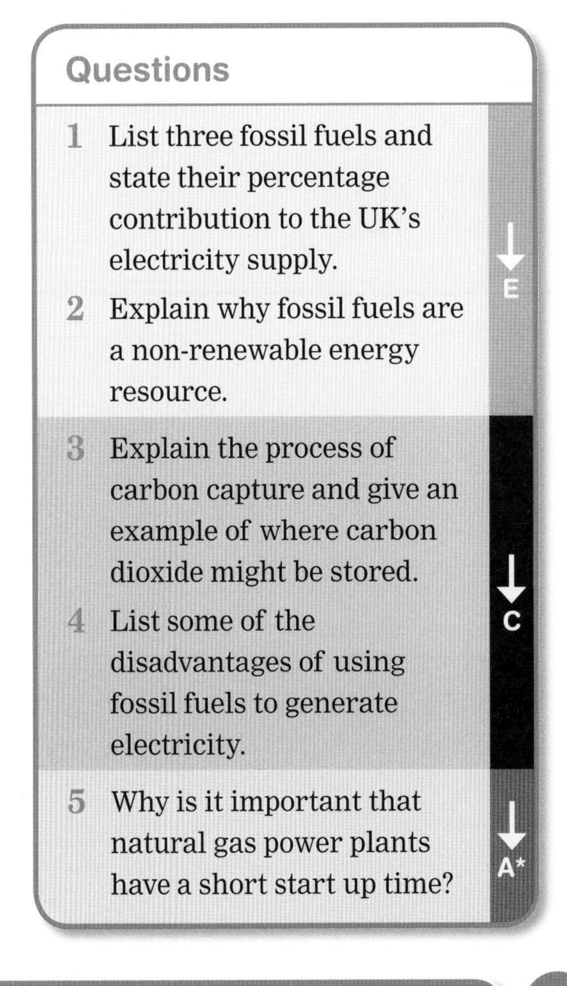

Questions

1. List three fossil fuels and state their percentage contribution to the UK's electricity supply.
2. Explain why fossil fuels are a non-renewable energy resource.

↓ E

3. Explain the process of carbon capture and give an example of where carbon dioxide might be stored.
4. List some of the disadvantages of using fossil fuels to generate electricity.

↓ C

5. Why is it important that natural gas power plants have a short start up time?

↓ A*

18: Nuclear power

Learning objectives

After studying this topic, you should be able to:

- describe how a nuclear reactor works
- outline some of the benefits and hazards of using nuclear power to generate electricity

Key words

nuclear reactor, nuclear fission, decommissioning, radioactive waste

A Name two fuels used inside a nuclear reactor.

Did you know...?

To produce the same amount of electricity you get from 1 kg of uranium, you would need over 15 000 kg of coal!

Inside a nuclear reactor

Nuclear power can be used to generate electricity. Inside a **nuclear reactor,** atoms of uranium or plutonium undergo **nuclear fission**. This releases a huge amount of energy in the form of heat. The heat is used to turn water into steam that drives turbines, as in thermal power plants. No burning is involved, so there is no release of carbon dioxide.

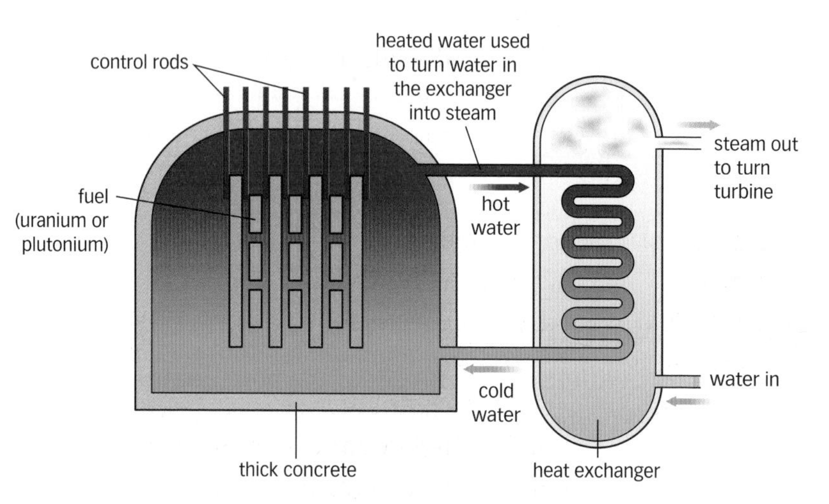

Controlled nuclear fission inside the nuclear reactor releases a huge amount of energy

Uranium is made into pellets that are inserted into a nuclear reactor

Arguments for and against nuclear power

Nuclear power is a controversial method of generating electricity. Currently around 13% of the electricity generated in the UK comes from nuclear power and there are plans to build several more nuclear reactors.

Nuclear power has some advantages when compared with other methods:

- Huge amounts of electricity can be generated for each kilogram of fuel used.
- No carbon dioxide is produced, so there is no contribution to global warming.
- The fuel is readily available and won't run out for thousands of years.

There are also disadvantages:

- Nuclear reactors produce highly radioactive nuclear waste. This remains dangerous for millions of years, and so has to be buried deep underground.
- The costs of building the plant and of taking it down when it has finished (called **decommissioning**) can be quite high. This means the electricity generated can be relatively expensive.
- Nuclear reactors have a very slow start up time. It takes a long time to increase or decrease the amount of electricity they are generating.
- There is always the risk of an accident that could release **radioactive waste** into the environment.

▲ Highly radioactive waste is stored in a pond until it can be processed

B How much of the UK's current electricity supply is generated by nuclear power?

Questions

1. List the advantages and disadvantages of using nuclear power. (E)
2. Describe how nuclear reactors can be used to generate electricity.
3. State the energy changes inside a nuclear power plant (ending with electrical energy from the generator). (C)
4. Is nuclear power a renewable or non-renewable energy resource? Explain your answer. (A*)

Exam tip

✔ Remember, one advantage of nuclear power is that a very large amount of electricity is generated per kilogram of fuel. It is not enough just to say that a lot of electricity can be generated, as that is also true for fossil fuels.

19: Biomass and biofuels

Learning objectives

After studying this topic, you should be able to:

- explain the meaning of the terms biomass and biofuel
- describe how biofuels may be used to generate electricity
- explain why biomass is considered to be carbon-neutral

Key words

biomass, renewable, biofuel

Biomass and biofuels

Biomass is a **renewable** energy resource. There are lots of different types, but all forms of biomass involve material produced by living organisms. Biomass used for burning is called a **biofuel**. These are similar to fossil fuels, but with fossil fuels the living organisms died millions of years ago.

▲ This biomass is ready for burning

Biofuels can be solids, liquids or gases. Some examples are:

- Wood and woodchips, from specially grown trees (new trees are planted when old ones are cut down).
- Alcohol fuels (such as ethanol), produced by fermenting sugar cane crops.
- Methane gas, given off by animal waste in storage tanks called sludge digesters, and also from other rotting waste (for example food waste from homes).
- Nutshells, that are a waste product from manufacturing cattle feed or other food.
- Vegetable oils, that can also be used to make biodiesel (for example, oil from oilseed rape crops). Biodiesel can also fuel cars, buses and even trains.
- Other crops, such as straw.

A Give three examples of different biofuels.

B Give an example of a use for biodiesel other than for generating electricity.

In the UK, biomass is used to generate more electricity than any other form of renewable energy. Over 40% of the electricity from renewable sources comes from biomass. The table below shows its advantages and disadvantages compared with other methods of generating electricity.

Advantages	Disadvantages
Uses products which might otherwise be wasted, so the fuel costs are very low	Releases atmospheric pollutants
Power stations can also supply hot water to local industry/ houses	In developing countries, land which could be used for food is now used to grow crops for biofuels, leading to food shortages
Carbon neutral	Power plants can be ugly to look at (visual pollution)

Large areas of forest are being cut down to provide land to grow biofuels

Carbon neutral

In a biomass power plant the biofuel is burnt and electricity is generated in a similar way to that in other thermal power stations (water to steam, etc). This releases carbon dioxide into the atmosphere. However, unlike fossil fuels, there is no overall increase in carbon dioxide as the amount released is the same as the plant absorbed while it was alive (as part of photosynthesis). This means biomass is considered to be carbon neutral.

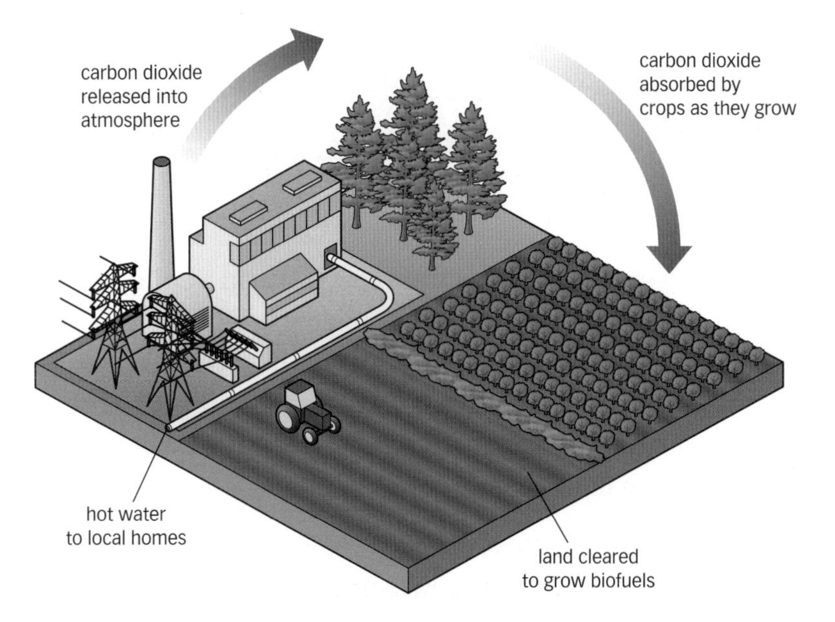

Using biomass is carbon neutral because there is no overall increase in carbon dioxide

Did you know...?

A large chicken and turkey farm in Norfolk uses bird waste to generate all its electricity. In fact, the birds produce so much of it that there is enough electricity left over to power another 5000 homes in the surrounding area. That's a lot of chicken poo!

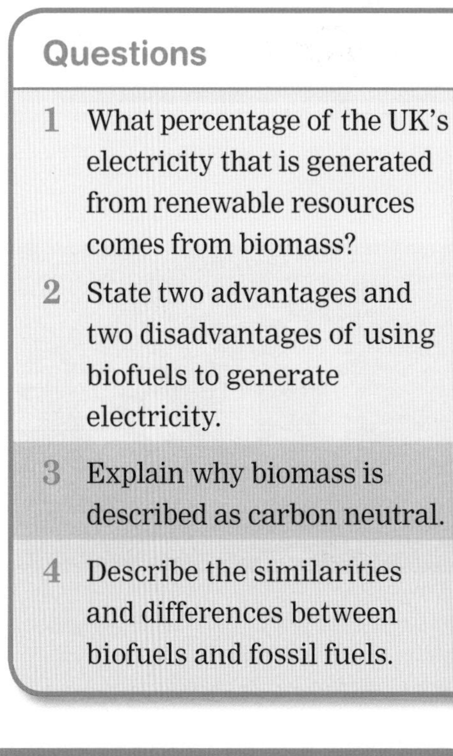

Questions

1. What percentage of the UK's electricity that is generated from renewable resources comes from biomass?
2. State two advantages and two disadvantages of using biofuels to generate electricity.

E

3. Explain why biomass is described as carbon neutral.

C

4. Describe the similarities and differences between biofuels and fossil fuels.

A*

20: Solar and wind power

Learning objectives

After studying this topic, you should be able to:

- describe how electricity can be generated using solar power
- describe how electricity can be generated using wind turbines

Energy from the Sun

Energy from the Sun can be used to generate electricity as well as heating water. **Solar cells** generate direct current when light energy is absorbed by them. They transfer light energy into electrical energy and produce direct current. The amount of electricity produced depends on the area of solar cell that has light shining on it.

Solar cells do not need much maintenance and there is no need for fuel. There is no need for power cables, so they can be used in remote locations. They do not produce any waste. They are a renewable energy resource. The disadvantages of solar cells are that a large area is needed to generate the same amount of electricity as from a thermal power station, and that less electricity is generated on cloudy days or at night.

▲ The mirrors reflect sunlight to the top of the solar thermal tower

▲ A solar cell power station

Another way of using energy from the Sun is in a **solar thermal tower**. A large number of mirrors are used to reflect light from the Sun into one spot at the top of a tower. The temperature can reach 500 °C. The energy is used to heat water into steam. The steam is then used to drive turbines in the same way as in a thermal power station.

▲ This solar stove works in the same way as a solar thermal tower. Light from the Sun is reflected on to the cooking pot.

A What energy transfer takes place in a solar cell?

Energy from the wind

Winds are convection currents set up in the Earth's atmosphere by energy from the Sun. The kinetic energy of the wind can be used to drive **wind turbines** directly.

Wind turbines are a renewable energy resource. They do not produce any waste, but many people think that they spoil the landscape, and they are also noisy.

Just like solar cells, they can be used in remote locations, but depend upon the wind blowing. They must also be built in open windy areas. A wind farm that could generate as much electricity as a thermal power station would take up a large area.

> **B** What energy transfer takes place in a wind turbine?

How a solar cell produces electricity

When infrared radiation is absorbed by a solar cell, electrons are knocked out of the atoms in the solar cell. These electrons are then free to flow as a direct current.

The amount of electricity produced by a solar cell also depends on the **intensity** of the light shining on it.

Questions

1. What is a renewable energy resource? (E)
2. What are the differences between a solar thermal tower and a thermal power station?
3. Draw up a table to summarise the advantages and disadvantages of each type of renewable energy described on these pages. (C)
4. Solar cells are used to provide electricity for garden lights. What is needed so that the lights can come on at night? (A*)

▲ Wind turbines

Key words

solar cell, solar thermal tower, wind turbine, intensity

Exam tip

- ✔ Remember the difference between a solar panel (spread P1.9) and a solar cell. A solar panel transfers energy to water, and a solar cell generates electricity directly.
- ✔ A solar cell cannot generate electricity at night.

21: Energy from water

Learning objectives

After studying this topic, you should be able to:

- explain how energy from waves, tides, and falling water can be used to drive turbines directly

Tidal stream turbine in Strangford Lough, Northern Ireland

A How can electricity be generated from tides?

Generating electricity from waves on Islay, Scotland

Tidal and wave power

Tides cause the sea level to move up and down. When the tide is high, the water can be held back by a **barrage** across a river estuary. After a few hours, the sea level on the other side of the barrage has fallen. Water is allowed to flow through turbines in the barrage to generate electricity.

Tidal barrages are a renewable energy resource – no fuel is needed. However, there are not many places where they can be built, and they can only generate power for 6–8 hours out of every day. Also, the time of high tide is not the same every day – power will be generated in the middle of the night.

The Rance tidal barrage in Brittany, France

Tides also cause tidal streams where seawater is moving from one place to another. In some places the tidal streams are very strong. Moving seawater has kinetic energy and it can be used to drive turbines directly. A **tidal stream turbine** in Strangford Lough in Northern Ireland generates enough electricity to supply around 2500 homes (1.2 MW). It works in the same way as a wind turbine. Tidal streams can be used to generate electricity for 18–20 hours every day. There are plans to install many more tidal stream turbines off the coast of Anglesey in North Wales.

Waves can be used to generate electricity, for example by forcing air up and down a tube. The moving air drives a turbine. The picture on the left shows a wave generator installation on the island of Islay in Scotland.

Hydroelectric power

A **hydroelectric** power station also transforms the kinetic energy in moving water into electrical energy. A dam is built in a hilly area to store water in a reservoir. Water then flows downhill in pipes to the power station. The flowing water drives turbines to generate electricity.

Hydroelectric power stations are a renewable energy resource. They do not need fuel. They can be started and stopped very quickly. However, a reservoir must be built, which can change the environment.

Small hydroelectric power stations can be used in remote areas with high rainfall.

▲ Turbine from a hydroelectric power station

Key words

barrage, tidal stream turbine, hydroelectric

Exam tip

✔ Remember the advantages and disadvantages of all the methods of generating electricity – you may need to know them in the exam.

B What is hydroelectric power?

Questions

1 Is fuel needed to generate electricity from moving water? (E)

2 Describe all the energy transfers that take place in a hydroelectric power station.

3 Draw up a table to summarise the advantages and disadvantages of all the renewable energy resources shown on these two pages.

4 How are all these methods of generating electricity similar to what happens in a thermal power station? (C)

5 Ellen says 'waves can always be used to generate electricity'. Is she correct? Explain your answer. (A*)

22: The National Grid

Learning objectives

After studying this topic, you should be able to:

- explain how electricity is transferred to consumers
- explain how transformers are used in the National Grid to minimise losses when energy is transmitted

An electricity substation

Distributing electricity

When electricity has been generated it needs to be distributed to consumers in homes, shops and offices. The electricity you use is transferred around the country by the **National Grid**. This is the **mains supply** that is available whenever you switch on an appliance at home.

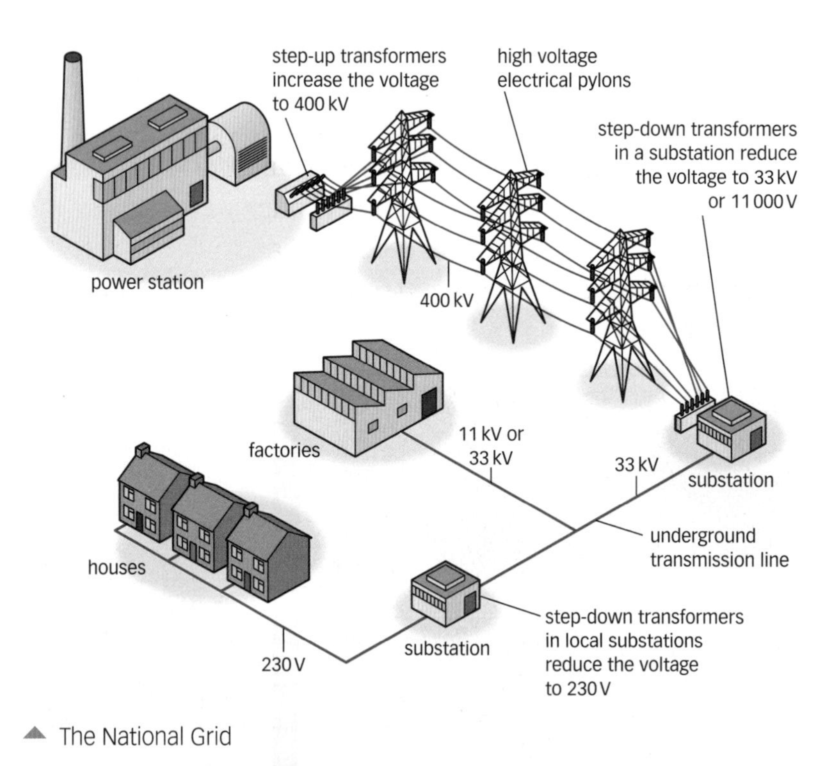

The National Grid

Transformers

The electricity goes from the power station to a **step-up transformer** where the **voltage** is increased. Electricity is transferred from the transformer by high-voltage transmission lines to an electricity **substation**. Here, the electricity goes through a **step-down transformer** to reduce the voltage to a safer level for use by consumers.

A step-up transformer increases the voltage but reduces the current. This minimises energy loss in transmission.

A What is the difference between a step-up and a step-down transformer?

B What happens in a substation?

Power lines

Large amounts of energy are transferred by the National Grid. The higher the current in the transmission cables, the greater the energy losses during transmission.

When a current flows through a wire, there is a heating effect and this dissipates some energy. The larger the current, the more heat is produced. This means that less energy is transferred when the current is higher.

Some power lines are carried overhead. Others are buried underground, so they cannot be seen and there is not the danger of tall vehicles or other objects tangling with them. However, it is much more expensive to bury power lines underground. They have to be insulated and waterproofed. Also, if they have to be repaired or renewed, the ground must be dug up.

▲ Repairing a power line

Key words

National Grid, **mains supply**, **step-up transformer**, **voltage**, **substation**, **step-down transformer**

Exam tip

- ✔ Remember all the parts of the National Grid – you may need to name them in the exam.

Questions

1. What is the National Grid? (E)
2. Explain how electricity is transferred from a power station to your home through the National Grid.
3. (a) What are the advantages of putting cables underground to distribute electricity?
   (b) What are the disadvantages?
4. Why is electricity stepped down before it enters the home? (C)
5. Why is electricity stepped up before being distributed by the National Grid? (A*)

23: Matching supply and demand

Learning objectives

After studying this topic, you should be able to:

- ✔ evaluate ways of matching supply with demand

Key words

pickup, **start up time**, **pumped storage**

Demand for electricity

The demand for electricity varies according to the time of year and the time of day. The demand can also change from minute to minute, depending on what is happening. If many people are watching a TV programme, the demand for electricity falls. When the programme ends, demand for electricity can suddenly increase because people turn on kettles or lights. This is called a **pickup**.

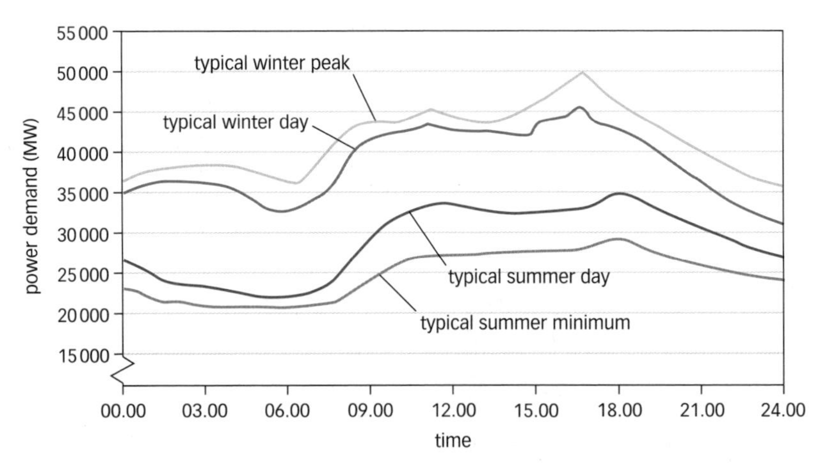

▲ Typical electricity demand for different days in the year

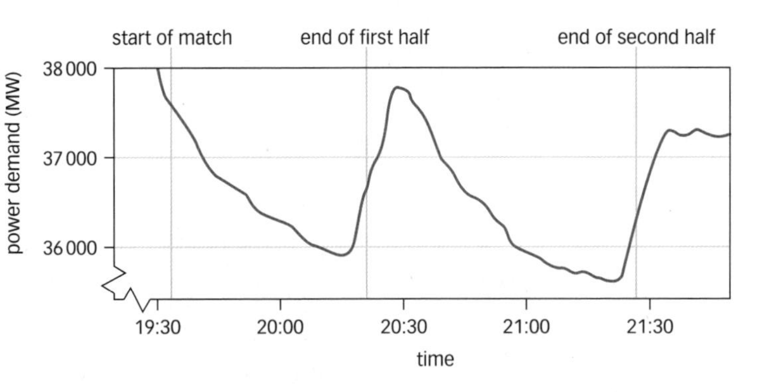

▲ How the electricity demand can vary when a football match is shown on TV

Did you know...?

The biggest TV pickup in the UK was in 1990 when England played Germany in the football World Cup semi-finals. The game went to a penalty shoot-out. When the programme finished, demand increased by 2800 MW.

It is wasteful to keep generating electricity at a high level. So power engineers stop some power stations from generating electricity when there is a low demand. They start them up again when there is a high demand.

Power engineers have to forecast when there is likely to be a high demand, so that there is the right amount of power available. They look at weather forecasts and TV schedules.

A What is a pickup in electricity demand?

B Why are hydroelectric power stations used to generate electricity during a pickup?

They keep some power stations generating electricity, even though not all of the electricity is needed. But they can also use hydroelectric power stations, because they can be switched on very quickly. It only takes a couple of minutes to start up a hydroelectric power station as they have a very short **start up time**.

It can take hours to start and stop fossil fuel power stations, so they are often left running. The extra electricity they generate is used in **pumped storage** hydroelectric power stations. At times of low demand, energy is stored at a hydroelectric station by pumping water from a lower reservoir to a higher one as shown in the diagram.

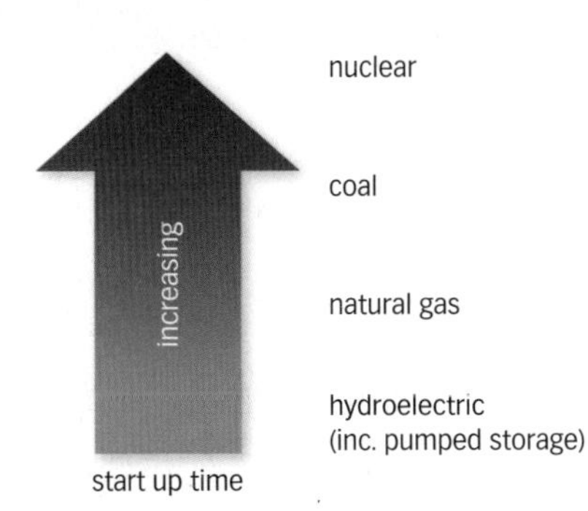

▲ Start up times vary for different power stations

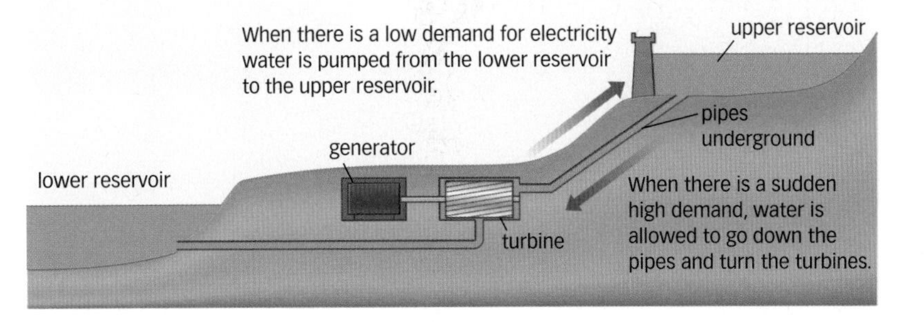

▲ A pumped storage power station

▲ Dinorwig in North Wales is an example of a pumped storage station found in the UK

Questions

1 What factors do power engineers have to think about when forecasting what the demand for electricity might be? (E)

2 More people are using air-conditioning. How do you think this will affect the electricity demand curves?

3 Draw a flow chart to show the energy transfers happening in a pumped storage power station when:
(a) it is pumping water from the lower reservoir to the higher reservoir
(b) it is generating electricity. (C)

4 When do pumped storage power stations pump water to the higher reservoir?

5 What events other than a football match could generate a big pickup in the demand for electricity? (A*)

Exam tip

✔ Remember that a pumped storage power station is a hydroelectric power station where water can also be pumped back up to the higher reservoir.

24: Waves all around us

Learning objectives

After studying this topic, you should be able to:

- describe how waves transfer energy from one place to another
- explain the terms frequency, wavelength and amplitude
- use the equation $v = f\lambda$

Key words

oscillation, vibration, energy, amplitude, wavelength, frequency, wave equation

What are waves?

Waves are around us all of the time. We see water waves on the surface of the sea, sound waves allow us to listen to our music – and just think what the world would be like without light.

Surfing some water waves

All waves are a series of **oscillations** (or **vibrations**) which travel from one place to another. Water waves make water molecules move up and down, and sound waves make air particles vibrate from side to side. In all cases the waves transfer **energy** from one place to another.

Energy is transferred from the speaker to your ears

A What do waves transfer from one place to another?

Did you know...?

Some waves can have very high frequencies measured in kHz (kilohertz; 1 kHz = 1000 Hz) and MHz (megahertz; 1 MHz = 1 000 000 Hz). An FM radio station might transmit waves with a frequency of 97.7 MHz – that's 97.7 million waves every second!

All waves have some key features.

Amplitude in metres (m)	Maximum height of the wave measured from the middle.
Wavelength in metres (m)	Shortest distance between a point on a wave and the same point on the next wave. For example, the distance from one peak to the next peak.
Frequency in hertz (Hz)	The number of waves passing a point per second. A frequency of 6 Hz would mean six waves pass a point every second.

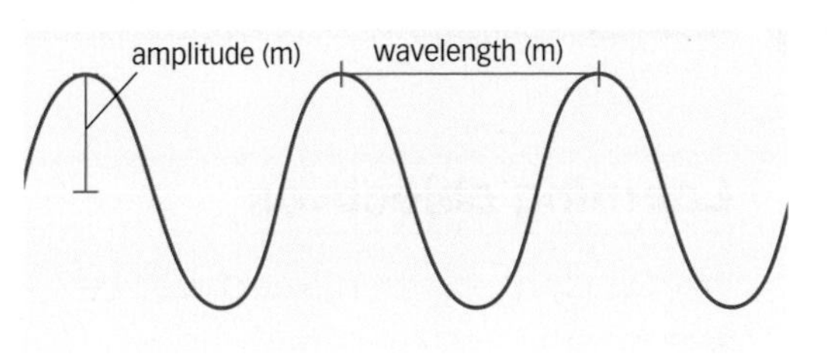

▲ Key features of a wave

The wave equation

The speed of a wave depends on its frequency and its wavelength. They are related in the **wave equation**:

$$\text{wave speed (metres/second, m/s)} = \text{frequency (hertz, Hz)} \times \text{wavelength (metres, m)}$$

If the wave speed is called v, the frequency f and the wavelength λ (a greek letter, pronounced lam-da), then $v = f\lambda$

Worked example 1

Dylan is standing on the end of a pier. He measures the water waves going past him. The wavelength of each wave is 1.3 m. He counts 2 waves every second. Find the wave speed.

wave speed = frequency × wavelength

$f = 2$ Hz (as there are two waves every second), and $\lambda = 1.3$ m, so

$v = 2 \times 1.3$

$v = 2.6$ m/s

Worked example 2

A flute produces a note with a wavelength of 75 cm. The speed of sound is 330 m/s. Find the frequency of the note.

wave speed = frequency × wavelength, to:

$$\text{frequency} = \frac{\text{wave speed}}{\text{wavelength}} \text{ or } f = \frac{v}{\lambda}$$

$f = \frac{330}{0.75}$

$f = 440$ Hz

Exam tip

✔ Remember to look carefully at the units when using the wave equation. Pay particular attention to wavelength. This must be measured in metres.

Questions

1. Define the terms wavelength, frequency and amplitude. (E)
2. Draw a wave with an amplitude of 3 cm and a wavelength of 8 cm.
3. Find the speed of a wave with a wavelength of 30 m and a frequency of 120 Hz. (C)
4. A speaker produces a sound at a frequency of 6.6 kHz. The wavelength of the sound wave is 5.0 cm. Use these values to show that the speed of sound is 330 m/s. (A*)
5. A radio station transmits waves with a frequency of 120 MHz. The radio waves travel at 3×10^8 m/s. Find the wavelength of the radio wave.

25: Transverse and longitudinal waves

Learning objectives

After studying this topic, you should be able to:

- describe the differences between transverse and longitudinal waves
- give several examples of transverse and longitudinal waves

Key words

transverse wave, longitudinal wave, compression, rarefaction

A Name the two types of wave.

B Give two examples of transverse waves.

Different types of wave

There are two different types of wave, **transverse waves** and **longitudinal waves**. Both types are made up of oscillations or vibrations, and they both transfer energy from one place to another. The oscillations which make up the waves are slightly different.

Transverse waves

If you were asked to sketch a wave, you would probably draw a transverse wave. They look like ripples on a pond, with peaks and troughs. In a transverse wave the oscillations are perpendicular (at right angles) to the direction of energy transfer.

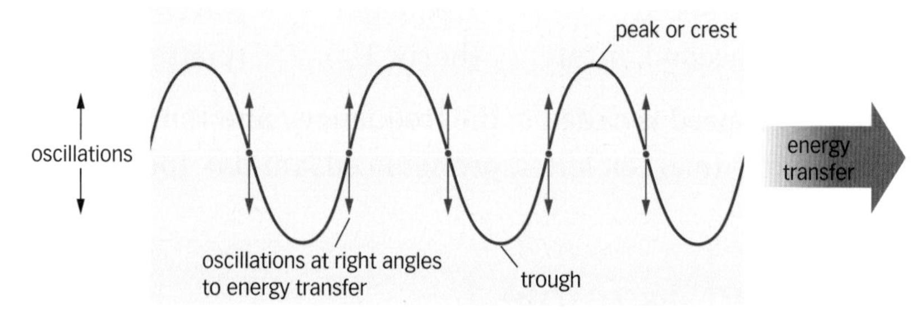

A transverse wave

Examples of transverse waves include:

- water waves
- light
- microwaves
- X-rays
- mechanical waves on strings (such as a plucked guitar string)
- some mechanical waves on springs.

Ripples on the surface of a pond are transverse waves

Longitudinal waves

In longitudinal waves the oscillations are to and fro along the same direction as the energy transfer. They are parallel to the direction of energy transfer.

Sound is an example of a longitudinal wave. If you look closely at a speaker, you can see the speaker cone moving in and out. This creates oscillations which travel through air in the same direction as the speaker movement. When the speaker moves out, it creates a **compression** as the air is bunched up. When it moves back in, it creates a **rarefaction**, where the air is more spread out. It is these compressions and rarefactions which make up a longitudinal wave.

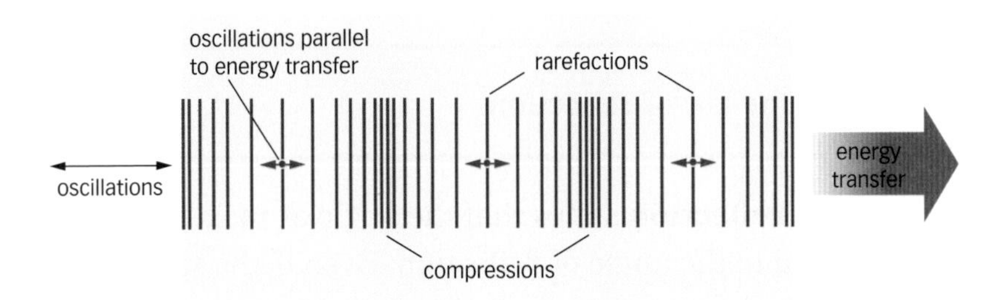

▲ A longitudinal wave

Other examples of longitudinal waves include:

- one type of seismic wave (in earthquakes)
- some mechanical waves on springs.

Questions

1 Sketch a diagram showing a transverse wave and one showing a longitudinal wave. Label the key features.
2 Give two examples of longitudinal waves.
3 Describe the differences between transverse and longitudinal waves.
4 Describe how a speaker produces compressions and rarefactions.
5 Describe how a spring could be used to demonstrate both transverse and longitudinal waves.

E C A*

Did you know...?

An earthquake produces both transverse (S) and longitudinal (P) waves. These travel at high speed through the Earth. Scientists have carefully recorded when these waves reach the surface. They have used this information to learn more about the structure of the Earth. Thanks to the differences between these types of wave we now know that the Earth has a solid iron inner core around 2000 km across, surrounded by a layer of molten iron 1000 km deep.

▲ A spring can be used to show a longitudinal wave

Exam tip

✔ When describing a transverse wave, you must say that the vibration is perpendicular, not that the wave vibrates 'up and down' – the movement could be side to side.

26: Experiments with waves

Learning objectives

After studying this topic, you should be able to:

- describe how waves can be reflected, refracted and diffracted
- use a normal line when drawing ray diagrams

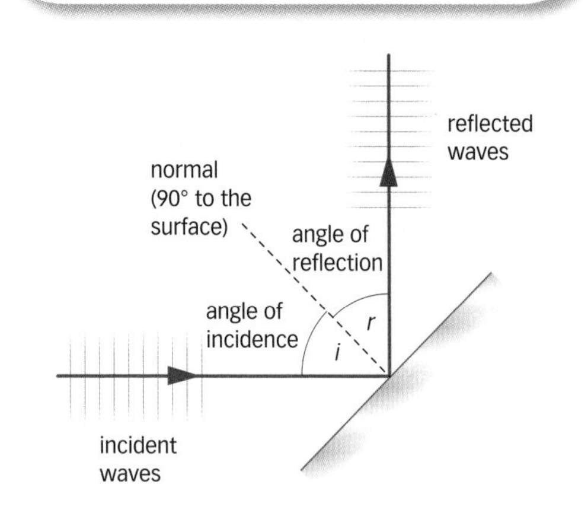

Reflection of a wave

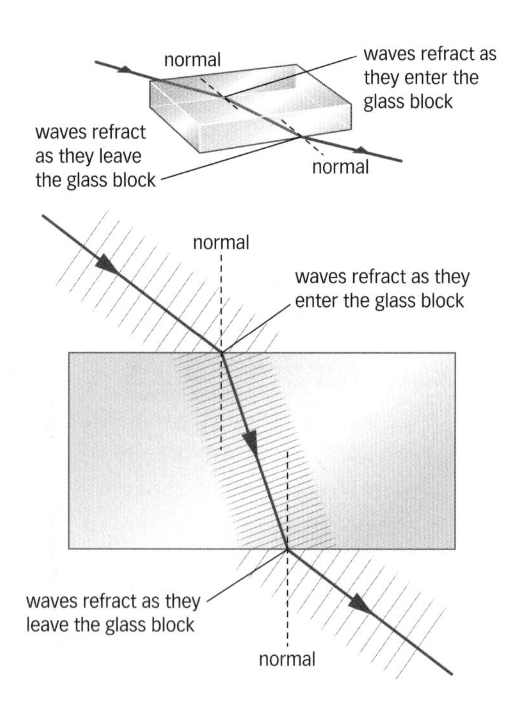

Light refracting as it passes through a glass block

Reflection

When you look in a mirror you see your reflection. But this doesn't only happen with light waves – all waves can be **reflected**. An echo is a reflection of sound waves, and radio waves are often reflected off buildings and interfere with your TV signal.

We can draw simple ray diagrams to show reflection. The rays show the direction of the energy transfer by the waves.

Whenever we draw ray diagrams we must include a **normal** line. This is a line at 90° to the surface. We always measure angles to the normal.

A State the law of reflection.

The **law of reflection** states that the angle of incidence is always equal to the angle of reflection. Even if the surface is really rough, the two angles are always the same.

Refraction

A material that a wave travels through is called a **medium**. When sound travels through air, the medium is air. When waves go from one medium to another they can be **refracted**. As they enter a different medium their speed changes and this causes them to change their direction. If the wave slows down it bends towards the normal, if it speeds up it bends away from the normal.

If the waves travel along the normal then, although their speed changes, they don't change their direction.

B Explain what is meant by the medium that waves travel through.

C Describe what happens to a ray of light when it enters a glass block.

Diffraction

Whenever waves pass through a gap or move around an obstacle, they spread out. This is called **diffraction**, and it happens with both transverse and longitudinal waves.

You might have noticed the effect with sound. If you have your door open and someone is talking outside, even though you can't see them, you can hear their voice. The sound waves diffract when they go through your open doorway, spreading out and filling the room. It might sound as though the sound wave actually comes from the doorway itself.

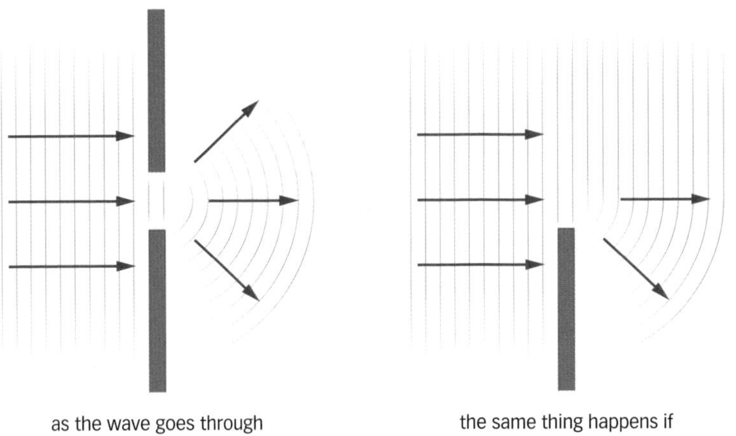

▲ Diffraction through a gap or around an obstacle

Like all waves, the radio waves used for transmitting radio and television signals diffract when passing over obstacles. This allows signals to reach the bottom of some valleys.

The size of the gap and wavelength of the wave affect how much diffraction takes place. The longer the wavelength, the greater the diffraction. The greatest diffraction takes place when the wavelength is around the same size as the gap.

▲ Water waves are diffracted when they enter a harbour. Radio waves may be diffracted by a hill or other obstacle.

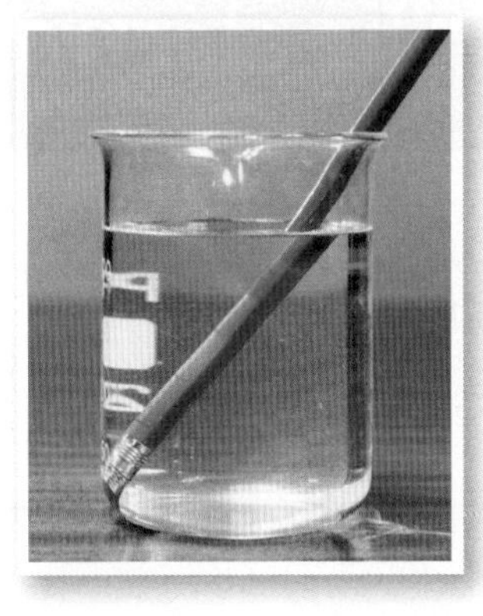

◀ Refraction can lead to some strange optical effects!

Did you know...?

Refraction can make swimming pools appear shallower than they actually are. A pencil can appear bent, and mirages in deserts are caused by refraction.

Key words

reflection, normal, law of reflection, medium, refraction, diffraction

Questions

1 Draw a diagram to show the law of reflection.

2 What types of waves can be diffracted?

↓ E

3 Draw two diagrams to show how waves can be diffracted.

4 Explain why waves are refracted.

5 Draw a diagram to show how light is refracted when it travels from water to air.

↓ C

6 Explain why in some valleys you can't get a TV signal, but you can pick up longer wavelength radio waves.

↓ A*

27: The electromagnetic spectrum

Learning objectives

After studying this topic, you should be able to:

- describe the key features of an electromagnetic wave
- list the order of waves within the electromagnetic spectrum in terms of wavelength, frequency, and energy
- describe some of the hazards and uses of the higher frequency waves in the spectrum

Key words

electromagnetic wave, vacuum, electromagnetic spectrum

Electromagnetic waves

Light, microwaves and X-rays are examples of **electromagnetic waves**. These are a special kind of transverse wave. They do not need a medium such as air or water to travel through. Electromagnetic (EM) waves can travel through a **vacuum** like space. This is how light and infrared waves reach the Earth from the Sun. If electromagnetic waves could not travel through a vacuum, then there would be no way to receive energy from the Sun. Life on Earth would not even exist.

A What makes electromagnetic waves special compared with other transverse waves?

The different electromagnetic waves form a family called the **electromagnetic spectrum**. This is a continuous spectrum with a very wide range of wavelengths. The longest electromagnetic waves are radio waves, which can be over 10,000 m long. Gamma rays have the shortest wavelength, as small as 10^{-15} m. That's a millionth of a billionth of a metre!

B List the electromagnetic waves from longest wavelength to shortest.

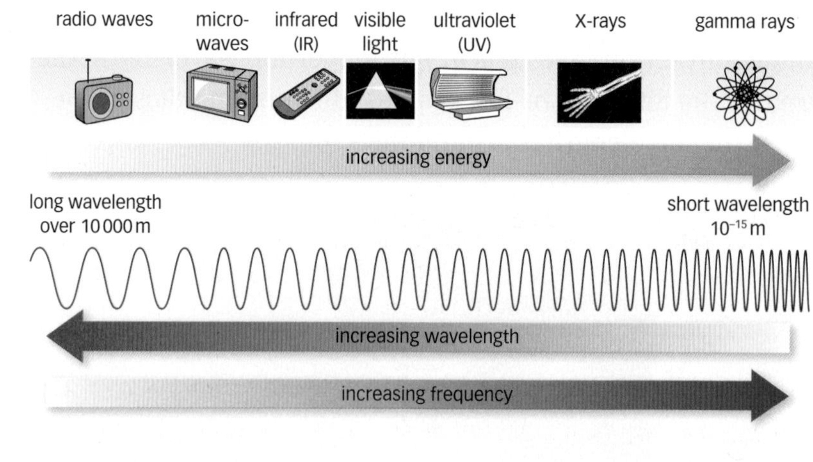

▲ The electromagnetic spectrum

Although they have different frequencies, energies and wavelengths, all electromagnetic waves travel at the same speed through space. This is the speed of light, or 300 000 000 m/s (3×10^8 m/s).

Did you know...?

According to Einstein's special theory of relativity, nothing can travel faster than the speed of light through a vacuum. Some science fiction shows on TV bend this rule and use 'warp drives', 'wormholes' or other fanciful future technology. However, the speed of light remains the ultimate speed limit.

Uses and hazards of electromagnetic waves

The different parts of the electromagnetic spectrum have many different uses, from cooking dinner to communicating with satellites. The properties of ultraviolet, X-rays, and gamma rays make them ideal for a variety of purposes.

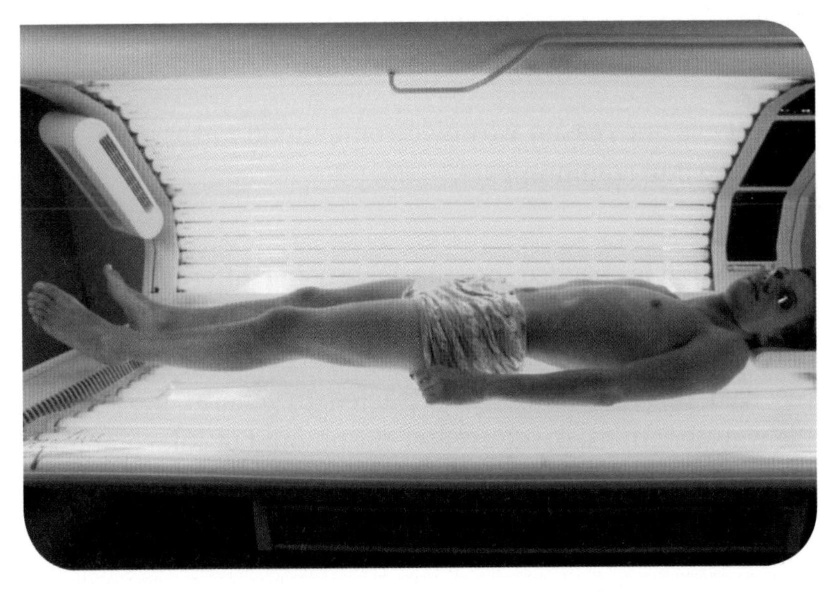

Sunbeds emit ultraviolet, giving you a tan, but they can be hazardous

Electromagnetic wave	Some uses
ultraviolet	detecting fake banknotes, sun tanning, sterilising drinking water
X-rays	detecting broken bones, looking for defects in metal products
gamma rays	sterilising medical equipment, killing cancerous cells

There are also hazards associated with the different parts of the electromagnetic spectrum. In particular, the three parts of the spectrum mentioned above can be very hazardous. These electromagnetic waves have very short wavelengths and transfer the most energy. They can travel through many materials and potentially damage or kill human cells.

Exam tip 

- Don't be tempted to say that higher energy waves such as gamma rays, travel faster – it's not true.

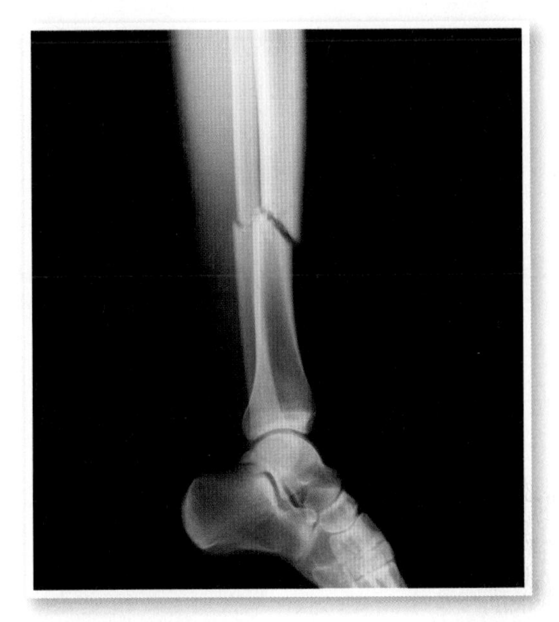

X-ray images help diagnose broken bones

Questions

1. Give three examples of electromagnetic waves. (E)
2. List the electromagnetic spectrum from highest frequency to lowest frequency.
3. Waves from which part of the electromagnetic spectrum travel the fastest? Explain your answer. (C)
4. State two uses for gamma rays.
5. Why are X-rays potentially hazardous and what measures, can be taken to reduce the risks when using them? (A*)

28: Communications and optics

Learning objectives

After studying this topic, you should be able to:

- describe examples of how electromagnetic waves are used to communicate
- outline the possible risks involved in using a mobile phone
- show how a mirror may be used to produce an image

Mobile phones use microwaves to communicate

Did you know...?

Mobile phones transmit and receive microwaves. Research has shown these waves can cause a small heating effect in the brain. Scientists are not sure whether using mobiles from a young age will cause any long-term damage. The current advice from the NHS is that mobiles should not be used regularly by younger children.

Using electromagnetic waves to communicate

As electromagnetic waves travel so fast, they are very useful for communications. Every time you listen to the radio, watch TV or chat to a friend on your mobile, you are using electromagnetic waves to communicate.

A Give one reason why electromagnetic waves are used to communicate.

Information is encoded into the wave using different techniques. It is then sent from a transmitter to a receiver (such as a mobile phone or TV aerial). When the wave is received, this information is extracted. In general, the shorter the wavelength used, the greater the amount of information that can be sent per second. The table shows some examples of how we use electromagnetic waves to communicate.

radio waves	TV, radio and wireless communications (like Bluetooth and WiFi)
microwaves	mobile phones and satellite TV
infrared	remote controls and some cable internet connections
light	photography and some cable internet connections

Drivers must use hands-free wireless devices such as Bluetooth headsets if they want to talk and drive at the same time

B Give three examples of electromagnetic waves used in communications.

Images

Visible light only forms a very small part of the electromagnetic spectrum. Our eyes are able to detect this tiny region of electromagnetic waves. This makes light very useful for communicating. Our brain forms images of the world around us when the light reflected off objects enters our eyes.

Our brain also forms an image when light is reflected off a flat, or plane, mirror. But the brain assumes that because light travels in a straight line, the light has come from somewhere in the mirror. This is why mirrors seem to have depth – it looks to us as though there is an image inside the mirror! This image is not really there, and so it is called a **virtual image**.

C What is the name of the type of image seen inside a mirror?

Images seen in flat mirrors stay the same way up as the object and the mirror (they are **upright**).

When light reflects off a mirror, there is another effect. You've probably noticed when you look at your reflection it looks as though you have been flipped horizontally. Close your left eye and the reflection in the mirror closes its eye on its right. As a result, the images formed in plane mirrors are described as being **laterally inverted**.

Questions

1 Name the type of electromagnetic wave used by mobile phones. (E)

2 Explain why electromagnetic waves are useful for communication.

3 Carefully draw a diagram to show how an image forms in a mirror. (C)

4 Which three terms describe the image formed by mirrors?

5 Explain why TV signals are generally a shorter wavelength than radio signals. (A*)

Key words

virtual image, upright, laterally inverted

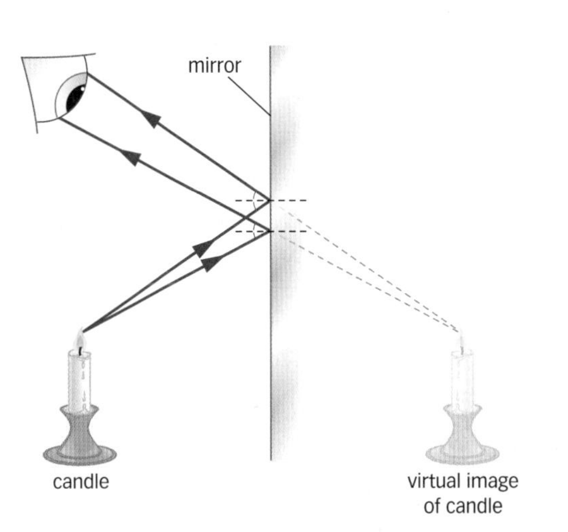

▲ A plane mirror produces an upright, laterally inverted virtual image

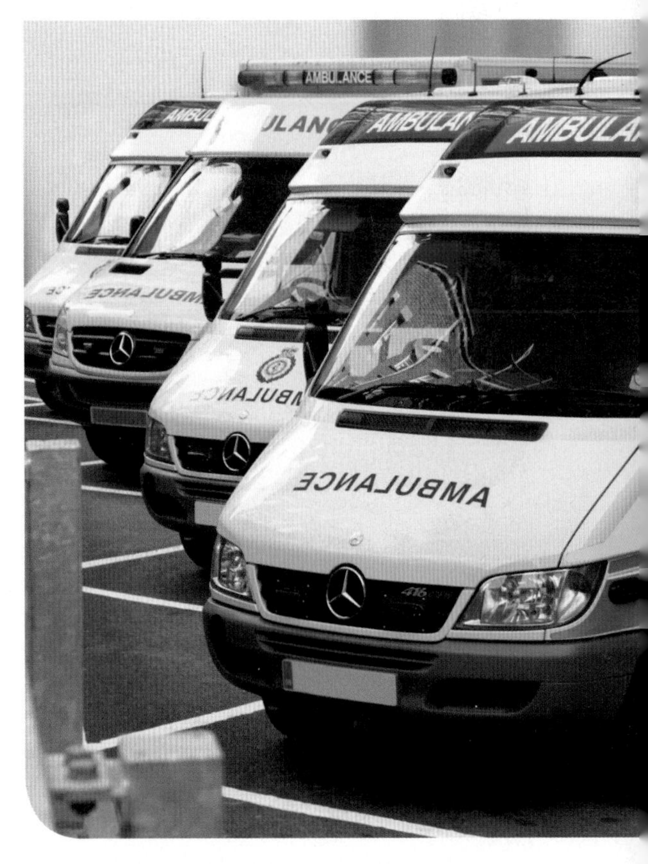

▲ Emergency vehicles have reverse writing on the front, so drivers see it the right way round in their rear-view mirrors

29: Sound waves

Learning objectives

After studying this topic, you should be able to:

- describe sound waves as longitudinal waves
- understand how the pitch of a sound is determined by its frequency, and how the volume is determined by its amplitude

Key words

sound waves, pitch, volume, ultrasound

Did you know...?

As you get older, you lose your ability to hear higher frequency sounds. By the time you are 20 you will probably only be able to hear up to around 16 000 Hz; this falls to 13 000 Hz at 30. But it is not just age that affects your hearing. Listening to music that is too loud, particularly through headphones, does permanent damage.

Listening to sound

Sound waves are created whenever an object vibrates. When you talk, your vocal chords vibrate. You can feel them if you gently press the front of your throat whilst talking. Your headphones contain tiny little speakers which vibrate in and out when they receive an electric current from your MP3 player. If you play a musical instrument, it is often a string, a column of air or a reed that is made to vibrate. These vibrations make the air around the instrument vibrate. These vibrations travel through the air as a sound wave.

▲ As the drumskin vibrates, it creates a sound

When vibrations reach our ear, they make our eardrum vibrate and we detect this as a sound.

A Describe how sound waves are formed.

Sound can travel through other media too, not just air. Sound can travel through all solids, liquids and gases. The denser the material the sound travels through, the faster the sound. It travels at 330 m/s through air, but 1500 m/s through water and even over 5000 m/s through metals like steel.

B Other than air, give three examples of materials that sound can travel through.

Although it can travel through a number of different materials, sound can't travel through a vacuum. There are no particles in the vacuum to vibrate. Just like the tagline of the film says, in space no-one can hear you scream!

Pitch and frequency, volume and amplitude

The faster something vibrates, the greater the number of vibrations every second. This means more sound waves are created each second, and so the frequency of the sound wave increases. The higher the frequency of a sound wave, the higher its **pitch**. A high-pitched sound has a high frequency. The loudness of the sound (its **volume**) depends on the amplitude of the sound wave. Louder sounds have much larger amplitudes.

A special piece of laboratory equipment called an oscilloscope can be used to produce a picture of a sound wave. If the sound increases in pitch, the waves get closer together on the screen. Their frequency has increased.

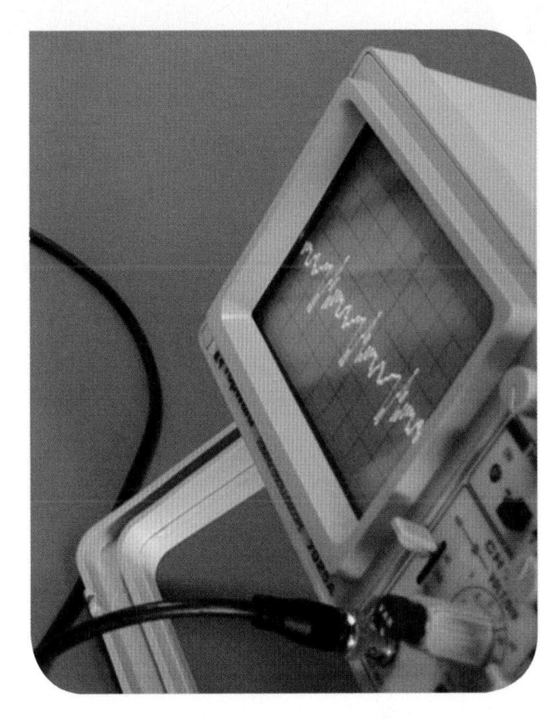

▲ An oscilloscope can produce an image of a sound wave

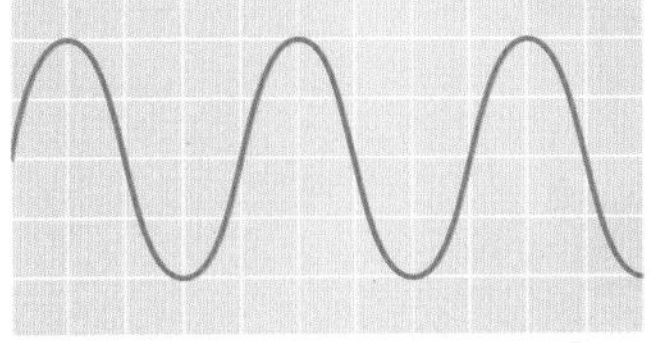

a low pitch, low frequency, loud sound

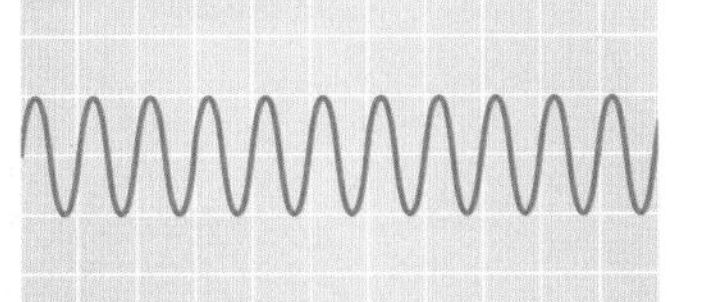

a higher pitch, higher frequency, quieter sound

▲ Sound waves with different pitch shown on an oscilloscope screen

An oscilloscope produces a picture of a sound wave that is easier to understand. This looks like a transverse wave, but remember that sound waves are longitudinal waves.

Humans can hear sounds with a frequency up to 20 000 Hz. Sound waves above this frequency are called **ultrasound**. Different animals can hear a range of different frequencies of sound. The frequency of a dog whistle is too high-pitched for humans to hear. Dogs can hear much higher frequency sounds, and so the dog hears the whistle and comes running.

▲ The frequency from a dog whistle is too high for humans to hear, but it's fine for dogs

Questions

1. Explain why sound can't travel through a vacuum. (E)
2. Describe how frequency and amplitude affect the pitch and loudness of a sound wave.
3. Describe what happens to the speed of sound as it passes through denser materials. (C)
4. Sketch the oscilloscope traces to compare two waves, one at 200 Hz, the other at 100 Hz. (A*)

Exam tip

✔ We often draw sound waves to look like transverse waves. This makes them easier to draw, but remember that they are actually longitudinal waves.

Learning objectives

After studying this topic, you should be able to:

- describe the Doppler effect
- explain why a change in wavelength and frequency is observed when a wave source is moving

Key words

Doppler effect, wave source

Moving wave sources

If you've ever heard an F1 car race past, you will have heard it make that distinctive *neeeeeaaaawwww* sound. This is an example of the **Doppler effect**. The sound comes from the engine of the car. At top speed, it produces a sound at a constant pitch. However, because the source of the sound (the car engine) is moving, the note sounds different depending on whether the car is moving towards or away from you. There is a change in the pitch of the sound as the car races past.

You can clearly hear the Doppler effect as an F1 car races past

When the car is approaching you, it makes the *neeeee* sound. The pitch sounds higher and higher; the waves have a higher frequency. When the car has gone past, it makes the *aaaawwww* sound. The sound gets lower pitched; the sound waves have a lower frequency.

Anything that emits waves is called a **wave source**. A car engine is a wave source, emitting sound waves. A light bulb is a wave source, emitting light. The Doppler effect happens with all waves, not just sound. Where the wave source is moving towards an observer, there is an increase in frequency. Where it is moving away, the waves have a lower frequency.

The same thing happens if the wave source is stationary and the observer moves towards or away from it. But you have to be moving really fast to notice the effect.

A What is the name given to the effect that causes a change in pitch when an F1 car travels past a stationary observer?

B If the sound has a higher pitch, does the sound wave have a lower or higher frequency?

Why does this happen?

When the source of waves is moving towards the observer, the waves it emits are all bunched up. They are compressed together, and this gives them a shorter wavelength. As a result there are more waves per second and so the wave has a higher frequency.

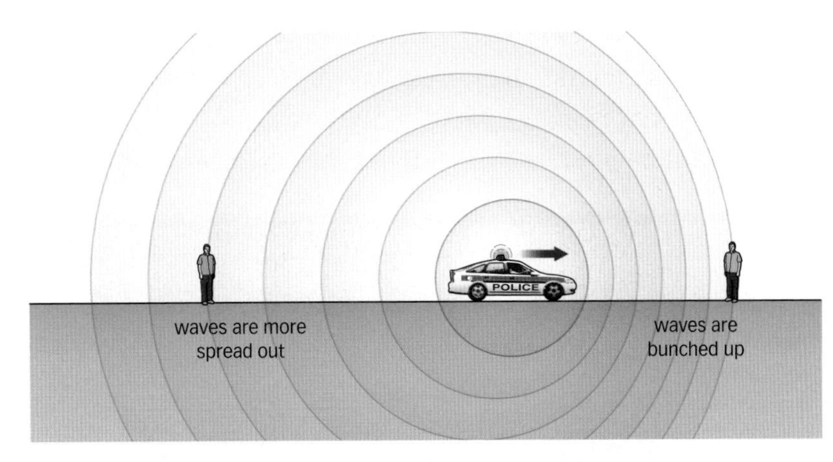

▲ The Doppler effect can be heard when a police car passes you with its siren sounding

When the wave source is moving away, the wavelength is stretched, the waves are much more spread out. Fewer waves arrive at the observer each second, and so the frequency is lower.

▲ The Doppler effect is used to measure vehicle speeds

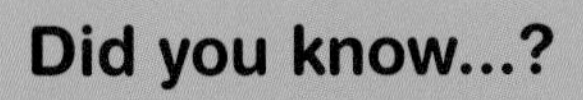

Did you know...?

The Doppler effect can be used to calculate an object's speed. Radar guns are used to measure the speed of many things, including the speed of a tennis ball after a serve and the speed of a football struck into the back of the net. They are also found in speed cameras used to monitor vehicle speeds. The guns emit radio waves which reflect off the vehicle being measured. If the vehicle is moving towards the camera, the reflected waves are more bunched up than when they were fired. Their wavelength has gone down due to the Doppler effect. The faster the vehicle is moving, the greater the change in wavelength, and so its speed can be determined very accurately.

Questions

1. Explain what is meant by wave source, and give two examples. (E)
2. Describe what causes the Doppler effect (C)
3. Explain what would happen to the change in pitch if the F1 car was moving much faster.
4. Describe what the driver hears as they race along a straight. Explain your answer. (A*)

31: Red-shift

Learning objectives

After studying this topic, you should be able to:

- ✔ know the meaning of the term red-shift
- ✔ describe how red-shift is observed in light from distant galaxies

Key words

red-shift

What is red-shift?

The Doppler effect happens to all waves, not just sound. The police rely on the Doppler effect from radio waves or microwaves to detect speeding motorists. Even wavelength of light changes if the light source moves away from, or towards you.

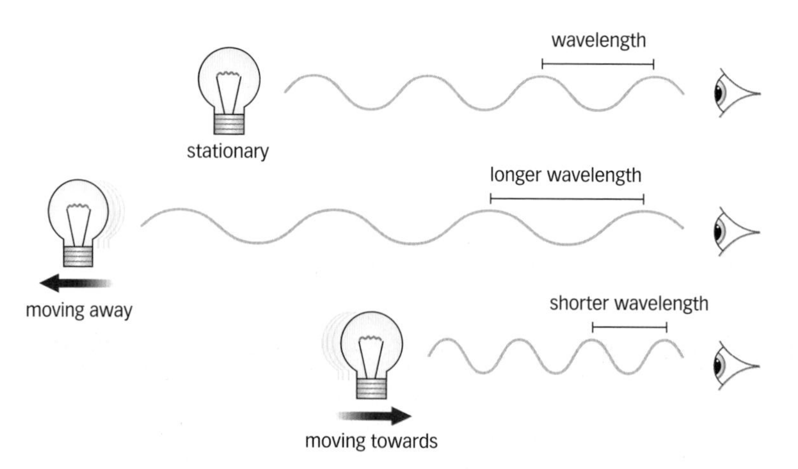

▲ Light wavelength shifting

If a light source moves away from you, the light from it gets stretched. Its wavelength increases, the waves get longer. Red light has the longest wavelength of any colour, so this effect is called **red-shift**. The light is shifted towards the red end of the visible spectrum.

The opposite happens if the light sources moves towards you.

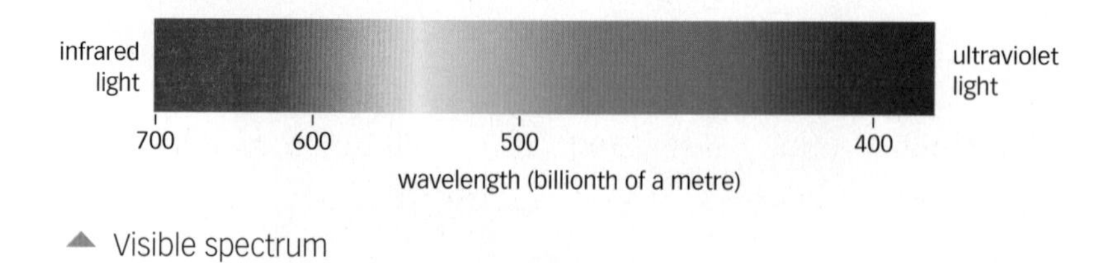

▲ Visible spectrum

Unlike sound, you don't notice the change. Light travels too fast; an object would have to be moving at close to the speed of light before it looked any different to the naked eye. It would look really odd if a car's headlights looked blue when it came towards you, but red when it moved away!

A What colour in the visible spectrum has the longest wavelength?

Galaxies and red-shift

It turns out when you look at the light from most of the other galaxies in the Universe, it is red-shifted. This must mean most galaxies are moving away from us.

B Are most galaxies moving towards or away from us?

When scientists carefully analysed this light they noticed another very important pattern. The galaxies furthest away from us showed a greater red-shift. These galaxies must be moving faster.

These observations on red shift lead to the conclusion that the further away a galaxy is from us, the faster it is moving.

▲ Hubble Space Telescope

Questions

1 When a galaxy is moving away from us, in which direction is the light said to be shifted? (E)

2 Describe what scientists have noticed about the red-shift from galaxies further away from us and explain what this means about those galaxies. (C)

3 Why don't you notice a change in colour like you notice a change in pitch when race cars go past?

4 What would happen to the light observed from a galaxy that was moving towards us? Suggest a name for this effect. (A*)

▲ A galaxy

Did you know...?

It was the American astronomer Edwin Hubble who first explained the red-shift in distant galaxies. The fact that more distant galaxies are moving faster is called Hubble's law. For such an important contribution to astrophysics, the Hubble Space Telescope is named in his honour.

Exam tip

✔ It's important to remember not only that observations on red-shift show us that galaxies are moving away from us, but also that the galaxies furthest away are moving fastest.

32: The Big Bang theory

Learning objectives

After studying this topic, you should be able to:

- outline the Big Bang theory, including the evidence supporting it
- describe the origin of cosmic microwave background radiation

A Explain how the Universe was formed, according to the Big Bang theory, and describe what it was like in the past.

It all started with the Big Bang

The **Big Bang** is at present the most widely recognised scientific theory on how the Universe began. It states that the Universe began from a very small, very dense and very hot initial point. It burst outwards in a giant explosion, and all matter and space was created in the Big Bang. It is even thought that this was the moment when time began.

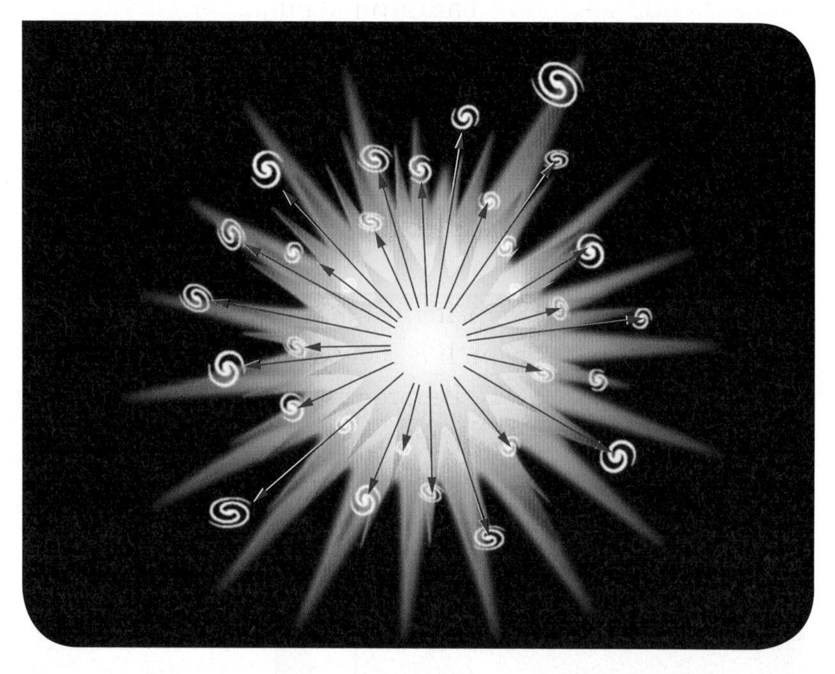

Most scientists think the Universe was created in the Big Bang

This theory may seem a little strange, but there is some good evidence to support it.

An exploding firework

Where's the evidence?

There are two key pieces of evidence for the Big Bang. The first is data collected on red-shift. You may remember that red-shift shows us that the galaxies which are furthest away from us are moving faster than galaxies closer to us. This suggests that the Universe is expanding outwards. The galaxies are a bit like the coloured sparks from an exploding firework. When the firework explodes, the sparks moving fastest travel the furthest.

If you could run time backwards, you would see the sparks coming back closer together and all starting in one point. The same is true for galaxies. They all started out at one point in space and then exploded outwards.

B What does red-shift suggest is happening to the Universe?

The second piece of evidence is even more compelling. In the 1960s, two scientists called Wilson and Penzias noticed that a form of microwave radiation was affecting their readings. No matter where they pointed their special telescope, they always detected the same background hum. This electromagnetic radiation was everywhere.

It is now called **cosmic microwave background radiation** (CMBR). Their explanations of this won Wilson and Penzias the Nobel prize for physics. They explained that CMBR must be the heat left over from the Big Bang. As the small, hot Universe expanded, it cooled and the radiation was stretched out. Today this radiation is in the microwave region of the electromagnetic spectrum. It is the same everywhere you look, because it fills the Universe. The Big Bang theory is currently the only theory than can explain the existence of this CMBR.

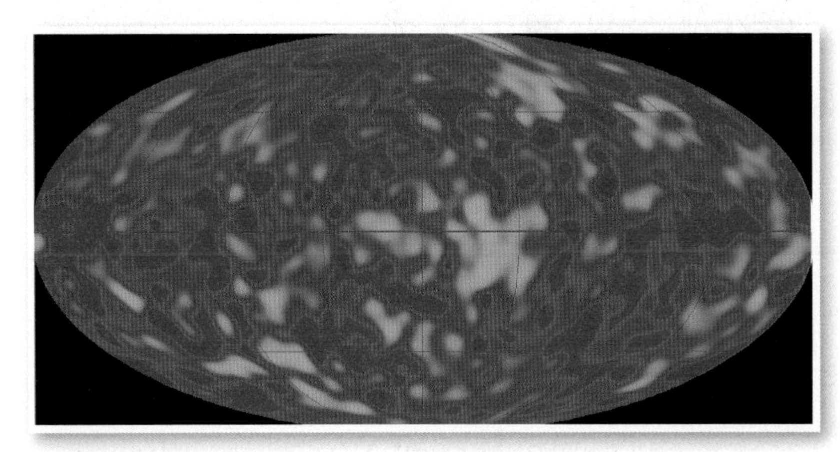

▲ Recent measurements show the distribution of cosmic microwave background radiation (CMBR)

Despite this **evidence** some scientists still disagree about the origin of the Universe. This is still one of the areas where future scientists will continue to explore other ideas. Who knows, maybe in a few years we will have a different theory. However, any change in the theory must explain the origin of CMBR.

Key words

Big Bang, cosmic microwave background radiation, evidence

Did you know...?

There are still a great number of unanswered questions in physics. For example, the scientific theories do not explain how or why the Universe began. Other recent observations suggest some very strange events at the edge of the Universe – it is not expanding in the way the Big Bang predicts. Scientists have come up with a number of temporary theories, including dark matter and dark energy, to try to explain these observations.

Questions

1. What are the two key pieces of evidence in support of the Big Bang theory? **E**
2. How is red-shift used to support the Big Bang theory?
3. Explain the origin of cosmic microwave background radiation. **C**
4. Explain why the Big Bang theory is at present the most widely accepted theory on the origin of the Universe. **A***

Course catch-up

Revision checklist

- Some power stations use fossil fuels, biomass, and nuclear fuels to heat water. Steam then turns a turbine that drives an electrical generator.
- Water in hydroelectric generators drives turbines directly. So does hot water and steam from geothermal areas. Solar cells convert the Sun's radiation energy directly into electricity.
- Different energy sources affect the environment differently; releasing substances into the atmosphere, producing noise and visual pollution, producing waste, or destroying habitats.
- Electricity is distributed from power stations to homes using the National Grid, using step-up and step-down transformers and high-voltage cables.
- For a given power, increasing the voltage reduces the current required and this reduces the energy losses in the cables.
- Waves transfer energy and can be either transverse or longitudinal. Waves have frequency, wavelength, and amplitude. They can be reflected, refracted, and diffracted.
- Electromagnetic waves are transverse, sound waves are longitudinal. Mechanical waves can be either transverse or longitudinal.
- Electromagnetic waves form a continuous spectrum. All types of electromagnetic waves travel at the same speed through a vacuum.
- Radio waves, microwaves, infrared, and visible light can all be used for communication.
- Longitudinal waves show areas of compression and rarefaction.
- The normal is a line perpendicular to the reflecting surface at the point of incidence. The angle of incidence is equal to the angle of reflection.
- The image produced in a plane mirror is virtual and upright.
- Sound waves are longitudinal and cause vibrations in a medium, which are detected as sound. The pitch of a sound determines its frequency. Echoes are reflections of sounds.
- The Doppler effect explains the change in wavelength and frequency of a wave seen by an observer. The Doppler effect on light from distant galaxies is called red-shift and is evidence for the Big Bang and expanding Universe.
- The Big Bang is the only theory that currently explains cosmic microwave background radiation (CMBR).

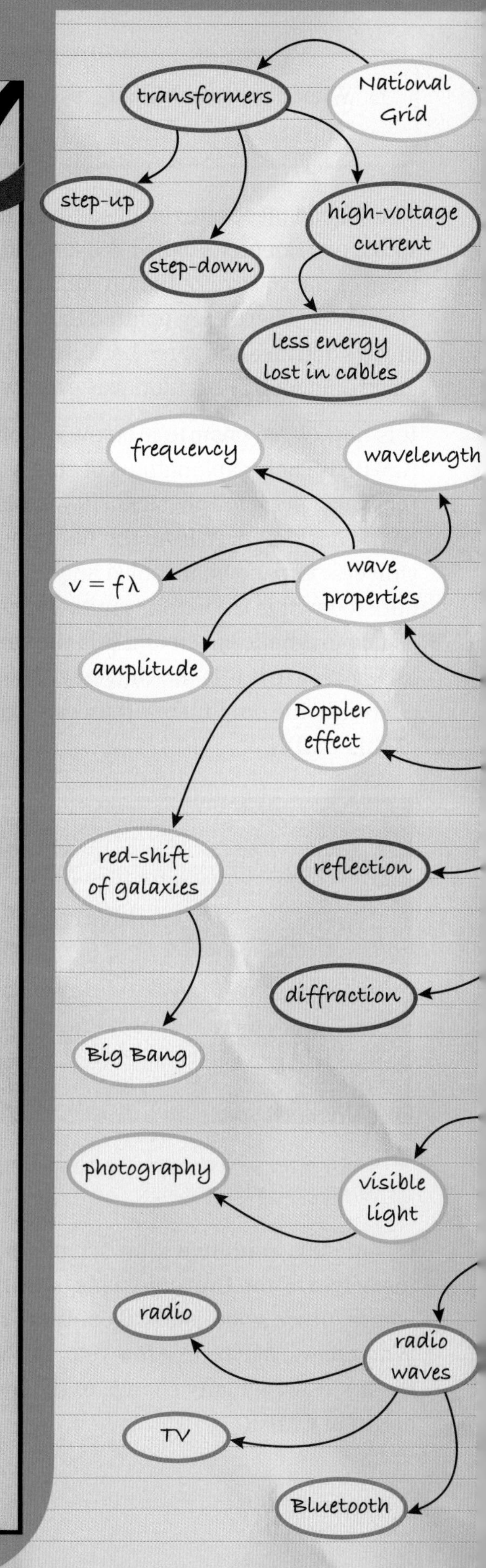

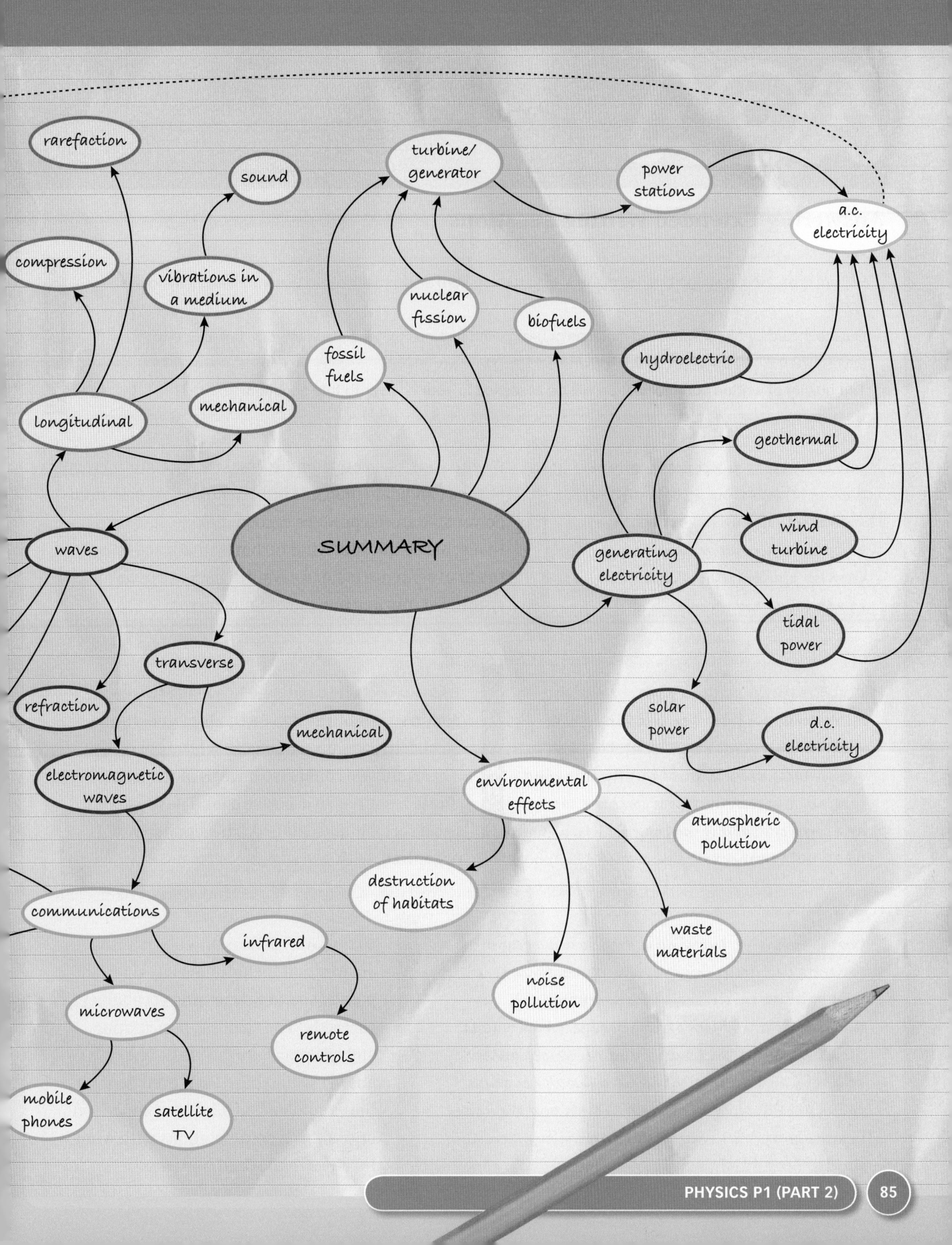

rarefaction
sound
turbine/ generator
power stations
a.c. electricity
compression
vibrations in a medium
nuclear fission
biofuels
fossil fuels
hydroelectric
longitudinal
mechanical
geothermal
waves
SUMMARY
generating electricity
wind turbine
tidal power
transverse
refraction
mechanical
solar power
d.c. electricity
electromagnetic waves
environmental effects
atmospheric pollution
destruction of habitats
communications
infrared
waste materials
noise pollution
microwaves
remote controls
mobile phones
satellite TV

Answering Extended Writing questions

QUESTION

William is an African schoolboy. He made a wind turbine from scrap materials to generate electricity for his home.

Outline the advantages and disadvantages of using wind turbines to generate electricity, compared to generating electricity in coal-fired power stations.

The quality of written communication will be assessed in your answer to this question.

Winde turbines do not make carbone dioxide, so there is no global warming from them My nan says that winde turbines in her village ruin the view from her bedroom. She is in a campaign against them.

G–E

Examiner: The points in the first sentence are correct, but the candidate then needs to mention that coal-fired power stations do produce carbon dioxide. It is true that some people find wind turbines unattractive, but the final sentence is not relevant. There are several spelling errors and one punctuation error.

Coal-fired power stations produce electrisity all the time, not just when it is windy, like wind turbines. Coal will run out, but wind is a reneweable resource. Birds can fly into wind turbines and then their lives come to a sad and sorry end, and if they pair for life, like doves, their partner will pine away.

D–C

Examiner: The candidate has made two good comparisons of the two methods of generating electricity, and used a scientific term correctly. The candidate could also have used the term 'finite resource' to describe coal. The final point is correct, but too detailed. The answer is well-organised, with two spelling errors.

Coal is expensive, but wind is free, so it is cheaper to generate electricity from wind turbines (once you have built the turbines and the power station). Burning coal produces carbon dioxide, which causes global warming, and sulfur dioxide, which causes acid rain and asthma. Wind turbines do not cause pollution. But wind turbines kill birds and the noise is annoying. They only produce electricity when it is windy.

B–A*

Examiner: This is a well organised answer that includes the key scientific points, and uses scientific vocabulary correctly. The spelling, punctuation, and grammar are faultless. The answer would have been even better if the candidate had added a comparison of coal-fired power stations with wind turbines in the final sentence.

Exam-style questions

1 A01 State whether each of the following is a renewable or a non-renewable source of energy.

hydroelectricity
natural gas
solar
nuclear
biomass
coal
wind
tidal
oil
wave

2 A01 The illustration represents a gas-fired power station.

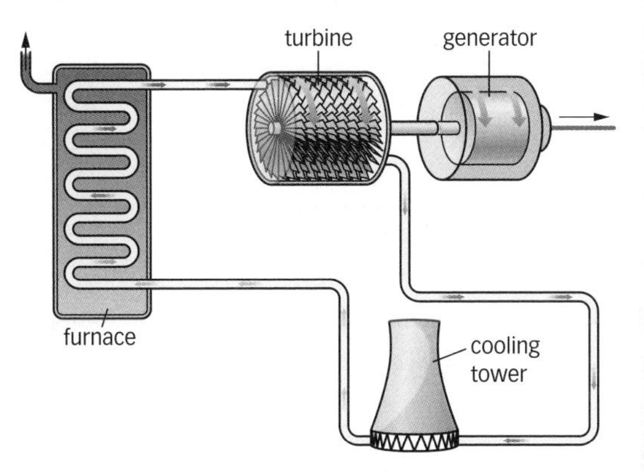

Explain what happens at each of the four stages illustrated above.

G–E

D–C

3 A02 Light and radio waves travel at 3×10^8 m/s, and sound travels in air at 330 m/s.

a Green light has wavelength 5×10^{-7} m. What is its frequency?

b The note middle C on the piano has frequency 256 Hz. What is its wavelength?

c What is the wavelength of an FM radio wave broadcasting at 100 MHz?

d A starting pistol is fired 200 m from where you are sitting. About how long is the interval between you seeing the smoke and hearing the bang?

B–A*

Extended Writing

4 A01 Describe some of the ways in which we use energy that reaches the Earth from the Sun.

G–E

5 A01 Explain what is meant by the Big Bang theory, and the main points of evidence that support it.

D–C

6 A01 There are many nuclear power stations operating successfully worldwide. Explain the process involved; and discuss some of the arguments for and against building more nuclear power stations.

B–A*

A01 Recall the science
A02 Apply your knowledge
A03 Evaluate and analyse the evidence

P2 Part 1

Forces and motion

AQA specification match

P2.1	2.1	Forces
P2.1	2.2	Force, mass, and acceleration
P2.1	2.3	Speed and velocity
P2.1	2.4	Distance–time graphs
P2.1	2.5	Acceleration
P2.1	2.6	Velocity–time graphs
P2.1	2.7	Resistive forces
P2.1	2.8	Stopping distances
P2.1	2.9	Motion under gravity
P2.1	2.10	Terminal velocity
P2.1	2.11	Hooke's law
P2.2	2.12	Work done and energy transferred
P2.2	2.13	Power
P2.2	2.14	Gravitational potential energy and kinetic energy
P2.2	2.15	Momentum
Course catch-up		
AQA Upgrade		

Why study this unit?

You can use physics to describe the motion of objects, and you can also use it to predict what will happen to an object in many different conditions. When objects move, energy transfers take place, for example from gravitational potential energy to kinetic energy when you drop an object and it falls to the floor. When engineers are designing cars, they need to be able to predict what will happen to the car and its occupants in order to minimise possible injuries if the car is in a crash.

In this unit you will look at how the acceleration of an object is linked to the force acting on it, and how this can change its motion. You will learn how the motion of the object can be represented in graphs, and how the object's motion is affected by air resistance. You will also learn about the distance it takes to stop a car, and how this distance is affected by different conditions and by the state of the driver. You will learn about momentum, and how this influences the design of car safety features.

You should remember

1. When a force acts on an object it can cause it to move.
2. Friction is a force that tries to stop things from moving.
3. When an object is moving, air resistance tries to slow it down.
4. Energy cannot be created or destroyed.
5. Energy can exist in different forms, such as kinetic energy and gravitational potential energy.

The world's fastest rollercoaster is the Ring Racer at the Nürburgring race track in Germany. It has a top speed of 217 km/h and accelerates from 0 to 217 km/h in 2.5 seconds. It has been designed to simulate the speed of a Formula 1 racing car. It does not have any loops or banked turns and has been designed simply to travel at high speeds. Engineers will have considered the forces acting on the rollercoaster and riders to give the acceleration they need to reach the top speed. They will also have considered how to keep the riders safe by including many safety features.

The Ring Racer operated briefly in 2009 until it was damaged by an explosion in the control system. It is now scheduled to open to the public in 2011. There is an even faster roller coaster under construction in Dubai – its top speed will be 240 km/h.

Learning objectives

After studying this topic, you should be able to:

- describe how when two objects interact, the forces they exert on each other are equal and opposite
- explain the meaning of the term resultant force
- understand effects of forces on an object

On take-off the Space Shuttle's engines provide a thrust that pushes against the pull of gravity

Pushing against a wall creates a pair of forces that are equal and opposite

What are forces?

Forces are all the different pushes or pulls which are around us all of the time. Everyday forces include air resistance, gravity and **friction**. All forces are measured in **newtons.** A teenager might be able to lift a weight of 800 N and a large jet engine might produce a thrust of 600 000 N.

A Give three examples of forces.

Pairs of forces between two objects

Whenever two objects interact, they exert forces on each other. They push onto each other and this produces a **pair of forces**. These pairs of forces are always **equal** in size and **opposite** in direction. You can feel this if you push down on the desk. It pushes back up at you with an equal and opposite force.

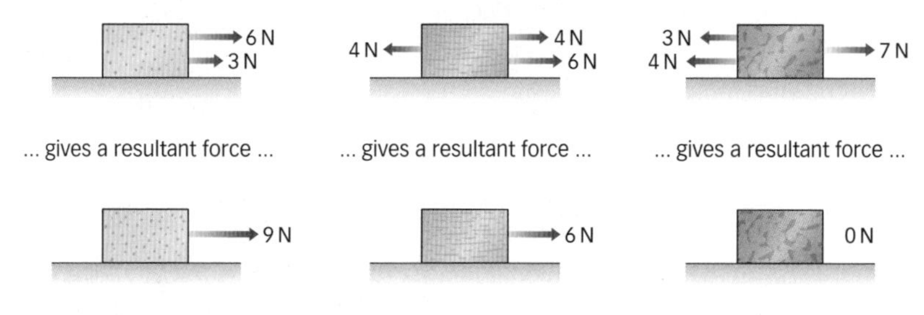

Calculating the resultant force acting on an object

Resultant force

There is often more than one force acting on an object. When a number of different forces act at a point on an object, they add up to a **resultant force**. This is the single force that has the same effect as all the original forces combined.

To work out the resultant force in a direction, you add up the forces acting in a straight line in that direction.

Any forces acting in the opposite direction need to be subtracted from the total, and so a number of forces might cancel out. This might give a resultant force of zero.

B Explain what is meant by resultant force.

What do forces do?

A resultant force is not needed to keep an object moving. When a car travels at a steady speed along a straight motorway, the force from the engine is equal to the resistive forces (air resistance and friction). There is no resultant force, it is zero, yet the car continues to move.

A resultant force *changes* the way something moves. If an object is not moving (stationary), then a resultant force makes it begin to move, and if it is already moving, a resultant force changes the motion of the object. If the driver takes their foot off the accelerator then there is a resultant force (backwards) and the car slows down. If the driver pushes their foot down then the force from the engine is greater than the resistive forces, there is a resultant force forwards and the car gets faster.

A resultant force makes an object **accelerate**, it will make it speed up, slow down or change direction.

	Zero resultant force	Resultant force
Stationary object	Remains stationary	Starts moving (accelerates)
Moving object	Continues moving at a steady speed in a straight line	Accelerates in the direction of the resultant force (this might mean slowing down if the force is in the opposite direction to the motion of the object)

Questions

1 List two things forces can do to objects.

2 Which two words describe the pair of forces produced when two objects interact? Explain what they mean.

3 Describe what happens to an object if there is resultant force acting on it.

4 Draw diagrams to show the forces acting on:

(a) A cyclist getting faster

(b) A cyclist travelling at a steady speed

(c) A cyclist getting slower.

E → C → A*

Key words

force, friction, newton, pair of forces, equal and opposite, resultant force, accelerate

Did you know...?

On Earth when you hit a tennis ball both gravity and air resistance act on the ball. This causes the ball to slow down and change direction (bend towards the Earth). A tennis ball that was hit in deep space would continue in a straight line at a steady speed. There are no forces to slow it down or change its direction.

▲ Skilled tennis players have mastered the forces acting on the ball

Exam tip

✔ Remember, forces don't make objects move. If there is a resultant force on an object, the object will change the way it is moving (accelerate).

Learning objectives

After studying this topic, you should be able to:

- ✔ understand how forces change the motion of objects
- ✔ use the equation resultant force = mass × acceleration

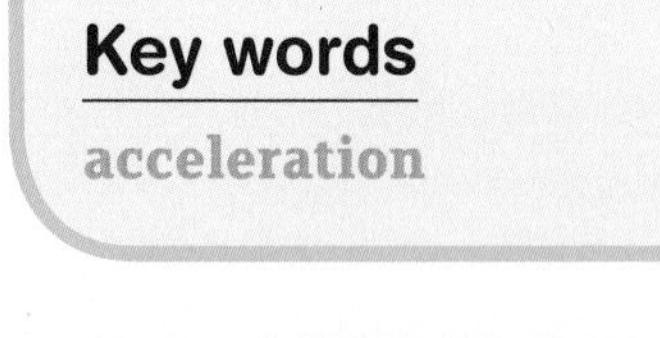

Key words

acceleration

▲ At the sound of the starter's gun, this sprinter's legs provide a force that accelerates him out of the blocks

A What is acceleration?

B Describe the relationship of acceleration and mass.

Forces and acceleration

You have already learnt that a resultant force will make an object accelerate. Forces such as friction or the thrust from an engine change the way an object moves. We will look at **acceleration** in more detail later, but you can think of acceleration as a change in speed or a change in direction.

If you double the resultant force acting on an object, its acceleration will double (as long as you keep the mass of the object the same). Applying a resultant force of 12 N might make an object accelerate at 3 m/s^2. Increasing this force to 24 N will make the same object accelerate at 6 m/s^2.

If you apply the same force to different objects, they may accelerate at different rates. The greater the mass of the object, the lower its acceleration. Imagine the engines in the vehicles shown below can all provide the same force. The bike will accelerate at the greatest rate. This is because the bike has the lowest mass. The lorry has the greatest mass and so it will accelerate at the lowest rate.

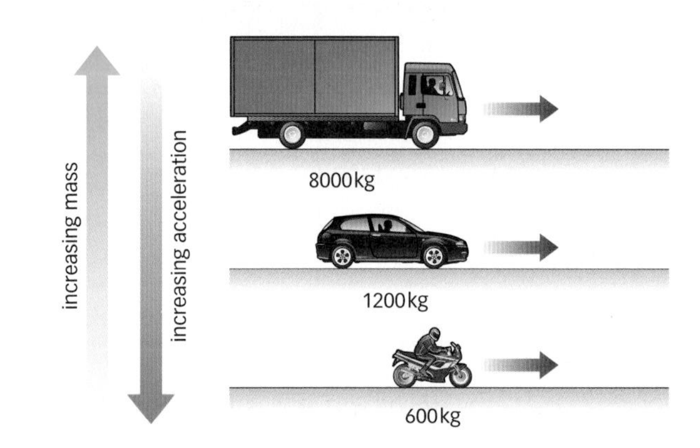

▲ The greater the mass of an object, the lower its acceleration for a particular force

Force = mass × acceleration

The resultant force acting on an object, its mass, and its acceleration are related in this equation:

resultant force (newtons, N)	=	mass (kilograms, kg)	×	acceleration (metres per second^2, m/s^2)

If the resultant force is called F, the mass m, and the acceleration a, then:

$$a = \frac{F}{m} \quad \text{or} \quad F = m \times a$$

Worked example 1

A sprinter accelerates out of the blocks at 3.0 m/s². She has a mass of 70 kg. What is the resultant force from her legs?

resultant force = mass × acceleration

$$F = m \times a$$

m = 70 kg and a = 3 m/s²

resultant force = 70 kg × 3 m/s²

= 210 N

Worked example 2

A passenger jet has mass of 320 000 kg and it has 200 000 N of thrust from each of its four engines. Calculate its acceleration.

resultant force = mass × acceleration, so

$$\text{acceleration} = \frac{\text{resultant force}}{\text{mass}} \quad \text{or} \quad a = \frac{F}{m}$$

mass = 320 000 kg

total resultant force = 4 × 200 000 N = 800 000 N

$$\text{acceleration} = \frac{800\,000\text{ N}}{320\,000\text{ kg}}$$

= 2.5 m/s²

Questions

1. State the equation that links resultant force, mass and acceleration, including units for each term. **E**
2. A charging rhino has a mass of 1400 kg and accelerates at 1.5 m/s². Calculate the resultant force providing the acceleration.
3. Sketch a graph of resultant force against the acceleration of an object. **C**
4. A football with a mass of 400 g is kicked with a force of 700 N. Calculate the ball's acceleration.
5. A small rocket has a weight of 200 N and a mass of 20 kg. When launched it accelerates at 5 m/s². Find: **A***
   (a) the resultant force
   (b) the thrust from the engine.

Did you know...?

▲ The mass of the flea is so small that the force from its legs provides a gigantic acceleration

Acceleration can be measured in 'G'. An acceleration of 2 G would be twice the acceleration due to gravity. Racing drivers often experience large accelerations of 4 G or 5 G when they take tight bends. When a flea jumps, it accelerates at over 1200 m/s². That's 120 G. Even the most experienced fighter pilots would pass out at around 8 G. The flea's legs produce only a small force, but the acceleration is huge because the flea has such a tiny mass.

Exam tip

✔ When using the equation $F = m \times a$, you must remember to use the resultant force. You may need to calculate this first.

3: Speed and velocity

Learning objectives

After studying this topic, you should be able to:

- ✔ calculate the speed of an object
- ✔ explain the difference between speed and velocity
- ✔ describe how cameras are used to measure speed

Worked example

Usain Bolt ran 100 metres in 9.58 seconds. On average, how fast did he run?

$$\text{average speed} = \frac{\text{distance}}{\text{time}}$$

distance = 100 m and time = 9.58 s, so

$$\text{average speed} = \frac{100\text{ m}}{9.58\text{ s}} = 10.4\text{ m/s}$$

This man is using a trundle wheel to measure distances

A A car travels 1000 metres in 40 seconds. What is the speed of the car?

B Usain Bolt ran 200 metres in 19.19 seconds. How fast did he run?

Speed

Speed is a measure of how fast someone or something is moving. It is the distance moved in a certain time. It is calculated using the equation:

$$\text{average speed (metres/second, m/s)} = \frac{\text{distance (metres, m)}}{\text{time (seconds, s)}}$$

For example, in a sprint race the athletes run a measured distance, and the time they take to run the distance is also measured. So you can work out their speed.

We can work out the speeds of these athletes because we know how long it took them to run a certain distance

Both the distance and the time need to be measured accurately to get an accurate measure of speed. You can measure distances using a surveyor's tape or a trundle wheel. The length of the circumference of the wheel is known, and the number of times the wheel rotates is counted. The time taken to move the measured distance can be measured using a stopwatch.

Speeds can also be measured in other units as well as in seconds. Speeds of cars and other vehicles are often measured in miles per hour (mph) or kilometres per hour (km/h).

Speed cameras

Speed cameras are used to measure the speeds of vehicles that are travelling faster than the speed limit.

Some speed cameras are used together with lines painted on the road, as shown in the picture on the next page. As the car passes over the lines, the camera takes two pictures 0.2 seconds apart. The distance travelled by the vehicle in that time is found by looking at the two photos. The speed is then calculated using the equation, speed = distance/time.

A speed camera. The lines painted on the road are usually 1 m apart. The speed of the car is worked out by measuring the distance the car travels in a certain time.

A pair of cameras can also be used to work out the average speed of a car. The time when the car passes each camera is recorded. The distance between the cameras is known, so the car's speed can be worked out.

Velocity

Speed tells you how fast something is moving, but it does not tell you what direction it is moving in. **Velocity** tells you the direction an object is travelling in as well as its speed. For example, you might say that a car was moving north at 30 km/h.

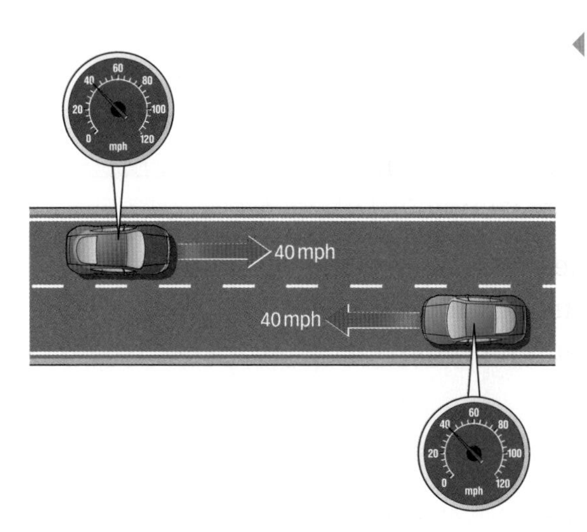

These cars may be travelling at the same speed, but they have different velocities

The two cars in the picture are both travelling at 40 mph; their speeds are the same. But they are moving in different directions, so their velocities are different.

We might say that the red car has a velocity of +40 mph. The blue car is travelling at 40 mph in exactly the opposite direction. We would say that it has a velocity of –40 mph.

Key words

speed, speed camera, velocity

Exam tip

- ✔ When calculating speed, take care with the units.
- ✔ Remember that speed and velocity are not the same thing. Velocity gives the direction of travel as well as the size of the speed.

C A car travels 4 metres in 0.2 seconds. How fast is the car travelling?

Questions

1. A cyclist covers a distance of 100 metres in 20 seconds. What is his speed in metres/second? (E)
2. A car travels 240 km in 3 hours. What is the speed of the car in km/h?
3. One car is travelling north at 70 mph. Another car is travelling eastwards at 70 mph. Do the cars have the same velocity? Explain your answer.
4. Explain how a speed camera and painted lines on the road are used to find the speed of a car. (C)
5. The car in Question C is in an area where the speed limit is 60 km/h. Is it breaking the speed limit? (A*)

4: Distance–time graphs

Learning objectives

After studying this topic, you should be able to:

- ✔ draw distance–time graphs
- ✔ understand that the gradient of a distance–time graph represents speed
- ✔ calculate the speed of an object from a distance–time graph

Time (s)	Distance (m)
0	0
50	1500
100	3000
150	4500
200	6000
250	7500
300	9000

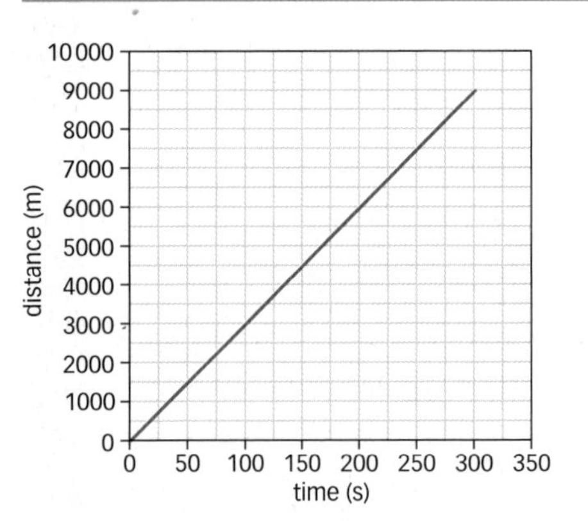

▲ Distance–time graph for the car on the motorway, using the data in the table

- ✔ When calculating speed from a distance–time graph, remember to find the change (difference) in distance and the change in time.

Recording distance and time

You can record the distances that an object travels and the time taken to travel those distances. The table in the margin shows the distance a car has travelled along a motorway. The distance and time are measured from where and when the car started.

You can plot this data on a **distance–time graph**. Time is usually plotted on the x-axis and distance on the y-axis.

Gradient

You can tell how fast something is moving by looking at the slope of the line. If the car is moving faster, it goes a greater distance in every 50 seconds and the slope of the line is steeper. If the car is slower, it moves a smaller distance every 50 seconds and the slope is less steep.

We call the slope the **gradient** of the graph. The gradient of a distance–time graph represents speed.

If a distance–time graph has a straight slope, this tells you that the object is moving at a constant speed. Where the line in a distance–time graph is horizontal, the object has not moved any distance – it is **stationary**.

A The table shows the time and distance measurements for a contestant in an athletics race. Draw a distance–time graph for the data shown in the table.

Time taken (s)	0	20	40	60	80	100	120
Distance run (m)	0	130	260	390	520	650	780

B Look at this distance–time graph for two runners in a race. Which runner is faster?

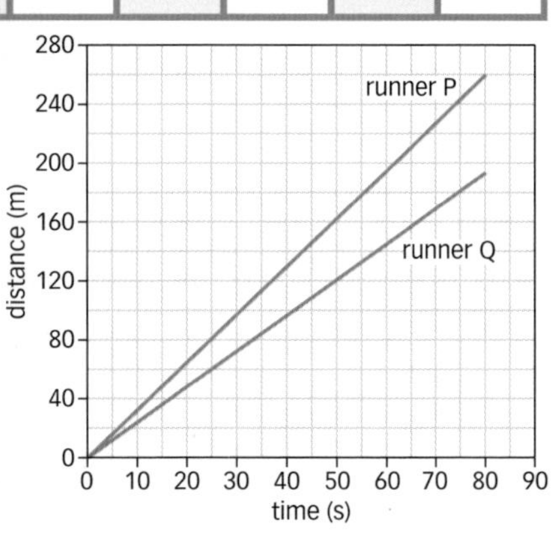

Calculating speed from a distance–time graph

You can calculate the speed of an object by finding the gradient of a distance–time graph. You can work out the gradient using the equation:

$$\text{gradient (speed)} = \frac{\text{difference in distance}}{\text{difference in time}}$$

Worked example

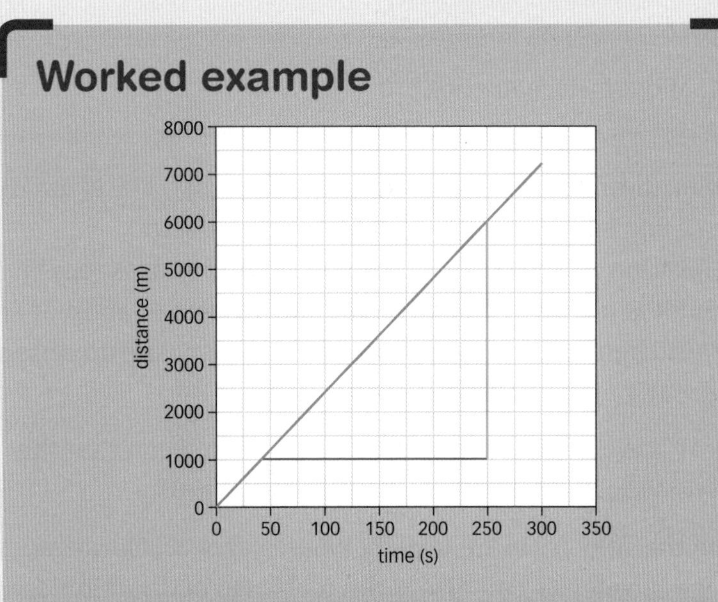

What is the speed of the car shown on the distance–time graph (blue line)?

Draw a right-angled triangle under the graph line.

difference in distance (green line): 6000 m – 1000 m = 5000 m

difference in time (red line): 250 s – 40 s = 210 s

$$\text{gradient (speed)} = \frac{\text{difference in distance}}{\text{difference in time}}$$

$$= \frac{5000\text{ m}}{210\text{ s}}$$

$$= 23.8\text{ m/s}$$

Did you know...?

All new lorries are fitted with a tachograph. This records the speed at which a lorry has travelled over time. They can be used to check that lorry drivers have taken the breaks that are required by law and to make sure that they have kept to the speed limit.

Key words

distance–time graph, **gradient**, **stationary**

Questions

1. What does the gradient of a distance–time graph tell you? (E)
2. The line of a distance–time graph slopes steeply up from left to right. What does this tell you about the motion of the object?
3. What does a horizontal line on a distance–time graph tell you?
4. A cyclist sets out on a straight road. After 50 seconds she has travelled 200 m. She stops for 100 seconds to adjust her bike. She then travels 1000 m in 200 seconds. Draw the distance–time graph for her journey. (C)
5. Find the speeds of the runners as shown in the graph in Question B. (A*)

P2 5: Acceleration

Learning objectives

After studying this topic, you should be able to:

- ✔ calculate the acceleration of an object

▲ A cheetah can speed up from rest to 20 m/s in less than 2 seconds. That's a greater acceleration than most cars are capable of.

▲ This car is accelerating because it is going round a bend

- ✔ Remember that slowing down is also acceleration in scientific language.
- ✔ When you are working out acceleration, don't forget to calculate the change in velocity – don't just use the final velocity.

Speeding up and slowing down

A moving object might speed up or slow down. This change in speed is called acceleration. The change can be negative as well as positive. When something is slowing down, it will have a negative acceleration. In everyday language negative acceleration is called **deceleration**.

A What is deceleration?

Acceleration also has a direction, like velocity. For example, when a car is pulling away from traffic lights, the acceleration is in the same direction that the car is moving in, and it is positive. When the car slows down at another set of traffic lights, the acceleration is in the opposite direction to the car's motion and it is negative.

If a car has a negative acceleration this can mean that the car is slowing down or even moving backwards.

When the velocity of an object changes, it is accelerating. A change in velocity can also mean a change in direction. Even if the speed stays the same, but the direction changes, the object is being accelerated.

B Why can we say that the car in the photo is accelerating?

Calculating acceleration

Acceleration is the rate at which velocity changes. It depends on how much the velocity changes and the time taken for the change of velocity.

$$\text{acceleration (metres per second squared, m/s}^2\text{)} = \frac{\text{change in velocity (metres per second, m/s)}}{\text{time taken for change (seconds, s)}}$$

The acceleration is called a, the initial velocity is called u, the final velocity is called v, and the time taken for the change is called t.

Change in velocity is final velocity – initial velocity = $v - u$, so the acceleration is:

$$a = \frac{v - u}{t}$$

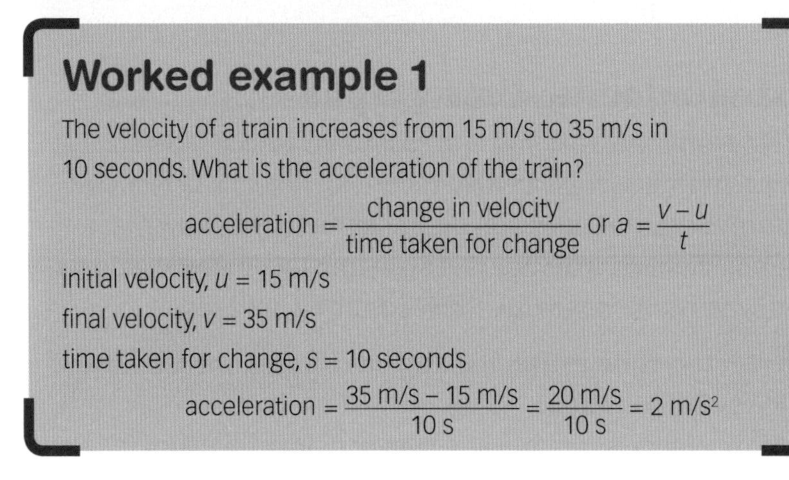

Worked example 1

The velocity of a train increases from 15 m/s to 35 m/s in 10 seconds. What is the acceleration of the train?

$$\text{acceleration} = \frac{\text{change in velocity}}{\text{time taken for change}} \text{ or } a = \frac{v - u}{t}$$

initial velocity, u = 15 m/s

final velocity, v = 35 m/s

time taken for change, s = 10 seconds

$$\text{acceleration} = \frac{35 \text{ m/s} - 15 \text{ m/s}}{10 \text{ s}} = \frac{20 \text{ m/s}}{10 \text{ s}} = 2 \text{ m/s}^2$$

C A cheetah accelerates from 3 m/s to 18 m/s in 3 seconds. What is its acceleration?

Acceleration is negative when an object is slowing down or if it starts to move backwards.

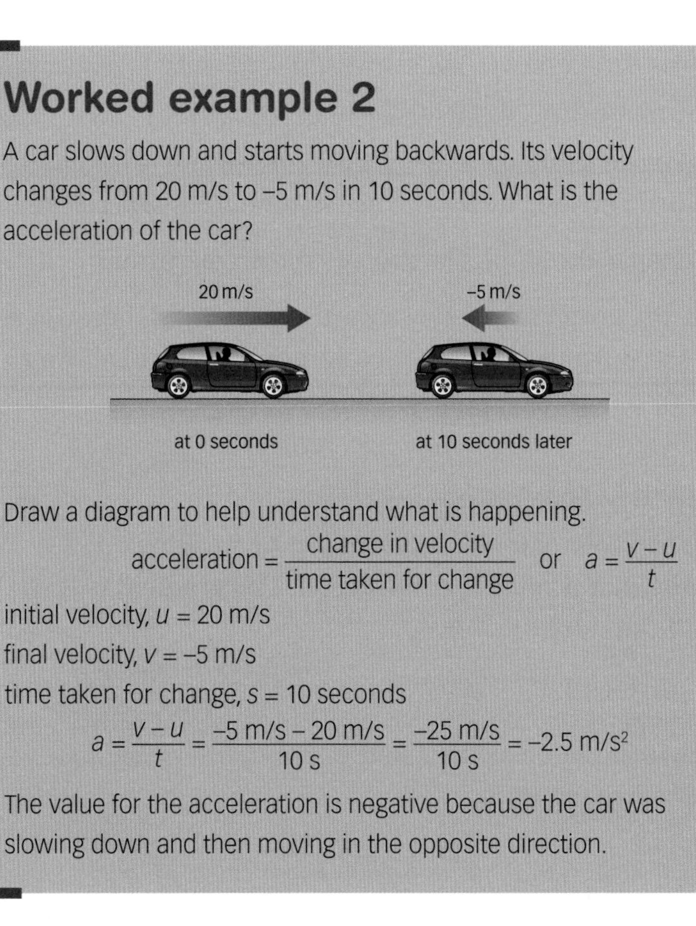

Worked example 2

A car slows down and starts moving backwards. Its velocity changes from 20 m/s to –5 m/s in 10 seconds. What is the acceleration of the car?

Draw a diagram to help understand what is happening.

$$\text{acceleration} = \frac{\text{change in velocity}}{\text{time taken for change}} \quad \text{or} \quad a = \frac{v - u}{t}$$

initial velocity, u = 20 m/s

final velocity, v = –5 m/s

time taken for change, s = 10 seconds

$$a = \frac{v - u}{t} = \frac{-5 \text{ m/s} - 20 \text{ m/s}}{10 \text{ s}} = \frac{-25 \text{ m/s}}{10 \text{ s}} = -2.5 \text{ m/s}^2$$

The value for the acceleration is negative because the car was slowing down and then moving in the opposite direction.

Key words

deceleration

Did you know...?

It is the effects of acceleration that provide the thrills on a roller coaster.

▲ This roller coaster will accelerate as it moves down the track

Questions

1. What is acceleration? (E)
2. What does it mean when an object has negative acceleration?
3. The speed of a train increases from 5 m/s to 55 m/s in 20 seconds. What is the acceleration of the train?
4. When an aircraft lands, its speed is 65 m/s. The speed decreases to 10 m/s in 11 seconds. What is the acceleration of the aircraft? (C)
5. A force of –5 N acts on an object of mass 2 kg moving at 10 m/s. After how many seconds will the object start to move backwards? (A*)

6: Velocity–time graphs

Learning objectives

After studying this topic, you should be able to:

- draw a velocity–time graph
- explain that the gradient of a velocity–time graph represents acceleration
- calculate the acceleration by using the gradient of a velocity–time graph
- use a velocity–time graph to calculate the distance travelled

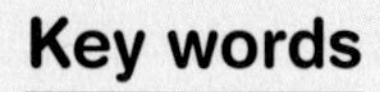

Key words

velocity–time graph

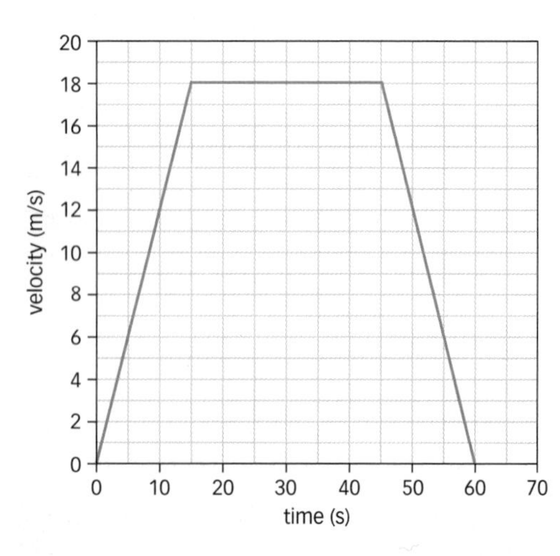

Velocity–time graph for a train travelling in a straight line between two stations

A What does a velocity–time graph show?

B What does a horizontal line on a velocity–time graph mean?

Using velocity–time graphs

In the same way that you can record an object's distance at different times, you can also record an object's velocity at different times. The graph on the left tells you how fast a train is moving, whether it is speeding up or slowing down, and whether it is moving forwards or backwards.

A **velocity–time graph** usually has time on the x-axis and velocity on the y-axis:

- If an object is moving at a steady (constant) velocity, the line on the graph is horizontal; the velocity is not changing.
- If the object is going steadily faster, the velocity is steadily increasing, and the graph shows a straight line sloping upwards.
- If the object is slowing down steadily, the velocity is steadily decreasing, and the graph shows a straight line sloping downwards.
- If the object moves backwards, the velocity is negative, and so those graph points will be plotted below the x-axis.

The slope or gradient of a velocity–time graph is the change in velocity divided by the change in time. This is acceleration.

The steeper the slope, the greater the acceleration.

When the graph slopes downwards to the right, because the object is slowing down, the acceleration is negative. We also say that the gradient is negative.

Calculating acceleration

You can calculate the acceleration by working out the gradient of a velocity–time graph. The gradient is given by the equation:

$$\text{gradient (acceleration)} = \frac{\text{change in velocity}}{\text{time taken for change}}$$

This is the same as the equation for acceleration that you met on spread P2.5:

$$a = \frac{v - u}{t}$$

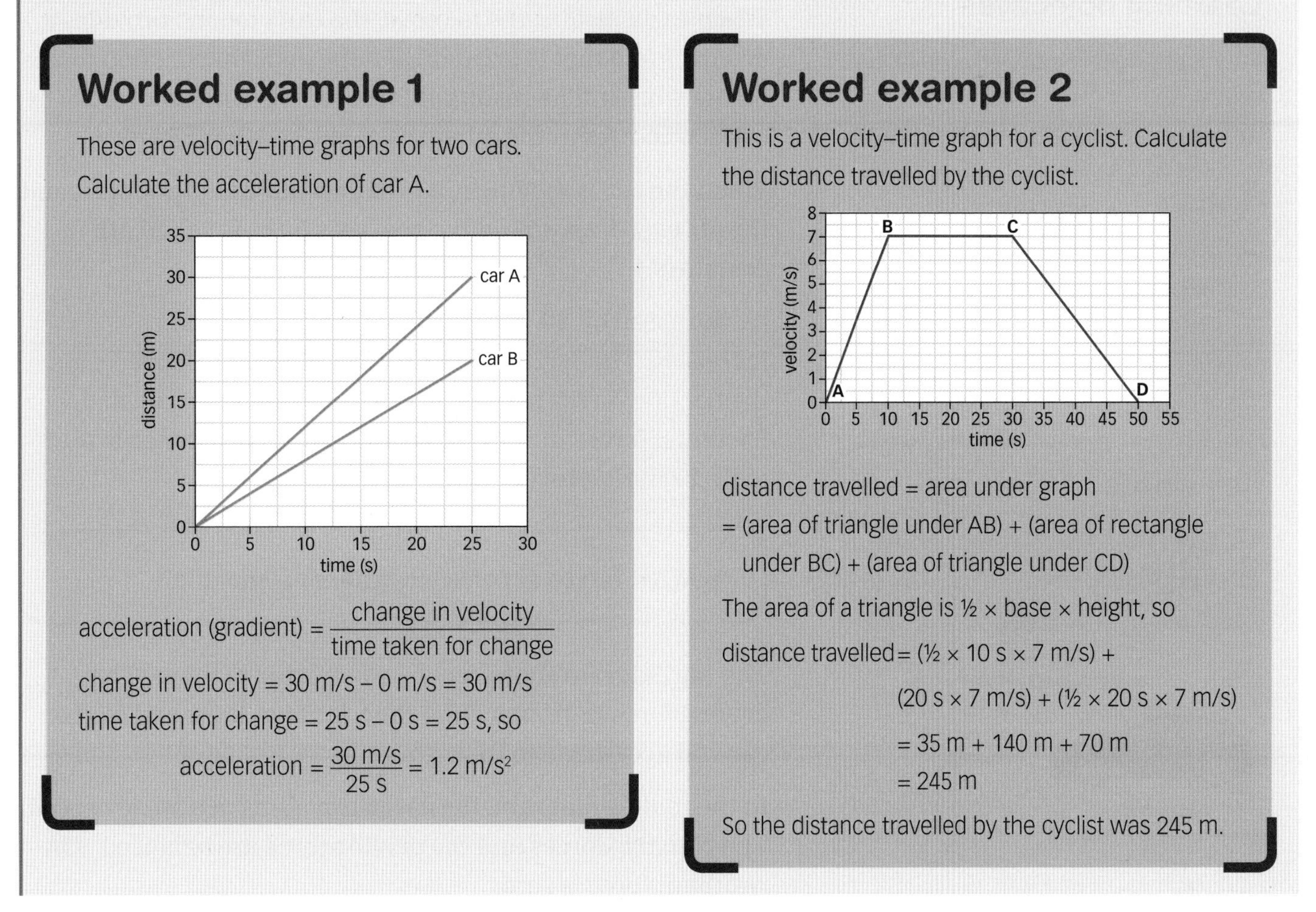

Distance travelled

The area under a velocity–time graph shows the distance travelled.

Worked example 1

These are velocity–time graphs for two cars. Calculate the acceleration of car A.

$$\text{acceleration (gradient)} = \frac{\text{change in velocity}}{\text{time taken for change}}$$

change in velocity = 30 m/s – 0 m/s = 30 m/s

time taken for change = 25 s – 0 s = 25 s, so

$$\text{acceleration} = \frac{30\text{ m/s}}{25\text{ s}} = 1.2\text{ m/s}^2$$

Worked example 2

This is a velocity–time graph for a cyclist. Calculate the distance travelled by the cyclist.

distance travelled = area under graph

= (area of triangle under AB) + (area of rectangle under BC) + (area of triangle under CD)

The area of a triangle is ½ × base × height, so

distance travelled = (½ × 10 s × 7 m/s) + (20 s × 7 m/s) + (½ × 20 s × 7 m/s)

= 35 m + 140 m + 70 m

= 245 m

So the distance travelled by the cyclist was 245 m.

Questions

1 What does the gradient of a velocity–time graph show?

E

2 Look at the velocity–time graphs for the two cars in the worked example. Does car A or car B have the greater acceleration? Explain your answer.

3 Draw a velocity–time graph for:

(a) a person walking at a constant velocity of 1 m/s for 10 seconds

(b) an aircraft accelerating from 0 m/s to 60 m/s over 20 seconds.

C

4 Look at the velocity–time graph for the train on the previous page. Calculate the acceleration of the train as shown in each part of the graph.

5 Look at the velocity–time graph for the car that stops and then reverses. Calculate the distance travelled by the car between 0 s and 4 s.

A*

7: Resistive forces

Learning objectives

After studying this topic, you should be able to:

- understand how the forces acting on a vehicle are balanced when it travels at a steady velocity
- explain that the braking force needed to stop a vehicle in a certain distance increases as the speed increases

Key words

air resistance, resistive force, braking force

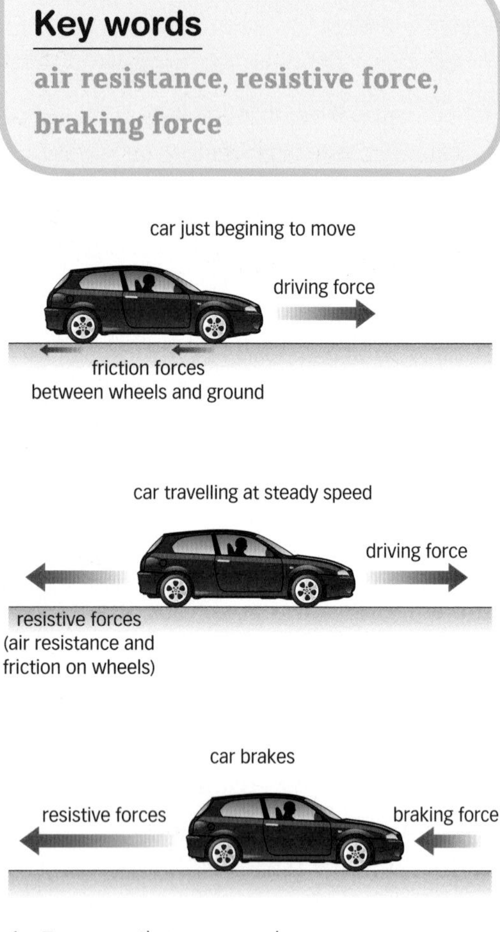

▲ Forces acting on a moving car

Forces on a car

When a car just begins to move, the car's engine provides a driving force that is greater than the opposing friction forces on the wheels. The resultant force on the car is not zero. The car accelerates in the direction of the resultant force.

When the car is moving it has to push air out of the way, and the air exerts a force on the car. This force is called **air resistance**.

As the speed of the car increases, the air resistance increases. The driving force of the car must still overcome the friction forces between the wheels and the road, and frictional forces in the engine, but the air resistance is the main **resistive force**.

A What forces act on a car when it is moving?

The top diagram shows a car just starting to move. The driving force is greater than the frictional force and so the car will accelerate to the right.

In the middle diagram, the car is travelling at a steady speed. The resistive forces and the driving force are balanced. The resultant force is zero.

In the bottom diagram, the resultant force is backwards. The car is slowing down. The resultant backwards force comes from removing the driving force and applying a **braking force**.

A car engine has a maximum force that it can provide. As the speed of a car increases, the resistive forces increase. When the car reaches a certain speed, the resistive forces will be equal to the driving force and the car will not be able to go any faster.

B What happens to the resistive forces as the speed increases?

C What is the direction of the resultant force when the car slows down?

Stopping a car

When a car brakes, a force is being applied to slow the car down. If the car has to stop in a certain distance, the force needs to be greater if the car is travelling faster. If the braking force is always the same (remains constant), then the distance needed to stop the vehicle increases as the speed of the vehicle increases.

▲ The shape of this train has been designed to reduce air resistance

negative acceleration

velocity (m/s)

10

20

40

acceleration (negative) needed to stop in a fixed distance

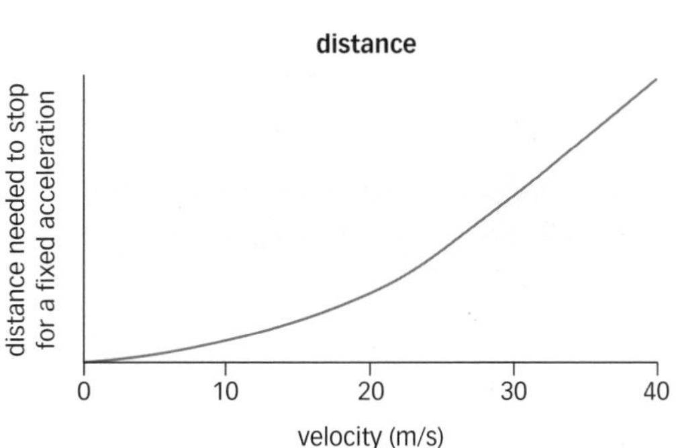

▲ The acceleration and distance needed to stop at different speeds. The acceleration is in the opposite direction to the motion of the car, so it is negative (in other words, a deceleration).

Questions

1. What is the resultant force on a car that is moving at a constant speed? (E)
2. What is the braking force on a car that is moving at a constant speed?
3. A car must stop in a certain distance. Explain the link between the speed of the car and the braking force needed.
4. The same constant braking force is used to stop a car, whatever speed it is travelling at to start with. Explain what happens to the car's stopping distance for higher speeds. (C)
5. Explain how the shape of the train shown in the photo enables it to travel faster. (A*)

Did you know...?

As faster vehicles are designed, more streamlining is needed to reduce air resistance. The train in the photo was built in Japan in 2009 and is designed to travel at 320 km/h.

Exam tip

✔ Remember that the acceleration will always be in the same direction as the resultant force.

Learning objectives

After studying this topic, you should be able to:

- explain what the stopping distance of a vehicle is
- state what affects the stopping distance
- describe the energy transfers that take place when a car is braked

A What is stopping distance?

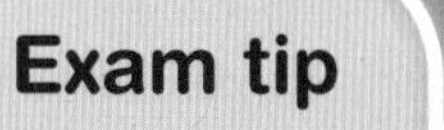

Exam tip AQA

- You don't need to remember actual stopping distances, but you do need to remember what can affect them, such as reaction time, condition of vehicle and road conditions.

Using a mobile phone in this way while driving can distract you. It is also illegal.

Key words

braking distance, reaction time, thinking distance, stopping distance

Stopping distances

The total distance needed to stop a car is not just the distance the car travels after the brakes have been applied, called the **braking distance**. There is also the time needed for the driver to react to seeing something. For example, the driver sees a red light and needs to move their foot onto the brake pedal. This is called the **reaction time**. During the reaction time, the car will have travelled a certain distance called the **thinking distance**.

So the total **stopping distance** is made up of the thinking distance and the braking distance.

The Highway Code gives stopping distances under normal conditions, as shown in the diagram. The calculations assume that the acceleration is $-6\ m/s^2$. (The minus sign means a negative acceleration, or deceleration. It is in the opposite direction to the direction of motion.)

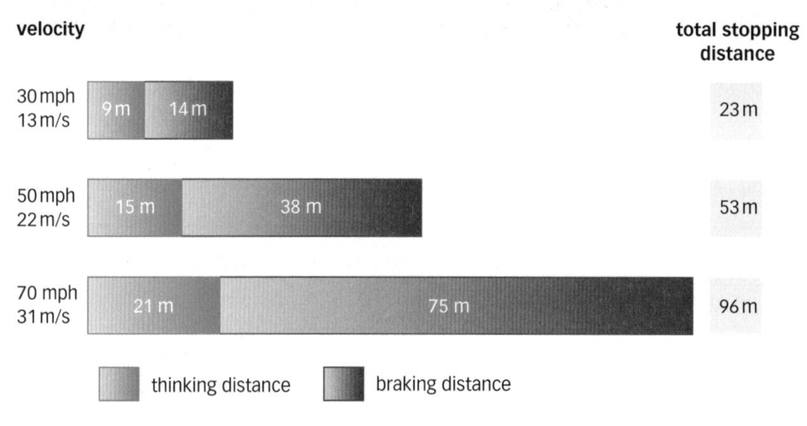

Stopping distances from the Highway Code

Thinking distance

There are many factors that can affect your reaction time and hence the thinking distance. When you are tired, you react more slowly. If you have used drugs such as alcohol or illegal drugs, your reactions are slower. Some drugs that are available over the counter or prescription drugs can also increase the time it takes you to react.

Distractions, such as listening to music, using a mobile phone or a satellite navigation system, or even talking to passengers, can increase reaction time. Also, people's reactions become slower as they get older.

Braking distance

The braking distance does not only depend on the speed of the car. When you press the brake pedal, the brakes are applied to the wheel. If the brakes are worn, this can reduce the force that they can apply. If too much force is applied, the wheels can lock and the car skids.

Road conditions can also affect the braking distance. If the road surface is icy, there is less friction between the tyres and the road and the tyres may slip. The braking distance will increase. On wet road surfaces the friction between the tyres and the road is also reduced.

▲ Stopping distances are longer in conditions like this

B What happens when you press the brake pedal?

C Why do brakes need to be in good condition?

Energy transfers

When you apply the brakes on a car, the frictional force is increased. Kinetic energy is transferred into heat energy in the brakes.

If the wheels of a vehicle lock when the vehicle is braking, the tyres will skid along the road. There is then much more friction between the tyre and the road surface, and much more kinetic energy is transferred into heat energy.

▲ The black marks on the runway are from the tyres of planes

Did you know...?

There are always skid marks where planes land at the end of airport runways. The plane's wheels are not moving at the moment that they touch the runway. When they touch the runway, they skid briefly. A small amount of smoke is produced – kinetic energy is transferred to heat energy, which burns a small amount of rubber.

Questions

1 How much (approximately) does the stopping distance increase when the speed increases from 30 mph to 50 mph?

E

2 Why can brakes overheat when a vehicle goes down a steep hill?

3 What factors can affect:
(a) thinking distance?
(b) braking distance?

C

4 Look at the stopping distances shown in the diagram on the previous page. About how long is reaction time?

A*

9: Motion under gravity

Learning objectives

After studying this topic, you should be able to:

- ✔ calculate the weight of an object
- ✔ describe the motion of an object falling under gravity

Key words

weight, mass, gravitational field strength, fluid

Worked example

A large bag of rice has a mass of 2 kg. What is its weight?

weight = mass × gravitational field strength

= 2 kg × 10 N/kg

= 20 N

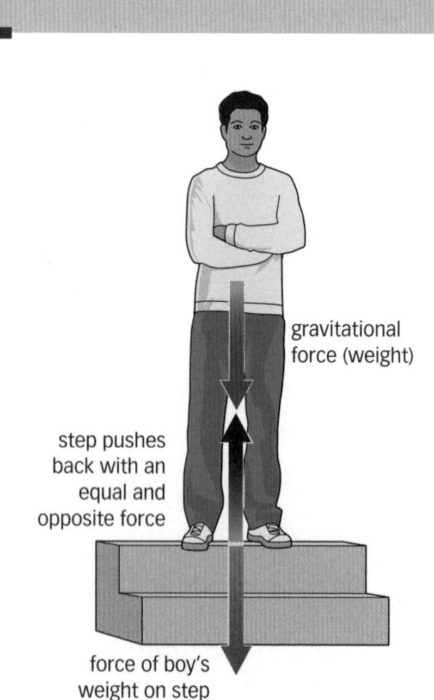

▲ The resultant force on the boy is zero

Weight

When you drop an object, it falls directly towards the ground. It falls because of the gravitational attraction between the object and the earth. It is pulled by the force of the Earth's gravitational field. This force is called **weight**.

The weight of an object depends on two things:

- its **mass** (the amount of matter) in kilograms (kg)
- the **gravitational field strength** in newtons/kilogram (N/kg).

The equation linking weight, mass, and gravitational field strength is:

$$W = m \times g$$

weight of object (newtons, N) = mass of object (kilograms, kg) × gravitational field strength (newtons per kilogram, N/kg)

The Earth's gravitational field strength is about 10 N/kg.

A Tom has a mass of 50 kg. What is his weight?

B A car has a mass of 1450 kg. What is its weight?

Falling under gravity

In the picture, the boy and the step are interacting objects that produce an equal and opposite pair of forces on each other.

To decide how the boy will move, we look only at the forces acting on the boy. There are two of them: the gravitational attraction downwards to the Earth, and the upward force from the step. These are the same size, the resultant force is zero and the boy is not accelerating. He is also not moving.

C What two forces are acting on the boy in the diagram?

If the boy moves forward off the step, there is no upward force acting on him, and his weight will cause him to accelerate downwards. The ball in the diagram on the next page does not have anything to stop it from falling towards the Earth. The force of gravity makes the ball accelerate downwards.

As the ball falls, its velocity continues to increase.

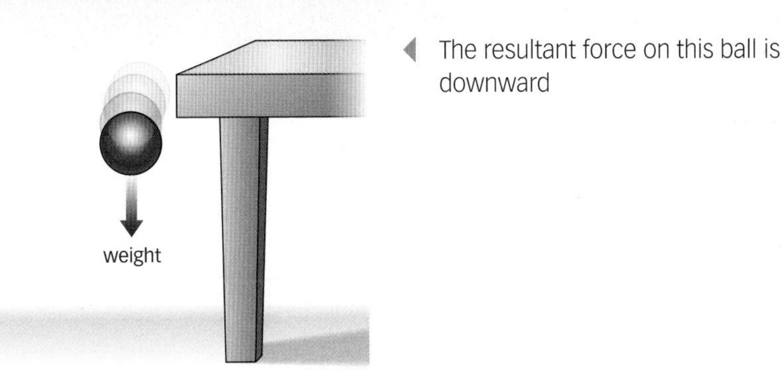

The resultant force on this ball is downward

Did you know...?

Gravitational attraction acts both ways. So when the leaf and the ball were falling downwards, they would actually be pulling the Earth a little bit upwards! But the effect would be much too small for us to notice.

Air resistance

When something is falling through the air, as its velocity increases, the size of the force in the opposite direction due to air resistance increases. The faster an object moves through air, or any other **fluid**, the greater the resistive force that acts on the object in the opposite direction.

The diagram on the right shows what happens to a falling leaf. When the leaf first begins to fall, the upward force due to air resistance is low. As the speed of the leaf increases, so does the air resistance. The downward force from the weight of the leaf stays the same.

The size of the air resistance also depends on the shape and the surface area of the object. The air resistance force is much greater for the leaf than for the ball shown earlier.

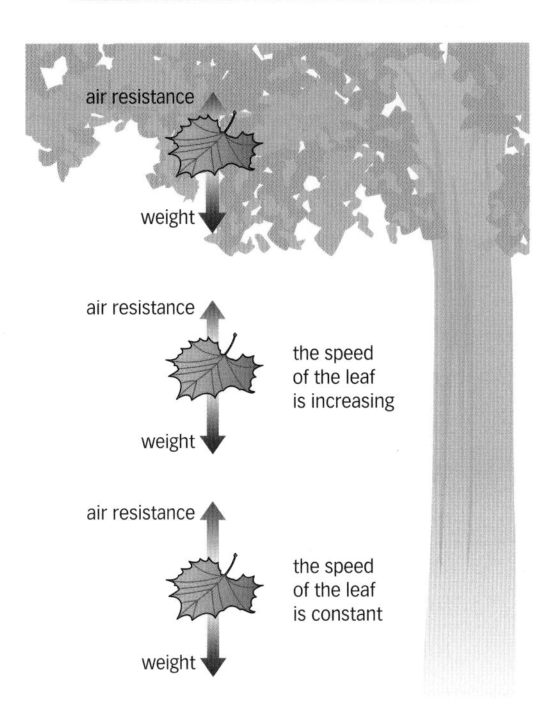

As the leaf falls more quickly (accelerates due to gravitational attraction), the air resistance gets bigger

D What happens to the force of air resistance when speed increases as an object falls under gravity?

Questions

1. What is the weight of an object? (E)
2. You drop a pen and it falls to the floor. Describe the motion of the pen.
3. Calculate the weight of each of the following:
   (a) a table with mass 25 kg
   (b) an elephant with mass 692 kg. (C)
4. Alex says that his weight is 65 kg. Explain why he is wrong.
5. An object has a weight of 12 N. What is its mass? (A*)

Exam tip

✔ In everyday language, the term 'weight' is used to mean 'mass' – when you talk about someone's weight, you usually mean their mass. Make sure you know the difference between the two and use them correctly.

Learning objectives

After studying this topic, you should be able to:

- understand how a falling object reaches a terminal velocity
- draw and interpret velocity–time graphs for objects that reach terminal velocity

Key words

terminal velocity

A What forces are acting on the skydiver in the pictures?

B What is the terminal velocity of an falling object?

C What can you say about the forces acting on anything moving at a terminal velocity?

Exam tip

- When the parachute opens, the skydiver's velocity decreases because there is an acceleration that acts upwards. The skydiver does not start to move upwards – they continue to fall, but at a lower velocity.

Terminal velocity

You already know that when something falls downwards due to gravity, it accelerates and its velocity increases. As its velocity increases, the upward force of air resistance increases, as shown for the skydiver in the diagram.

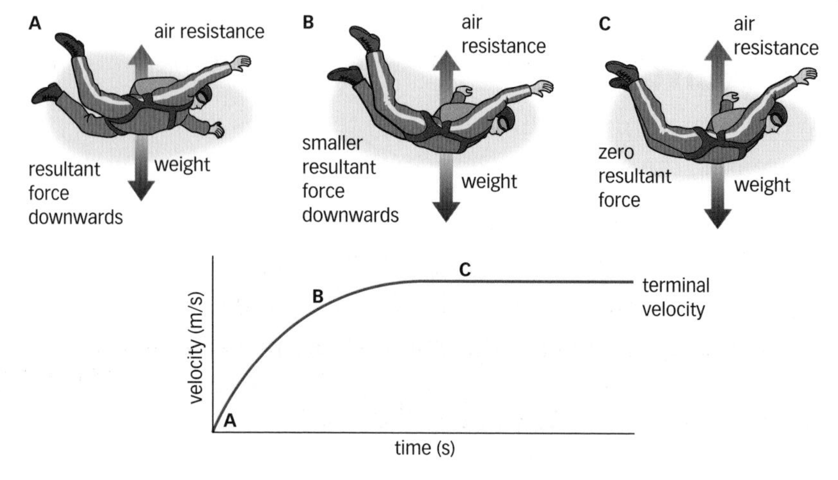

Skydiver reaching a terminal velocity: resultant forces and velocity–time graph

As the skydiver speeds up, the air resistance increases. The skydiver's weight stays constant.

A There is still a resultant force downwards, so there is still an acceleration downwards and the velocity is still increasing.

B Because the velocity is still increasing, the air resistance gets bigger. The resultant force downwards gets smaller. The acceleration downwards also decreases. There is still some acceleration though, so the skydiver's velocity is still getting bigger.

C The velocity, and so the air resistance, increase until the skydiver's weight is balanced by the upward force of air resistance. The resultant force is zero, and the skydiver will not accelerate any more. The velocity of the skydiver stays the same. This steady speed downwards is the **terminal velocity**.

Reducing the terminal velocity

D When the skydiver opens a parachute, the force of air resistance increases. This means that there is a resultant force upwards and an acceleration upwards. The velocity decreases.

As the velocity gets smaller, the air resistance force also decreases. The resultant force upwards is less and the acceleration upwards also decreases.

The resultant upward force and upward acceleration are still there though. The speed still gets less and so the force of air resistance also keeps decreasing.

E The air resistance balances the weight of the skydiver. The resultant force is zero again and the speed downwards now stays the same. This is a new terminal velocity. The new terminal velocity is much lower than the terminal velocity without a parachute.

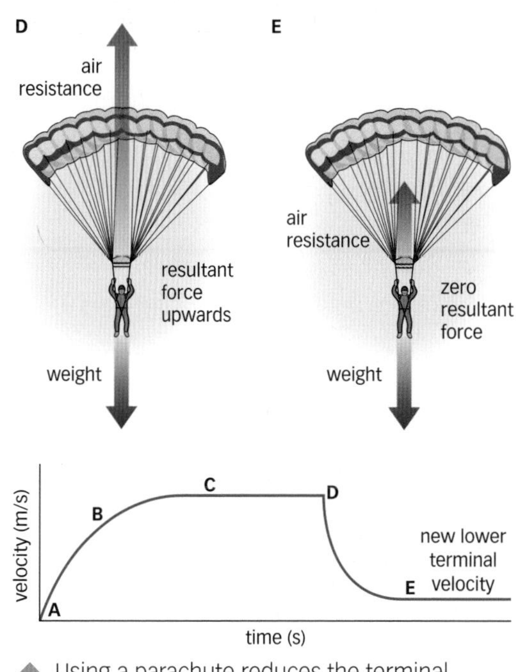

Using a parachute reduces the terminal velocity

Shape and terminal velocity

Cars and ships that move horizontally also have a terminal velocity when the driving force from their engine becomes balanced by air resistance.

Falling parachutists want their terminal velocity to be very low. A racing driver wants the terminal velocity of the car to be as high as possible.

The shape of an object affects its terminal velocity. Racing cars are designed to minimise the forces due to air resistance. They have a streamlined shape so that air can flow over them more easily. This means that they can reach a higher terminal velocity than a car that has the same thrust force but meets greater air resistance forces.

This ferry is streamlined

D How does shape affect terminal velocity?

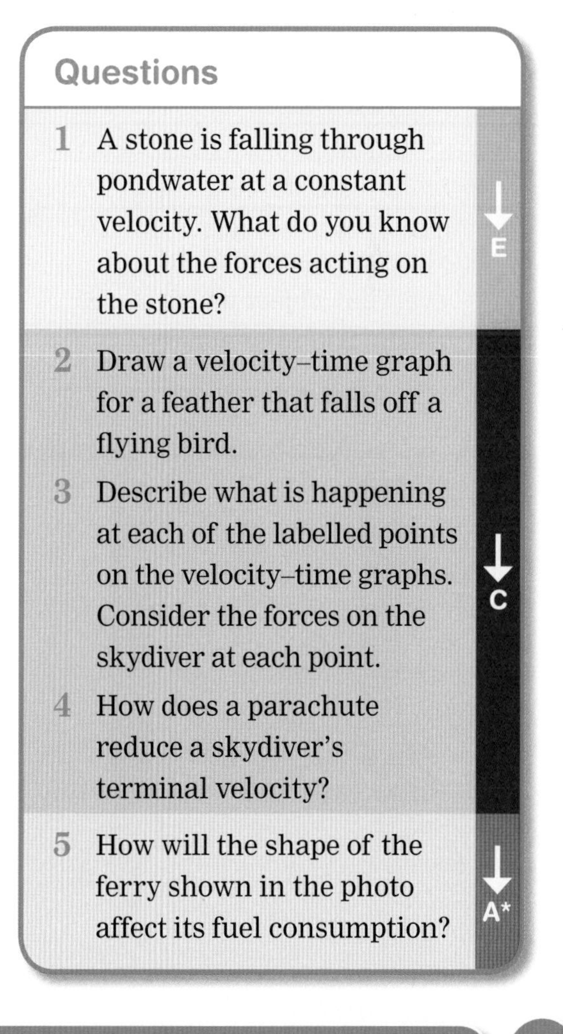

Questions

1. A stone is falling through pondwater at a constant velocity. What do you know about the forces acting on the stone? (E)
2. Draw a velocity–time graph for a feather that falls off a flying bird.
3. Describe what is happening at each of the labelled points on the velocity–time graphs. Consider the forces on the skydiver at each point.
4. How does a parachute reduce a skydiver's terminal velocity? (C)
5. How will the shape of the ferry shown in the photo affect its fuel consumption? (A*)

11: Hooke's law

Learning objectives

After studying this topic, you should be able to:

- describe how forces acting on an object may cause a change in its shape
- describe the energy changes when a force is used to change the shape of a spring
- state and use the relationship between the force applied and the extension of the spring

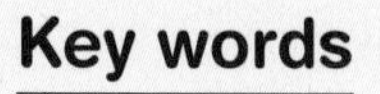

Key words

elastic potential energy, extension, directly proportional, limit of proportionality

A Give an example of an everyday object which stores elastic potential energy.

Stretching and compressing objects

Forces not only change the way objects are moving, but can also change the shape of objects. They can squash or stretch them.

When a spring is squashed or stretched by a force, energy is transferred into **elastic potential energy** and stored within the spring. When the force is removed, this energy is converted into other forms which return the spring to its original length.

The same thing happens when you pull back an elastic band. This process of storing elastic potential energy is useful for spring toys, wind-up clocks, and car suspension systems.

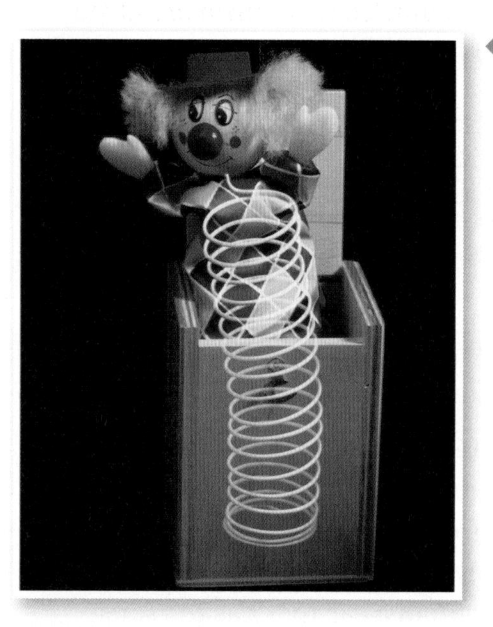

When this toy jumps, it converts elastic potential energy stored in the spring into kinetic energy

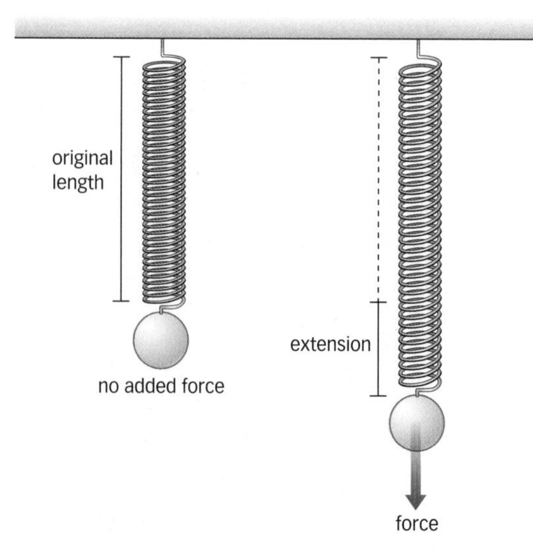

When a force is applied to a spring, it changes shape and extends

Hooke's law

In 1676 the British scientist Robert Hooke investigated the relation between the force applied to an elastic object and the change in its length (called the **extension**).

He found the force applied to an object was **directly proportional** to its extension, up to a certain point. When he doubled the force on a spring he found the extension would also double. If applying 20 N caused a spring to extend 0.05 m, then applying 40 N would cause it to extend 0.10 m.

You may have used a newtonmeter or weighing scales that make use of this relationship. The extension of a calibrated spring is used to determine the weight of an object.

This relationship continues up to the **limit of proportionality**. Above this point, the spring starts to deform – it will no longer return to its original length.

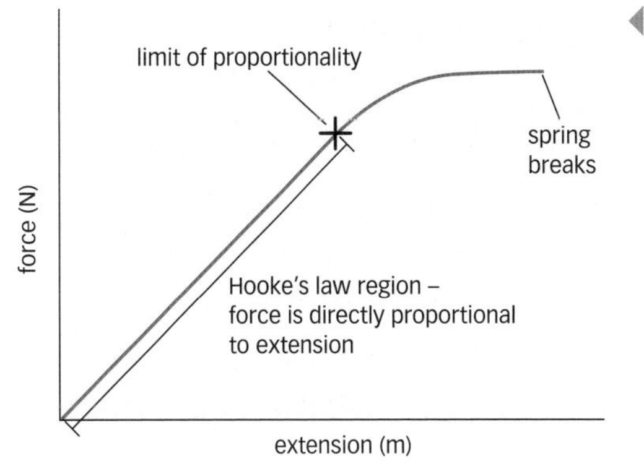

The graph obtained by applying different forces to a spring and measuring its extension

How much a certain spring extends depends on the force applied and the spring constant. This is a measure of the stiffness of the spring, and it is measured in newtons per metre (N/m).

The relationship between force, spring constant and extension is represented in the equation below.

force (newtons, N) = spring constant (newtons per metre, N/m) × extension (metres, m)

If the force is called F, the spring constant k, and the extension e, then:

$$F = k \times e$$

Questions

1 Describe the energy changes when you compress the spring toy shown in the picture on the left. (E)

2 A spring has a spring constant of 2000 N/m and extends 0.5 m when a force is applied. Calculate the value of the force and give the units.

3 Describe Hooke's law. (C)

4 A force of 600 N is applied to a spring which extends 20 cm. Find the value of the spring constant.

5 Sketch a graph of force against extension and label the key points. Then draw a second line showing the extension for a spring with a spring constant twice the size of the spring constant of the first spring. (A*)

Did you know...?

If the force applied to an object is too great, the object might change shape too much, causing it to break. The materials used to build bridges and buildings must be able to withstand the huge forces involved.

Worked example

A spring found in a wind-up toy has a spring constant of 10 N/m. Calculate the force needed to extend the spring by 20 cm.

spring constant = 10 N/m

extension = 20 cm = 0.2 m

force = spring constant × extension

= 10 N/m × 0.2 m

= 2 N

A much stiffer spring found in a car suspension system has a spring constant of 5000 N/m. Calculate the force needed to extend the spring by 20 cm.

force = spring constant × extension

= 5000 N/m × 0.2 m

= 1000 N

Exam tip

- Hooke's law relates the force applied to an object to its extension, not its length. Work out the extension of the object from its original length.
- 'Directly proportional' means that if one variable goes up by a factor of x, so does the other variable. If one doubles then the other doubles, or if one goes up by ×5 then the other goes up by ×5.

12: Work done and energy transferred

Learning objectives

After studying this topic, you should be able to:

- understand that work done is equal to energy transferred
- describe how when a force moves something through a distance, against another force such as gravity or friction, energy is being transferred
- use the equation linking the work done, the force and the distance moved in the direction of the force

Key words

work done, joule

Work is done when weights are lifted

A What is the relationship between work done and energy transferred?

B Give three examples of doing work.

Working hard?

In science, the term **work done** (or just 'work') has a very specific meaning. 'Work done' is another way of saying energy has been transferred. Lifting weights, as in the picture on the left, transfers energy to the weights. We can say work has been done on the weights.

Work done = energy transferred

When you lift up the weights you might transfer 100 J of energy to them. This means the work done on the weights is 100 J. Work done, just like energy, is measured in **joules**. If no energy is transferred then no work is done.

Other examples of doing work include climbing stairs, pushing a trolley at a supermarket, or pulling a sledge along the ground. Work is always done when a force is used to move something through a distance against an opposing force.

How much work?

The amount of work done on an object depends on the force applied and the distance moved.

$$\text{work done (joules, J)} = \text{force (newtons, N)} \times \text{distance moved in the direction of the force (metres, m)}$$

If W is the work done, F is the force, and d is the distance moved in the direction of the force, then:

$$W = F \times d$$

The distance moved must be in the same direction as the force. On the left of the diagram, the book has been lifted upwards against the gravitational force (weight), which acts downwards. On the right, the book has been lifted sideways and up, but it has been moved the same distance against the gravitational force of weight. The same amount of work has been done against the weight, so the same amount of energy has been transferred to the book.

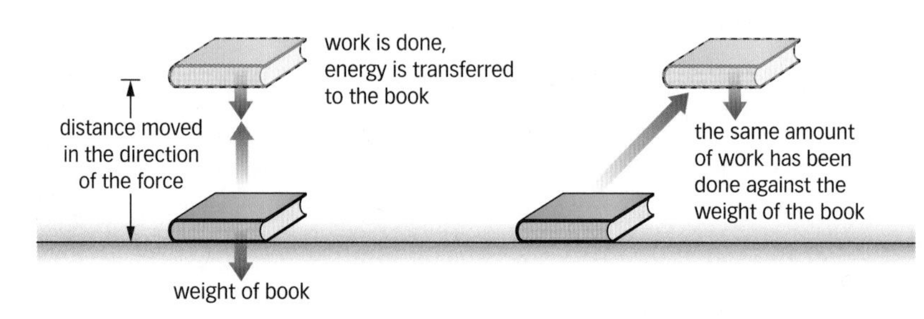

When calculating work done, you use the distance moved in the direction of the force

Worked example

A TV with a weight of 300 N is lifted onto a wall mount 1.2 m from the ground. Calculate the work done.

work done = force × distance moved in the direction of the force, or $W = F \times d$

force = 300 N

distance moved = 1.2 m

work done = 300 N × 1.2 m

= 360 J

Work done against frictional forces

When objects rub against each other, energy is transferred from kinetic energy into heat. We say that work is being done against friction.

This can be a problem because energy is transferred away from the moving object, but it can be very useful too. Most brakes use this principle to slow vehicles down. Pads or discs rub against moving parts and heat up to very high temperatures. Some kinetic energy is transferred into heat and the vehicle slows down. Next time you stop on your bike, if you put your hand near the brake pad you will be able to feel this heat being transferred to the surroundings.

Questions

E

1 What units are used to measure work?

C

2 Calculate the work done to pull a sledge 80 m against a frictional force of 6.0 N.

A*

3 Explain why you do more work pulling a sledge uphill than along the flat.

4 A delivery driver lifts 20 boxes, each with a mass of 3.0 kg, into his truck 1.5 m above the ground. Find:

(a) the work done on each box

(b) the energy transferred to each box

(c) the total work done lifting all the boxes into the truck.

Exam tip

- The distance moved must be in the direction of the force. For questions where objects are lifted, this is the vertical distance moved.
- Make sure you use the force in the 'work done' equations. You may need to calculate the weight of an object from its mass.

▲ Friction on a racing car's brakes can make them heat up so much that they glow

▲ The Space Shuttle's kinetic energy is transformed into heat on re-entry

Did you know...?

When the Space Shuttle re-enters the Earth's atmosphere it is travelling at a whopping 18 000 mph. At this speed, the air resistance causes the shuttle to heat up to over 1500 °C. Most of the shuttle's kinetic energy is transferred into heat and so it slows down. It lands at just over 200 mph.

13: Power

Learning objectives

After studying this topic, you should be able to:

- state that the unit of power is the watt
- describe power as the rate of doing work
- use the equation power = work done (or energy transferred) divided by time taken

PRO*line*
MOD.:ST44

	2450MHz
230V ~ 50Hz	MICROWAVE INPUT POWER : 1550 W
	MICROWAVE ENERGY OUTPUT : 950 W

SERIAL NO. 81000138
MADE IN KOREA
CE

WARNING – HIGH VOLTAGE

▲ Different microwaves have different power outputs. This one has a power of 950 W.

Did you know...?

Most humans can produce a power output of just over 100 W, peaking at over 1000 W for very short periods. An average horse can sustain a power output of around 750 W for much longer. Brake horse power (BHP) is often used as a measure of vehicle power output. A family car may have around 100 BHP, or the same power as 100 horses. A high performance car may produce over 500 BHP, that's the same as 500 horses or 375 000 W!

'Watt' is power?

Power, like work, means something different in scientific language from its everyday use. A politician may be a very powerful person in terms of governing a country, but they are likely to be much less powerful physically than a honed athlete.

In the scientific sense of transferring energy, a powerful person or machine can do a lot of work in a short space of time. Power is the work done in a given time (or the rate of doing work).

Boiling water

A more powerful kettle will do more work in a certain time. It will transfer more electrical energy to heat every second. This means the water will boil quicker. A more powerful car transfers more chemical energy (in the fuel) into kinetic energy per second. This means it can accelerate more rapidly to a high speed.

Power is measured in **watts** (W). One watt is one joule of work done (or energy transferred) every second. A 1500 W hairdryer transfers 1500 J of energy every second. A large TV may have a power output of 120 W, an average family car 75 000 W, and an express train a huge 12 000 000 W.

◀ Express trains have a power output that is many times greater than a typical family car

A Name the unit of power.

B How many times more powerful is an express train than a typical family car?

Horses can transfer more energy per second than a human being. They have an average power output of 750 W.

Key words

power, watt

Calculating power

If power is the work done in a given time (rate of doing work), then:

$$\text{power (watts, W)} = \frac{\text{work done (joules, J)}}{\text{time taken (seconds, s)}}$$

'Work done' is just another way of saying 'energy transferred', so this can be written as

$$\text{power (W)} = \frac{\text{energy transferred}}{\text{time taken}} \text{ (J/s)}$$

If power is P, the energy transferred is E, and the time taken for the energy transfer is t, then:

$$P = \frac{E}{t}$$

Worked example 1

A man pushing a wheelbarrow does 400 J of work in 5 s. Calculate the power that he develops.

$$\text{power} = \frac{\text{work done}}{\text{time taken}}$$

work done = 400 J

time taken = 5 s

$$\text{power} = \frac{400\text{ J}}{5\text{ s}} = 80\text{ W}$$

Worked example 2

An electric shower transfers 540 000 J of energy to the water in 1 minute. Calculate its power.

$$\text{power} = \frac{\text{energy transferred}}{\text{time taken}}$$

energy transferred = 540 000 J

time taken = 60 s (as it took 1 minute)

$$\text{power} = \frac{540\,000\text{ J}}{60\text{ s}} = 9000\text{ W or 9 kW}$$

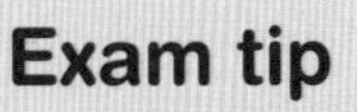

✔ When using the equations for power, you must make sure time is in seconds. If it is in minutes, you will need to convert it to seconds.

Questions

1. What is the equation for power?
2. A cyclist has a power meter on her bike to tell her about her performance. It is showing a steady reading of 200 W. How much work is she doing every second?
3. Calculate the power provided by a small solar panel on a satellite when it transfers 720 J of energy every minute.
4. A boy pulling a sledge does 6000 J of work in 2 minutes. Calculate the power he develops.
5. A 40 W bulb is left on. Calculate the energy transferred in:
   (a) 10 seconds
   (b) 30 minutes
   (c) 24 hours.

14: Gravitational potential energy and kinetic energy

Learning objectives

After studying this topic, you should be able to:

- understand what factors affect gravitational potential energy
- calculate changes in gravitational potential energy
- understand the factors affecting the kinetic energy of an object
- use the kinetic energy equation
- describe the benefits of regenerative braking

A Apart from gravitational field strength, which two factors affect the GPE of an object?

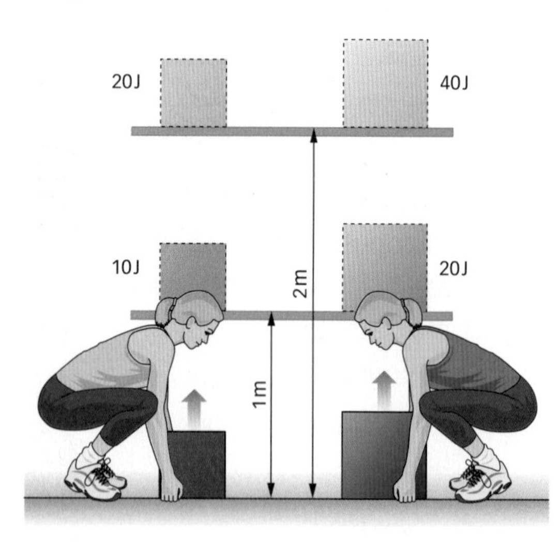

▲ The change in gravitational potential energy (GPE) depends on the mass of the object and the change in height

Gravitational potential energy

When you lift up a book and place it on a shelf, you are doing work on the book. The book gains **gravitational potential energy** (or GPE) as it is lifted away from the ground. The GPE of an object is the energy it has because of its position in a **gravitational field**, like the one around the Earth. This energy depends on

- the mass of the object
- its height above the ground
- the strength of the gravitational field.

The higher you lift an object above the ground, the greater its GPE. To calculate any change in GPE we can use the equation below.

change in GPE (joules, J)	=	mass (kilograms, kg)	×	gravitational field strength (newtons per kilogram, N/kg)	×	change in height (metres, m)

If E_p is the GPE, m the mass, g the gravitational field strength and h the change in height, then:

$$E_p = m \times g \times h$$

On Earth, the gravitational field strength is 10 N/kg.

Worked example 1

Find the change in GPE when a book of mass 1.2 kg is lifted 1.5 m and placed on a shelf.

change in GPE = mass × gravitational field strength × change in height

$$E_p = m \times g \times h$$

mass of book = 1.2 kg, gravitational field strength on Earth = 10 N/kg, and change in height = 1.5 m.

change in GPE = 1.2 kg × 10 N/kg × 1.5 m

= 18 J

B A person of mass 60 kg runs up a flight of stairs 3.0 m high. Calculate their change in GPE.

Kinetic energy

Any moving object has a **kinetic energy** (or KE). The size of the kinetic energy depends on

- the mass of the object
- the speed it is travelling at.

An object with a greater mass will have more kinetic energy. An object that is moving faster will have more kinetic energy. This is the equation that links kinetic energy, mass and speed:

kinetic energy (joules, J) = ½ × mass (kilograms, kg) × speed² (metres per second, m/s)²

If E_k is the kinetic energy, m the mass, and v the speed, then:

$$E_k = ½ \times m \times v^2$$

Worked example 2

Find the kinetic energy of a car of mass 1200 kg travelling at 20 m/s.

kinetic energy = ½ × mass × speed² or $E_k = ½ \times m \times v^2$

mass of the car, m = 1200 kg, and speed of the car, v = 20 m/s

kinetic energy = ½ × 1200 kg × (20 m/s)²

= ½ × 1200 × 400

= 240 000 J or 240 kJ

Energy transfers

The law of conservation of energy states that energy cannot be created or destroyed. However, it can be transferred. When you drop a ball, its GPE is transferred into kinetic energy as it falls. The same thing happens to bungee jumpers. When they jump, their GPE is converted into kinetic energy as they accelerate towards the ground.

This hybrid car converts kinetic energy into useful electrical energy

Some hybrid cars make use of a clever technology called **regenerative braking**. The brakes in these vehicles are specifically designed to convert some of the kinetic energy back into energy to be stored in the car's batteries. This makes the cars much more fuel-efficient and the brakes wear down less quickly.

Key words

gravitational potential energy, kinetic energy, regenerative braking

Exam tip

✔ When using the equation for kinetic energy, that is, kinetic energy = ½ × mass × speed², don't forget to square the speed!

Questions

1. State the equation for change in gravitational potential energy and give all the units. (E)
2. Calculate the change in gravitational potential energy when a crane lifts a 3000 kg concrete block 40 m into the air.
3. Describe an advantage of using regenerative braking. (C)
4. Find the kinetic energy of a football of mass 400 g travelling at 20 m/s
5. Calculate the speed of a horse of mass 600 kg with a kinetic energy of 43 200 J. (A*)

Learning objectives

After studying this topic, you should be able to:

- define and calculate momentum
- explain and apply the law of conservation of momentum
- explain the benefit of air bags, crumple zones and other safety devices in cars

The momentum of this runner depends on his velocity and his mass

A A woman with a mass of 60 kg is skydiving and has a velocity of 45 m/s. What is her momentum?

B A football has a mass of 450 g and is moving at 25 m/s. What is its momentum?

Momentum of an object

The **momentum** of a moving object depends on its mass and velocity. You can calculate momentum using the equation:

$$\underset{(\text{kg m/s})}{\text{momentum}} = \underset{(\text{kg})}{\text{mass}} \times \underset{(\text{m/s})}{\text{velocity}}$$

We can use the symbol p for momentum. Then if mass is m and velocity is v:

$$p = m \times v$$

As with velocity, you have to state the direction of momentum as well as how big it is.

If an object is not moving, its velocity is zero so it does not have any momentum.

Conservation of momentum

The total amount of momentum in a system of objects that interact only with each other always stays the same. For example, if two moving objects collide, the total momentum of the two objects does not change. This is the **law of conservation of momentum**.

Only the objects themselves must be involved. If any external forces act on the objects, such as friction, then momentum is not conserved.

You can use the law of conservation of momentum to solve problems involving collisions and explosions:
total momentum before collision (or explosion) = total momentum after collision (or explosion).

Worked example

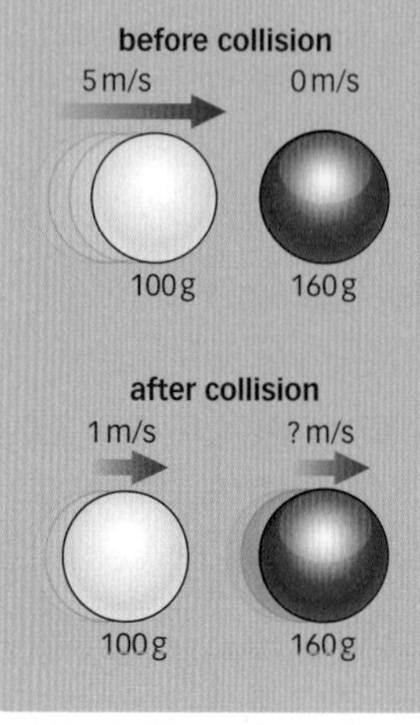

The diagram shows a cue ball of mass 100 g rolling at 5 m/s towards a stationary pool ball. The pool ball has a mass of 160 g. After the collision, the cue ball moves at 1 m/s. What is the velocity of the pool ball?

momentum = mass × velocity

$$\text{total momentum before collision} = \text{total momentum after collision}$$

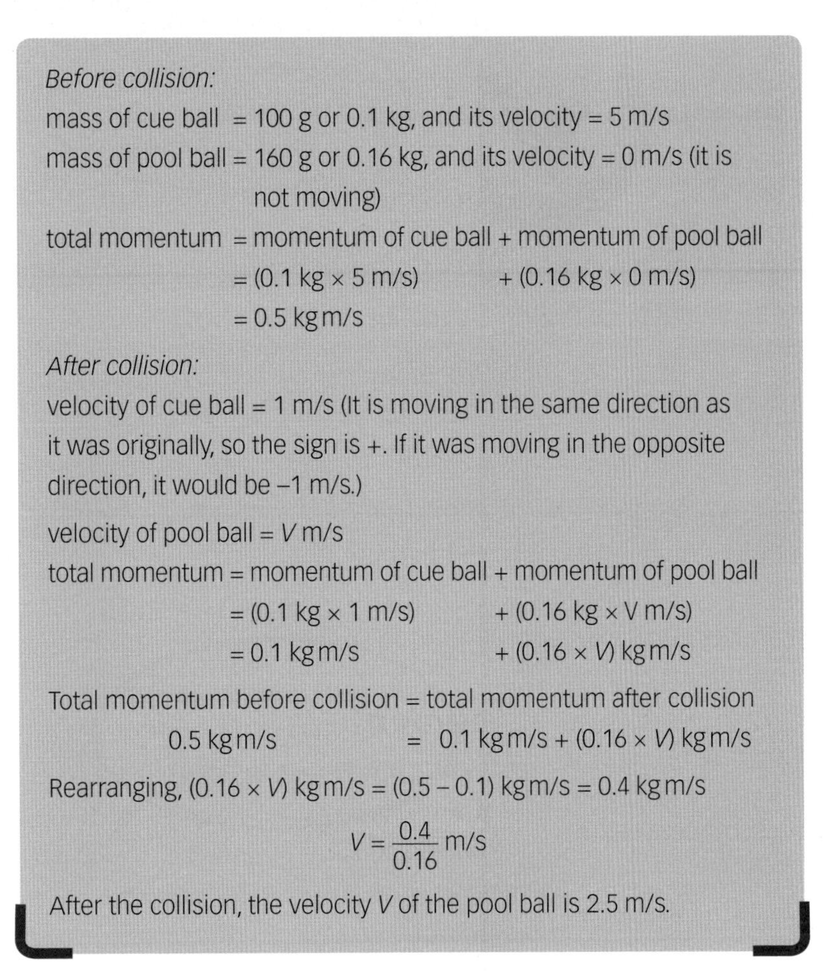

Before collision:

mass of cue ball = 100 g or 0.1 kg, and its velocity = 5 m/s

mass of pool ball = 160 g or 0.16 kg, and its velocity = 0 m/s (it is not moving)

total momentum = momentum of cue ball + momentum of pool ball

$= (0.1\ \text{kg} \times 5\ \text{m/s}) + (0.16\ \text{kg} \times 0\ \text{m/s})$

$= 0.5\ \text{kg m/s}$

After collision:

velocity of cue ball = 1 m/s (It is moving in the same direction as it was originally, so the sign is +. If it was moving in the opposite direction, it would be –1 m/s.)

velocity of pool ball = V m/s

total momentum = momentum of cue ball + momentum of pool ball

$= (0.1\ \text{kg} \times 1\ \text{m/s}) + (0.16\ \text{kg} \times V\ \text{m/s})$

$= 0.1\ \text{kg m/s} + (0.16 \times V)\ \text{kg m/s}$

Total momentum before collision = total momentum after collision

$0.5\ \text{kg m/s} = 0.1\ \text{kg m/s} + (0.16 \times V)\ \text{kg m/s}$

Rearranging, $(0.16 \times V)\ \text{kg m/s} = (0.5 - 0.1)\ \text{kg m/s} = 0.4\ \text{kg m/s}$

$$V = \frac{0.4}{0.16}\ \text{m/s}$$

After the collision, the velocity V of the pool ball is 2.5 m/s.

Car safety

In a crash, a car stops very quickly. Both the car and the occupants have kinetic energy and momentum. You will keep moving and will hit parts of the car such as the windscreen unless you are wearing a seatbelt.

The sudden stop means that a very large change in momentum happens very quickly. The longer the time that this takes to happen, the smaller the forces involved. Air bags, and stretching seat belts increase the time taken for you to stop and crumple zones mean the car takes longer to stop, significantly reducing the forces acting.

These features and others such as side impact bars also absorb some of the huge amounts of kinetic energy involved when they deform.

Key words

momentum, law of conservation of momentum

Exam tip

✓ Remember that mass must be in kilograms not grams.

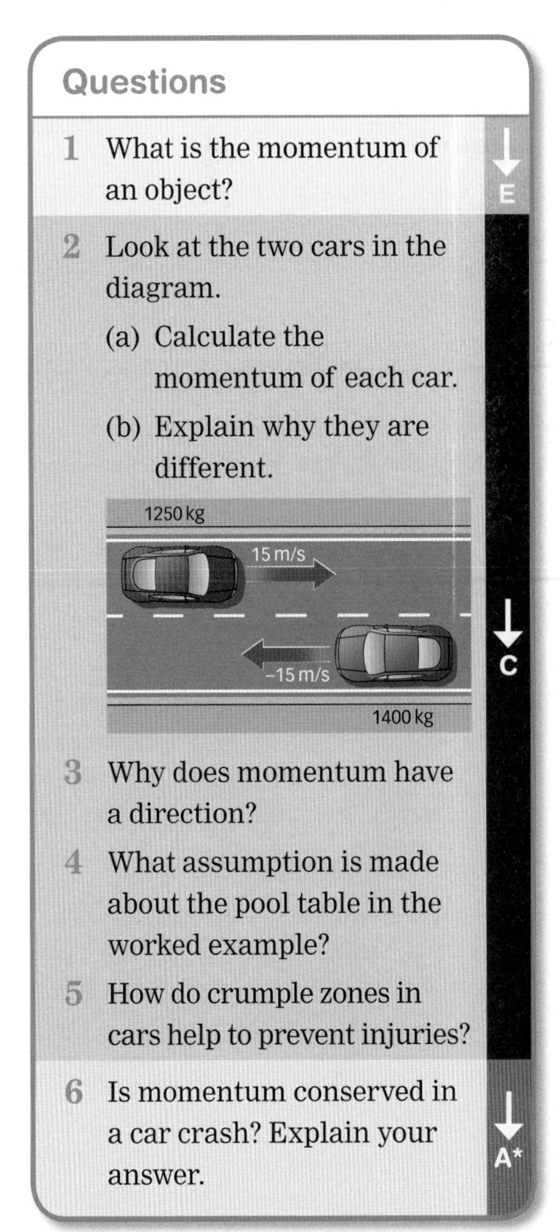

Questions

1. What is the momentum of an object?
2. Look at the two cars in the diagram.
   (a) Calculate the momentum of each car.
   (b) Explain why they are different.

3. Why does momentum have a direction?
4. What assumption is made about the pool table in the worked example?
5. How do crumple zones in cars help to prevent injuries?
6. Is momentum conserved in a car crash? Explain your answer.

Course catch-up

Revision checklist

- Forces are pushes and pulls. Objects exert equal and opposite forces on one another. Resultant force is the sum of all the forces acting on an object.
- Resultant forces change the motion of objects, causing them to accelerate. Resultant force = mass × acceleration.
- Speed is a measure of how fast something is moving. Speed = distance/time. This principle is used in speed cameras.
- Velocity describes direction as well as speed.
- The gradient of a distance–time graph represents speed. Acceleration is a change in speed, and can be positive (speeding up) or negative (slowing down).
- The gradient of a velocity–time graph represents acceleration. The area under a velocity–time graph represents distance travelled.
- A vehicle must overcome the forces of friction and air resistance in order to start moving. When these are balanced by the vehicle's driving force, the vehicle is travelling at a steady speed.
- Stopping distance (thinking distance plus braking distance) is affected by speed, road conditions, vehicle condition, and driver reaction time.
- Weight = mass × gravitational field strength.
- As an object falls under gravity, its velocity increases until its weight is balanced by the force of air resistance. The object then reaches a terminal velocity (steady speed).
- Forces can change the shape of objects. Squashing or stretching a spring gives it elastic potential energy. Hooke's law dictates that the force applied to an object is directly proportional to its extension, up to a certain point.
- Work done is calculated using the equation, work done = force × distance moved in the direction of the force.
- Power = work done/time taken.
- Gravitational potential energy (GPE) is defined by mass, height above the ground, and the Earth's gravitational field. GPE is transferred into kinetic energy (KE) when an object moves.
- A moving object's KE is defined by its mass and its speed.
- The momentum of a moving object is defined by its mass and its velocity.
- Car safety devices increase the time taken for the change in momentum experienced in a crash.

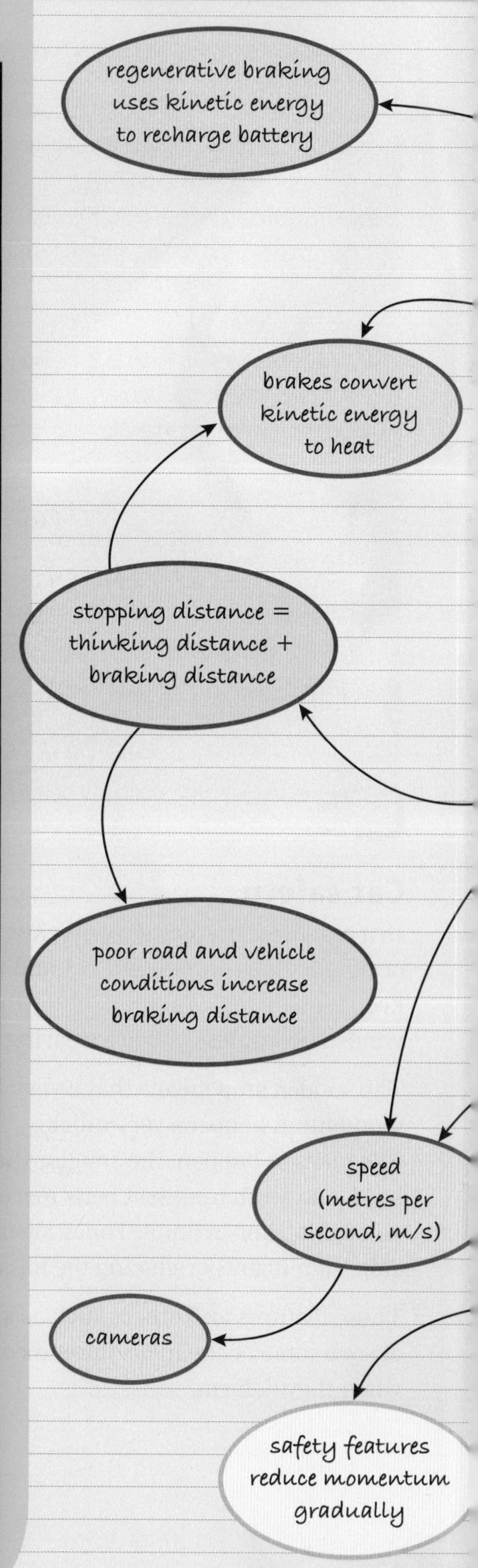

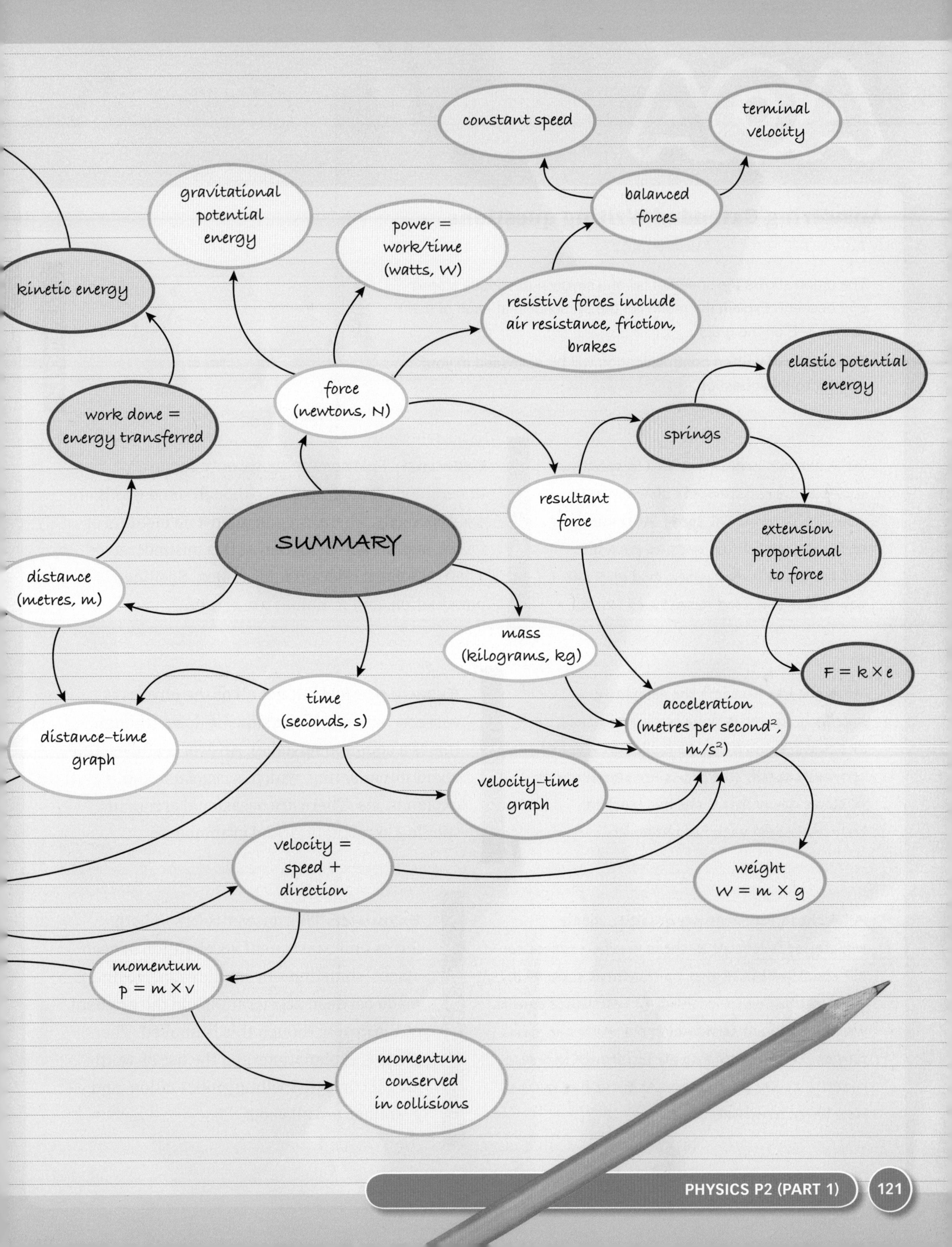
constant speed
terminal velocity
gravitational potential energy
power = work/time (watts, W)
balanced forces
kinetic energy
resistive forces include air resistance, friction, brakes
elastic potential energy
force (newtons, N)
work done = energy transferred
springs
resultant force
extension proportional to force
SUMMARY
distance (metres, m)
mass (kilograms, kg)
F = k × e
time (seconds, s)
acceleration (metres per second², m/s²)
distance–time graph
velocity–time graph
velocity = speed + direction
weight W = m × g
momentum p = m × v
momentum conserved in collisions

Answering Extended Writing questions

QUESTION

The graph shows the speed of fall of a skydiver at various stages of a descent. Explain the motion of the parachutist at each of the stages A–E on the graph.

The quality of written communication will be assessed in your answer to this question.

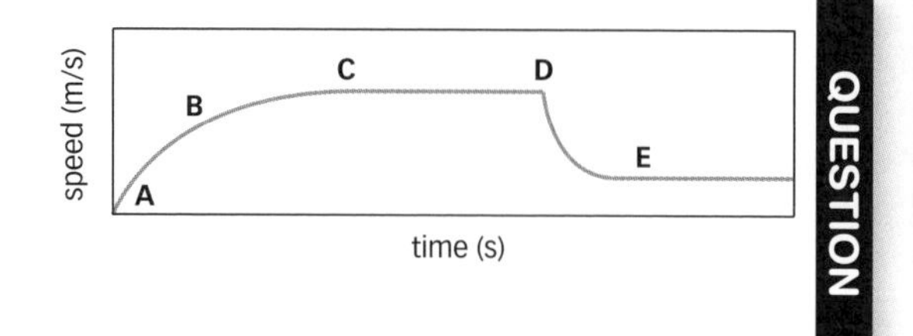

G–E

At A she has jumpd out and is speeding up At B are resistance is slowwing her down, C is terminal spede. At D she opens the parachute, goes up for a bit till her chute is fuly open and at E she comes down a good desent and lands.

Examiner: This answer shows only vague understanding of the physics, though some words are used correctly. There is again almost no mention of forces, and the actual motion at D is misunderstood. There is some irrelevant information. Spelling, punctuation, and grammar are erratic.

D–C

At A she has just left the aeroplane, speeding up. At B there is air resistence, so she slows down a bit. At C she has reached terminal speed. She opens her parachute at D, slows down. By E she has reached terminal speed with her parachute open.

Examiner: Most but not all of the physics is correct – at B she is not slowing down (though this is a common mistake). Answer includes little about forces acting, which is crucial to the explanations. There are occasional errors in spelling, punctuation, and grammar.

B–A*

At A she is accelerating quickly, as air resistance is low. At B she is moving faster, so air resistance is higher, so resultant force and acceleration are lower. At C air resistance equals weight, she is at terminal speed. At D she opens parachute – area and so air resistance increase greatly, so she decelerates. At E she has reached new lower terminal speed.

Examiner: This answer refers to both forces and consequent acceleration at each stage – sometimes by implication. No link is made between acceleration and the gradient of the graph, though this is implied. The physics explanations and the use of words are all correct. Spelling, punctuation, and grammar are all good.

Exam-style questions

1 Match these quantities with their units.

A01

force	m/s
mass	m
acceleration	J
velocity	m/s²
kinetic energy	W
distance	kg
power	N

2 This is a velocity–time graph for a tube train moving between stations.

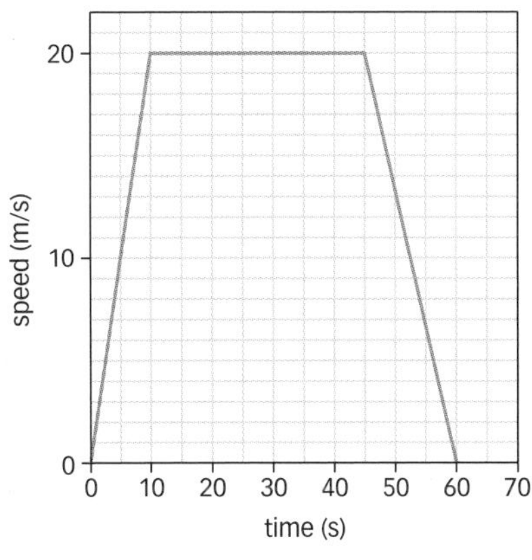

A02 **a** How long did the journey take altogether?

A02 **b** Calculate the acceleration in the time 0–10 s.

A02 **c** Calculate the acceleration in the time 45 s–60 s.

A02 **d** Describe what is happening in the time 45 s–60 s.

A02 **e** Calculate the distance the train moves while it is travelling at a steady speed.

G–E

D–C

3 A weightlifter heaves a load of 30 kg from the floor to a height of 2 m. He then drops the bar, which hits the ground.

A02 **a** Calculate the weight of the load.

A02 **b** How much gravitational potential energy does the load have at the top of the lift?

A02 **c** Calculate the speed at which the load hits the floor.

4 A model rocket has mass 5 kg. When it is fired vertically upwards, the initial thrust from the exhaust gas is 80 N.

A02 **a** What is the resultant upward force accelerating the rocket?

A02 **b** What is the initial acceleration of the rocket?

A02 **c** Assuming the thrust remains 80 N, explain why the acceleration of the rocket would increase.

B–A*

Extended Writing

5 A01 What is meant by the terms stopping distance, thinking distance, and braking distance?

A02 Explain why a car's braking distance is greater if the road is icy or wet.

G–E

6 A02 An archer fits an arrow to his bow, draws the bow back and then fires the arrow, which sticks into a target. Describe the energy changes that occur.

D–C

7 A02 Explain how airbags in a car help to protect you if a collision occurs.

B–A*

A01 Recall the science

A02 Apply your knowledge

A03 Evaluate and analyse the evidence

P2 Part 2

Electricity and radiation

AQA specification match	
P2.3	2.16 Electrostatics
P2.3	2.17 Current and potential difference
P2.3	2.18 Circuit diagrams
P2.3	2.19 Current–potential difference graphs
P2.3	2.20 Series circuits
P2.3	2.21 Parallel circuits
P2.3	2.22 Lamps and light-emitting diodes (LEDs)
P2.3	2.23 Light-dependent resistors (LDRs) and thermistors
P2.4	2.24 Direct current and alternating current
P2.4	2.25 Mains electricity in the home
P2.4	2.26 Current, charge, and power
P2.5	2.27 The atom and the nucleus
P2.5	2.28 Radioactive decay and half-life
P2.5	2.29 Alpha, beta, and gamma radiation
P2.5	2.30 Uses of radiation
P2.6	2.31 Fission and fusion
P2.6	2.32 Star life cycles
Course catch-up	
AQA Upgrade	

Why study this unit?

What would the world be like without electricity? A flow of tiny, negatively charged, sub-atomic particles is vital to the operation of every electrical appliance, from small touch-screen mobile phones to large 3D TVs. An understanding of electric current is essential to all scientists, engineers, and anyone interested in how things work. In this unit you will learn about electric circuits, the differences between static, current, and mains electricity, how to control an electric current, and some of the dangers of all forms of electricity.

You will also learn more about the atom, and how radiation from unstable nuclei breaking down surrounds us all of the time, continuously bombarding the cells in our bodies. Finally, you will learn how scientists have been able to split the atom to devastating effect, while they have yet to master fusing it back together. Looking up at the stars may yet offer solutions to this challenge, and provide a clean, cheap energy source for the future.

You should remember

1. Electricity can produce a variety of different effects, including heating.
2. There are two different types of electric circuit: series and parallel.
3. Electric circuits can be used to control an electric current.
4. Some materials are charged. Opposite charges attract, and like charges repel.
5. All materials are made up of atoms.

Lightning is one of nature's most impressive and terrifying uses of electricity. Each strike only lasts around 30 millionths of a second, but in that time it transfers a massive 5 billion joules of energy to its surroundings. The voltage in a strike can be as high as 100 million volts – that's the same as 66 million AA batteries.

Even more frightening is what happens to the air around this giant electric spark. The air is heated to almost 30 000 °C, which is five times hotter than the surface of the Sun. The heating takes place so rapidly that the air expands in a supersonic shock wave. We hear this explosion as thunder.

Learning objectives

After studying this topic, you should be able to:

- describe how some materials can become charged by rubbing them
- describe static electricity effects in terms of the transfer of electrons
- explain that like charges repel and opposite charges attract

Electrostatic charge

You have experienced an effect of **static electricity** if you have ever had a shock when touching a metal door knob, or getting out of a car.

When you rub certain types of **insulating materials** together, they can become charged. This is sometimes known as an **electrostatic charge**.

Insulating materials can be charged with positive charge or negative charge. The charge is caused by electrons, which have a negative charge, being transferred from one insulating material to another when they are rubbed together.

Positives and negatives

When an acetate rod is rubbed with a woollen cloth, electrons move from the acetate rod to the cloth. The acetate rod becomes positively charged.

When a polythene rod is rubbed with a woollen cloth, electrons move from the cloth to the polythene rod. The polythene rod becomes negatively charged.

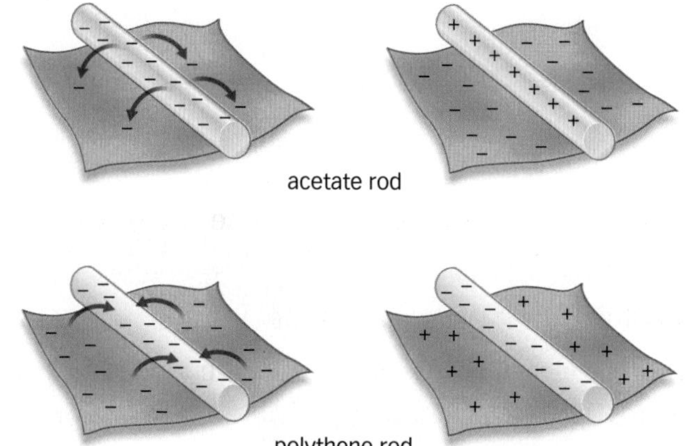

▲ The rods become positively or negatively charged when electrons are transferred

These effects are only seen with insulating materials, where the transferred charge will stay in a particular area. In materials such as metals, electrons can move freely through the material and can travel to earth, charge does not accumulate in one place.

A Why might rubbing an insulator cause it to become charged?

B Has a positively charged object gained or lost electrons?

Exam tip

- Remember that you can only charge something with static electricity by rubbing two insulators together.

Key words

static electricity, insulating material, electrostatic charge, repel, attract

Opposite charges attract, like charges repel

If two bodies are both positively charged, they will **repel** each other, or push each other away. If both bodies are negatively charged, they will also repel each other. But if the two bodies have opposite charges, they will **attract** each other.

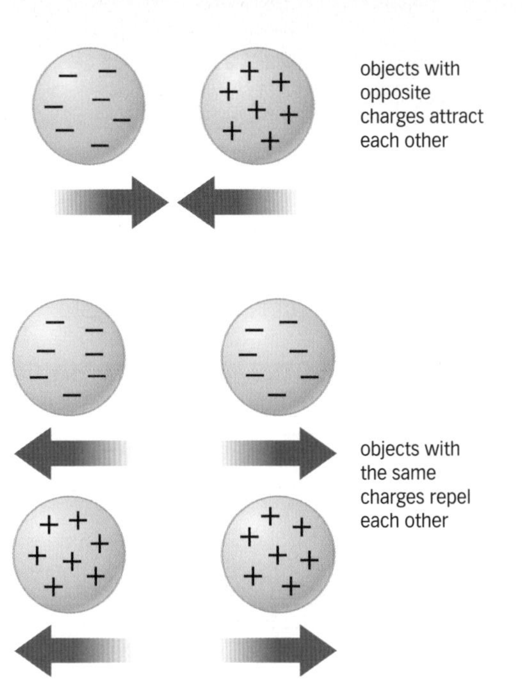

▲ Attraction and repulsion of charged particles

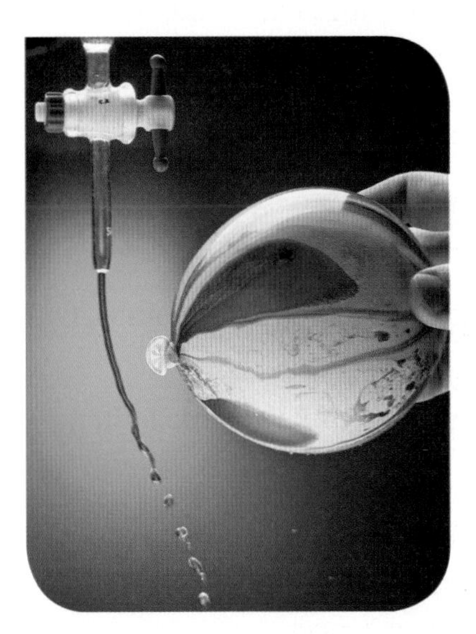

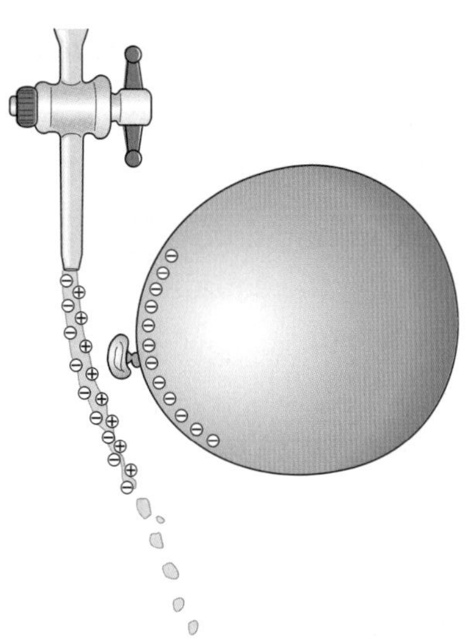

▲ The uncharged stream of water is attracted to the charged balloon

Did you know...?

Charged objects can attract other uncharged objects. The balloon in the photo on the left has been negatively charged by rubbing. The negative charge on the balloon is repelling away electrons in the stream of water. So the surface of the water nearest the balloon is now positively charged. The stream of water is attracted towards the balloon.

Questions

E

1 Name the two types of charge.

C

2 What kind of material can become charged when you rub it?
3 Do like charges attract or repel each other?
4 Explain how a plastic comb could be used to pick up small pieces of paper.

A*

5 (a) Explain how an acetate rod becomes charged when you rub it with a woollen cloth.
(b) A polythene rod is rubbed with metal foil. Will it become charged? Explain your answer.

▲ The boy has gained an electrostatic charge while sliding down. His hair is sticking out because the charges on the hairs are all the same type, so they are repelling one another.

Learning objectives

After studying this topic, you should be able to:

- understand that electric current is the rate of flow of charge
- calculate the size of an electric current
- explain what potential difference is
- explain how to connect cells in series to increase the potential difference

A What type of charge do electrons carry?

B A charge of 1950 C flows past a point in a kettle element in 150 seconds. What is the size of the current flowing through the kettle?

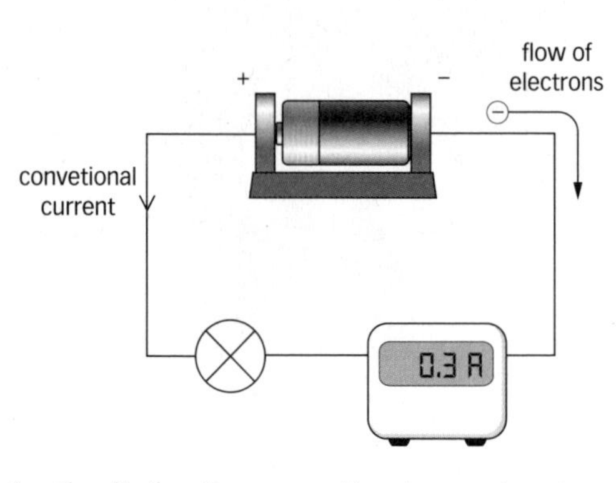

Circuit showing conventional current and electron flow

Current and charge

An electric **current** is a flow of **charge** around a circuit. Negatively charged electrons flow from the negative terminal of a **cell** round the circuit to the positive terminal.

The charge carried by each electron is very small, so electric charge is measured in a much larger unit called the **coulomb**, with the symbol C.

The size of the current is given by the amount of charge passing a point in a circuit each second.

You can calculate the size of the current flowing in a circuit by using the equation:

$$\text{current (amperes, A)} = \frac{\text{charge flowing past a point (coulombs, C)}}{\text{time (seconds, s)}}$$

If the current is called I, the charge Q, and the time t, then:

$$I = \frac{Q}{t}$$

Worked example 1

In 60 seconds a charge of 300 C flows past a point in a circuit. What is the current in the circuit?

$$\text{current} = \frac{\text{charge flowing past a point}}{\text{time}} \quad \text{or } I = \frac{Q}{t}$$

charge = 300 C and time = 60 s

$$\text{current} = \frac{300\text{ C}}{60\text{ s}}$$

$$= 5\text{ A}$$

Nowadays, we know that the negatively charged electrons move from the negative terminal of a cell through the circuit to the positive terminal. However, earlier scientists thought that charge moved from the positive terminal to the negative; they showed this **conventional current** direction on their circuit diagrams. We still use this notation today – circuit diagrams show current flowing from positive to negative.

Potential difference

As electrons pass through the cell in a circuit, they gain energy. They lose energy as they pass through the components of the circuit, when their electrical energy is transferred into other types of energy.

The difference between the energy carried by the current going into a component and the energy carried by the current leaving the component is measured using the **potential difference**.

The potential difference is the amount of energy transferred (or work done) for each coulomb of charge as it passes through the component. It is measured in volts.

$$\text{potential difference (volts, V)} = \frac{\text{energy transferred (or work done) (joules, J)}}{\text{charge (coulombs, C)}}$$

If the potential difference is called V, the work done (or energy transferred) W and the time t, then:

$$V = \frac{W}{Q}$$

Worked example 2

When a charge of 30 C flows through a DVD player, 6900 J of electrical energy are transferred. What is the potential difference across the DVD player?

$$\text{potential difference} = \frac{\text{energy transferred}}{\text{charge}}$$

energy transferred = 6900 J and charge = 30 C

$$\text{potential difference} = \frac{6900\text{ J}}{30\text{ C}}$$
$$= 230\text{ V}$$

Potential difference is also known as **voltage**. It can be measured using a voltmeter that is connected across a component.

Increasing potential difference

A cell usually supplies a potential difference or voltage of about 1.5 V. Many appliances need a higher potential difference than this. You can increase the potential difference supplied by connecting cells end to end in **series** to form a **battery**. We can roughly think of each cell providing a 'push' to the electrons.

You can find the total potential difference by adding up the potential differences of all the connected cells. But the cells must all be 'pushing' the same way – the positive terminal of one cell must be connected to the negative terminal of the next to increase the potential difference.

Key words

current, charge, cell, coulomb, conventional current, potential difference, voltage, series, battery

Did you know...?

It takes the charge on 6 241 509 750 000 000 000 electrons to provide one coulomb of charge!

C When a car engine is started, 600 J is transferred for a charge of 50 C. What is the potential difference across the engine?

Questions

E

1. What is an electric current?
2. A torch needs 4.5 V. How many 1.5 V cells do you need to use?
3. A charge of 1000 C flows past a point in a hairdryer circuit in 250 seconds. What is the current through the hairdryer?

C

4. A lamp transfers 12 000 J of energy as a charge of 500 C passes through it. What is the potential difference across the lamp?
5. A current of 0.25 A flows through a computer for 30 minutes. How much charge passes through the computer?

A*

18: Circuit diagrams

Learning objectives

After studying this topic, you should be able to:

- recognise standard circuit symbols
- draw and interpret circuit diagrams

Did you know...?

Circuit symbols are now an almost universal language. It means that when you draw a circuit diagram, most electricians around the world can understand it.

Circuit symbols

These are some circuit components and their symbols.

Symbol	Component	Photo
	open **switch** closed switch	
	cell	
	battery	
	diode	
	resistor	
	variable resistor	
	lamp	
	fuse	
	voltmeter ammeter	
	thermistor	
	light-dependent resistor (LDR)	
	light-emitting diode (LED)	

Exam tip

- You need to know all of the standard symbols shown in the table.

A Draw the symbols for a variable resistor, a light-emitting diode (LED), an ammeter, and a voltmeter.

Circuit diagrams

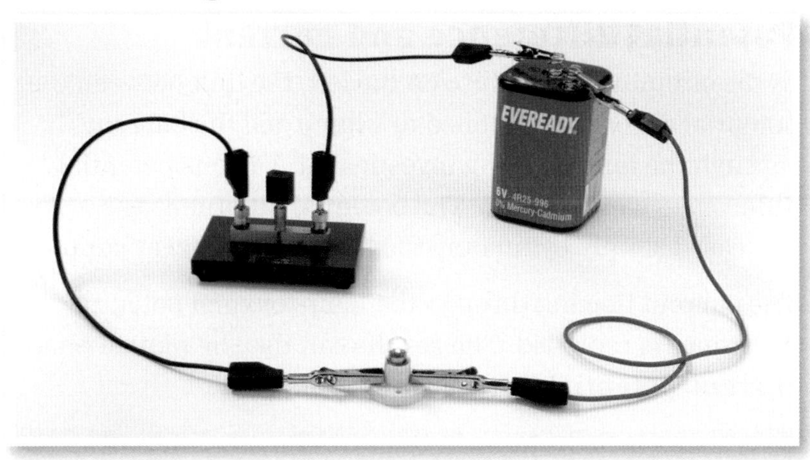

▲ A simple circuit with a battery, lamp, and switch

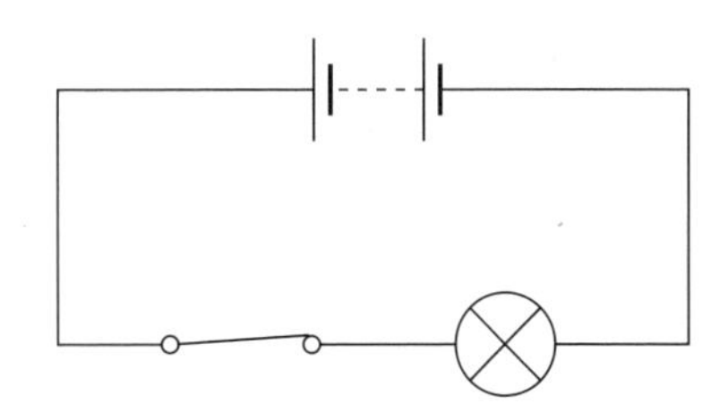

▲ The circuit diagram for the circuit shown in the photo

Look at the circuit shown in the photo. You can draw a circuit diagram using the standard symbols.

You can use circuit diagrams to set up or describe a circuit, or to interpret how a particular circuit might behave.

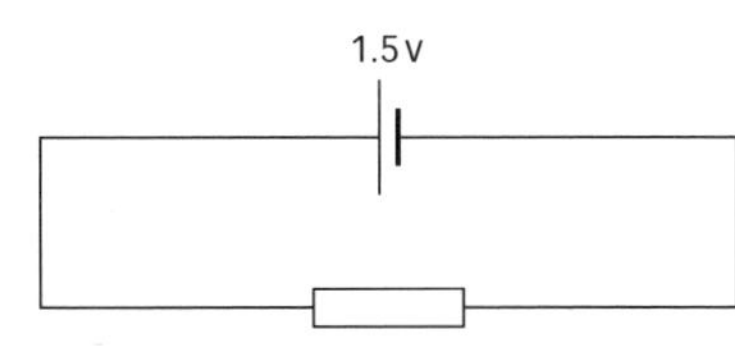

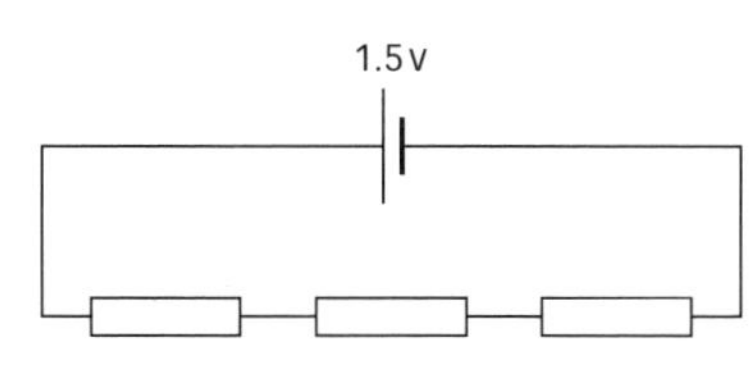

▲ Two simple circuits

All the resistors in the two circuit diagrams above are identical. So it will take charged electrons longer to move around the circuit with three resistors. The flow of charge is slower. In other words, the current is smaller.

B What components are shown in these two circuit diagrams?

C If all four resistors are identical, which circuit has the highest current?

Key words

switch, diode, resistor, variable resistor, fuse, thermistor, light-dependent resistor, light-emitting diode

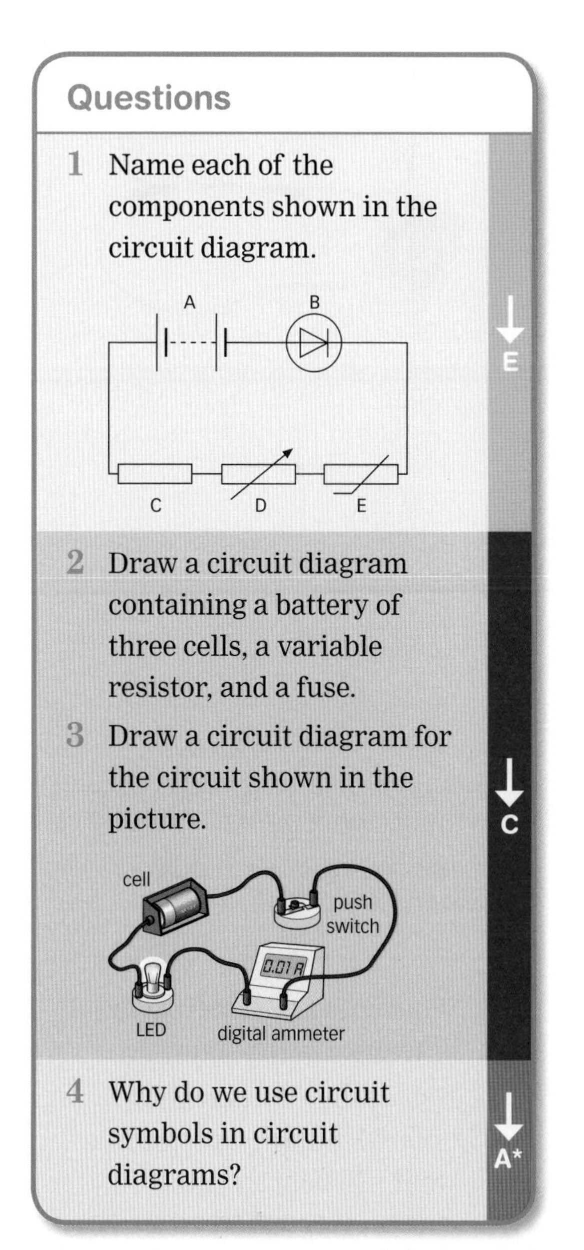

Questions

1 Name each of the components shown in the circuit diagram.

2 Draw a circuit diagram containing a battery of three cells, a variable resistor, and a fuse.

3 Draw a circuit diagram for the circuit shown in the picture.

4 Why do we use circuit symbols in circuit diagrams?

E
C
A*

Learning objectives

After studying this topic, you should be able to:

- ✔ explain what current–potential difference graphs show
- ✔ describe how to find the resistance of a component
- ✔ calculate current, potential difference, and resistance

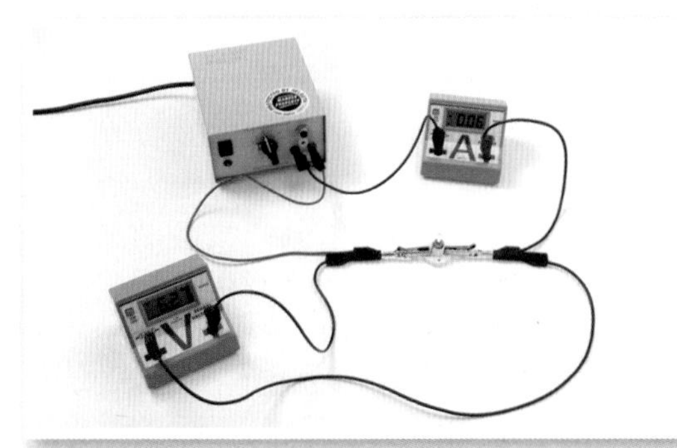

▲ A simple circuit with an ammeter and a voltmeter

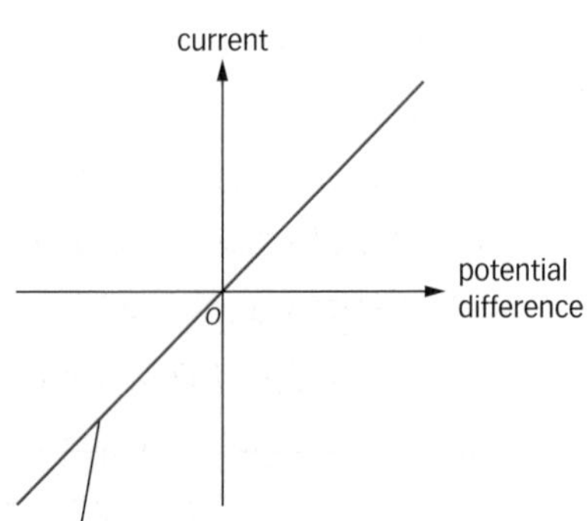

This part of the graph shows that if you reverse the potential difference, you also reverse the direction of the current

▲ Current–potential difference graph for a resistor (at constant temperature)

A What does a current–potential difference graph show?

B If you halve the potential difference across the resistor, what will happen to the current through it?

Potential difference and current

In the circuit in the picture on the left, the link between the potential difference supplied to a lamp and the current through the lamp is being investigated. Various potential differences can be set up. The potential difference is increased simply by turning the knob on the power supply.

The current flowing through the lamp for each potential difference is recorded. The results can then be plotted on a **current–potential difference graph**.

You can replace the lamp in this circuit with different components, and plot the graph of current versus potential difference for each one.

Current–potential difference graph for a resistor

In the graph on the left the line passes through zero and has a straight upward slope (or constant positive gradient). This shows that:

- The current through the resistor is directly proportional to the potential difference across the resistor.
- If you double the potential difference, the current will also double.
- Any change in the potential difference changes the current in the same proportion.

The graph for a resistor will only be a straight line if the temperature of the resistor stays the same.

Calculating resistance, current, and potential difference

If it is more difficult for electrons to pass through a component, we say that the component has a higher **resistance**. This value is shown by the symbol R, and the units are 'ohms' (symbol Ω).

Resistance is linked to current and potential difference by the equation:

$$\underset{\text{(volts, V)}}{\text{potential difference}} = \underset{\text{(amperes, A)}}{\text{current}} \times \underset{\text{(ohms, }\Omega\text{)}}{\text{resistance}}$$

If potential difference is V, current is I and resistance is R, then:

$$V = I \times R$$

Worked example

The potential difference across a lamp is 12 V, and the current flowing through it is 6 A. What is the resistance of the lamp?

potential difference = current × resistance or $V = IR$

$$\text{resistance } (\Omega) = \frac{\text{potential difference (V)}}{\text{current (A)}} = \frac{12\text{ V}}{6\text{ A}} = 2\ \Omega$$

Resistors often have resistance values that are thousands of ohms. So resistance is often given in kilo-ohms (kΩ), with 1 kΩ = 1000 Ω. When the resistance is large, the current will be small – much less than 1 A. When this happens, currents are given in milliamps (mA), with 1 A = 1000 mA.

Resistance and current

You can see from the current–potential difference graphs below that for a potential difference of 10 V, the current through resistor A is 0.3 A, but the current through resistor C is only 0.1 A. It is more difficult for charge to pass through resistor C: it has the highest resistance of the three resistors.

At a particular value of potential difference, the current through any component depends on the resistance of the component. If the resistance increases, the current through the component decreases.

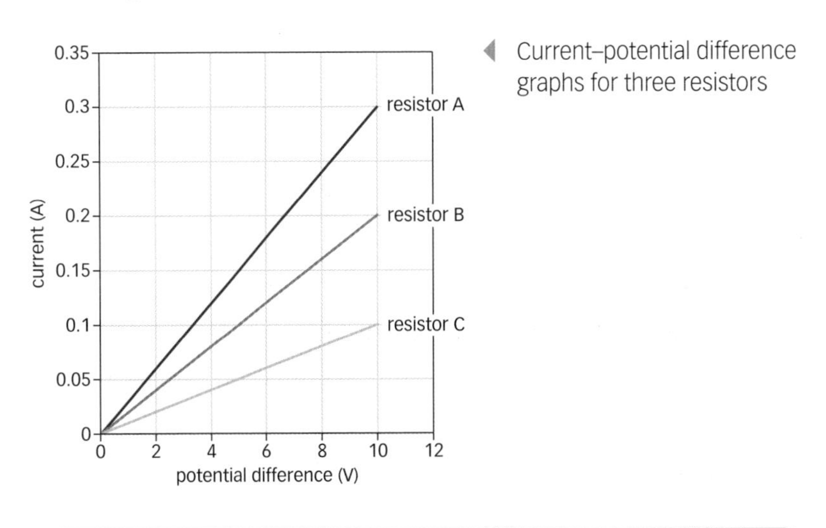

Current–potential difference graphs for three resistors

C How do you know that resistor C has the highest resistance?

Key words

current–potential difference graph, resistance

Exam tip

- ✔ Take care with the units of current and resistance when using them in calculations. Always make sure you convert kilo-ohms to ohms and milliamps to amps when needed.
- ✔ Remember that when resistance goes up, current comes down.

Questions

1. A current–potential difference graph is a straight line. What does this tell you? (E)
2. The current through a 8 Ω resistor is 1.5 A. What is the potential difference across the resistor?
3. Use points from the graph to calculate the resistances of resistors A, B, and C.
4. The potential difference across a variable resistor is 12 V. Calculate the current through the resistor for the following resistances:
   (a) 6 Ω
   (b) 24 Ω
   (c) 1.2 kΩ. (C)
5. What does the gradient of a current–potential difference graph at a particular point represent? (A*)

20: Series circuits

Learning objectives

After studying this topic, you should be able to:

- calculate the resistance of components connected in series
- explain that the size of the current is the same throughout a series circuit
- describe how the potential difference provided by the supply is shared between the components

Did you know...?

Some Christmas tree lights are connected in series – if one of the bulbs breaks, none of the lights will work.

Series connection

In the circuit shown in the picture, all the electrons moving round must pass through first one lamp then the other. There is only one possible path that the current can take. The lamps are connected in series.

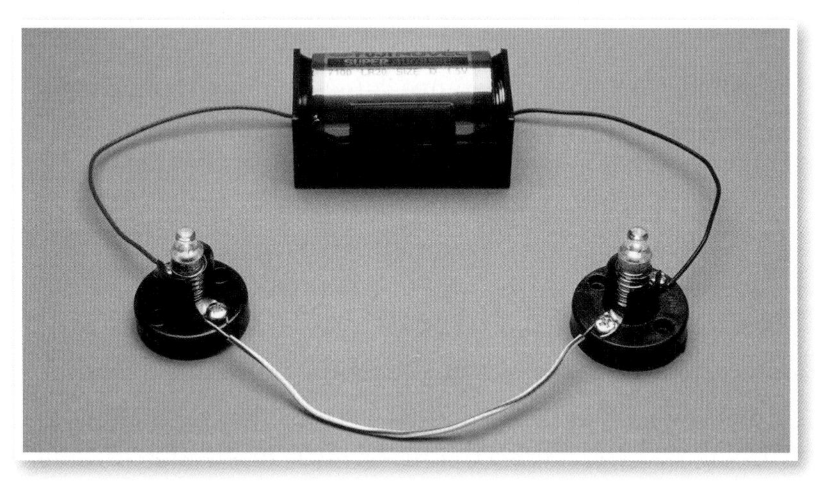

▲ These lamps are connected in series

Resistance in a series circuit

Often, several components can be connected in series in a circuit. It is harder for the current to flow through these circuits because electrons have to flow through each component in turn.

If you add more components, the resistance of the circuit will increase. All components have a resistance, including lamps, LDRs, and thermistors.

You can find the total resistance of the circuit by adding together the resistances of all the components.

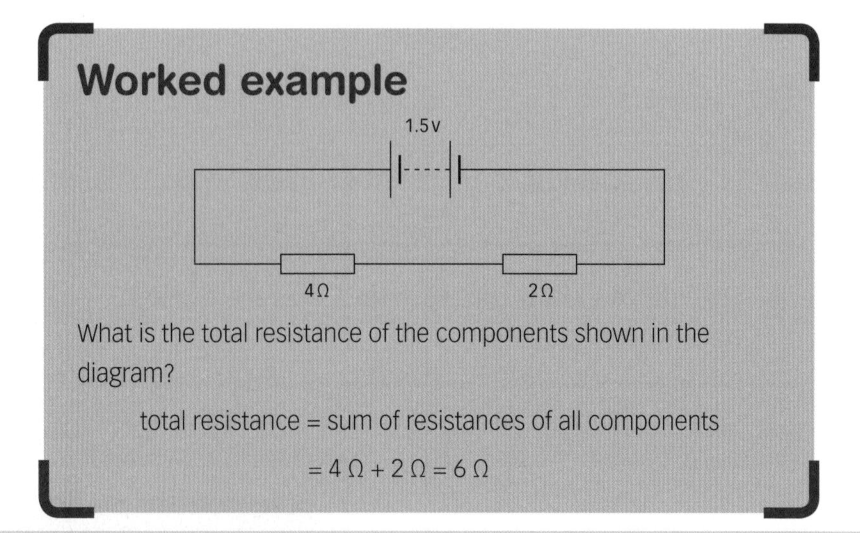

What is the total resistance of the components shown in the diagram?

total resistance = sum of resistances of all components

$= 4\ \Omega + 2\ \Omega = 6\ \Omega$

A Three resistors with resistances 5 Ω, 10 Ω, and 20 Ω are connected in series. What is their total resistance?

Current in a series circuit

The current in a series circuit is the same everywhere in the circuit. In the circuit shown here, all three ammeters show that the current flowing in the circuit is 0.3 A.

The current is the same throughout the circuit because the electrons have to flow at the same rate through all the components in the circuit.

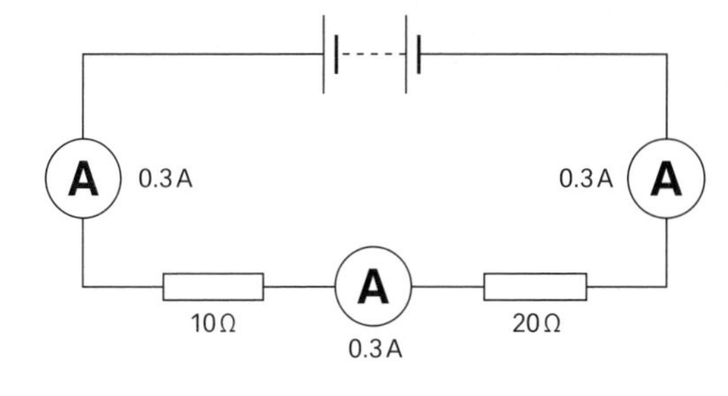

▲ The same current flows through every component in a series circuit

Potential difference in a series circuit

The potential difference of the supply is shared between the resistors in the circuit. The sum of the potential differences across each component will equal the potential difference of the supply.

The total potential difference across the two resistors is 9 V. The potential difference across the 10 Ω resistor is 3 V and the potential difference across the 20 Ω resistor is 6 V. The potential difference has been shared between the components in proportion to the size of the resistance.

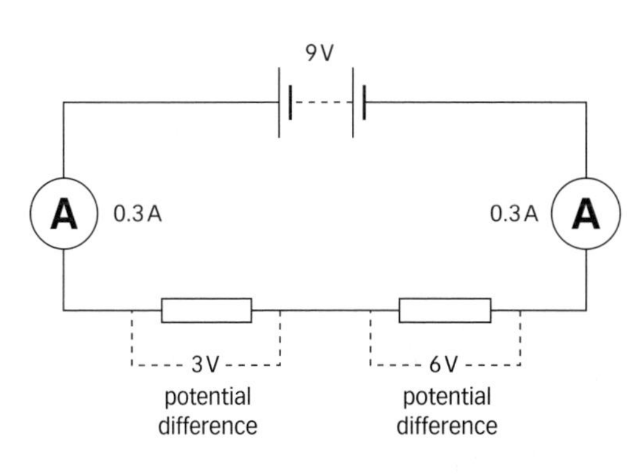

▲ The potential difference of the supply is shared between the resistors

Questions

1. Why is the current the same everywhere in a series circuit?
2. What can we say about the potential difference across components in a series circuit?

E

3. In the worked example circuit, what is the current if the potential difference of the supply is 3 V?
4. Look at the two circuit diagrams.

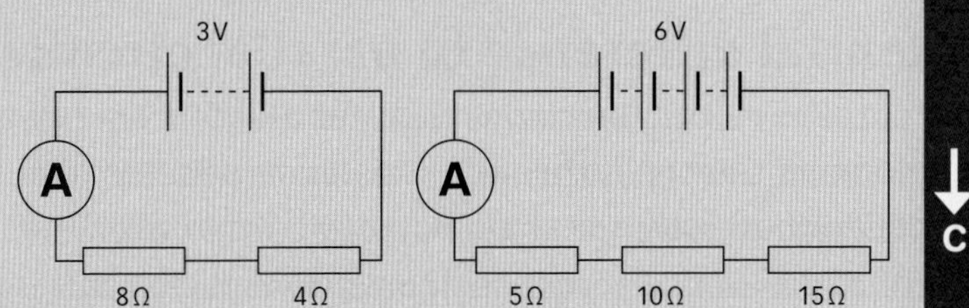

(a) What is the total resistance of each circuit?

(b) What is the current flowing in each circuit?

(c) Calculate the potential difference across each resistor.

C

5. Should an ammeter have a high or low resistance? Explain your answer.

A*

B What is the potential difference across R_2? Explain how you worked out your answer.

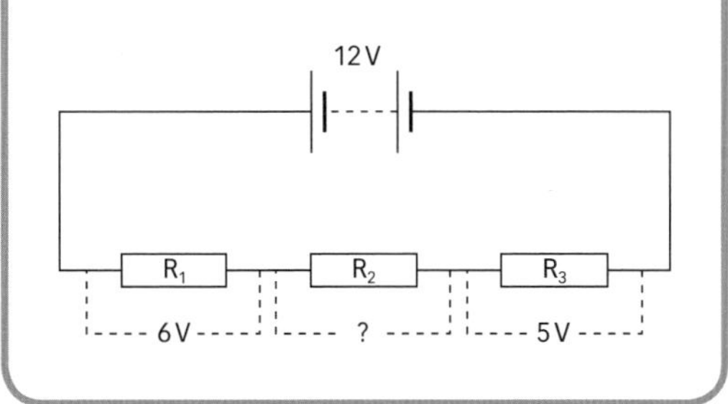

Exam tip AQA

✔ Remember that an ammeter is always connected in series.

P2 21: Parallel circuits

Learning objectives

After studying this topic, you should be able to:

- explain that the potential difference across components connected in parallel is the same
- describe how the current in a parallel circuit splits between the branches

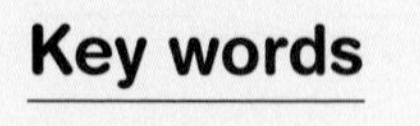

Key words

parallel circuit

Parallel connections

In the circuit shown below, there is more than one route that electrons can take when moving round the circuit. They can move through one lamp or the other. There are two branches that the current can flow through. The lamps are connected in parallel. A **parallel circuit** has two or more branches.

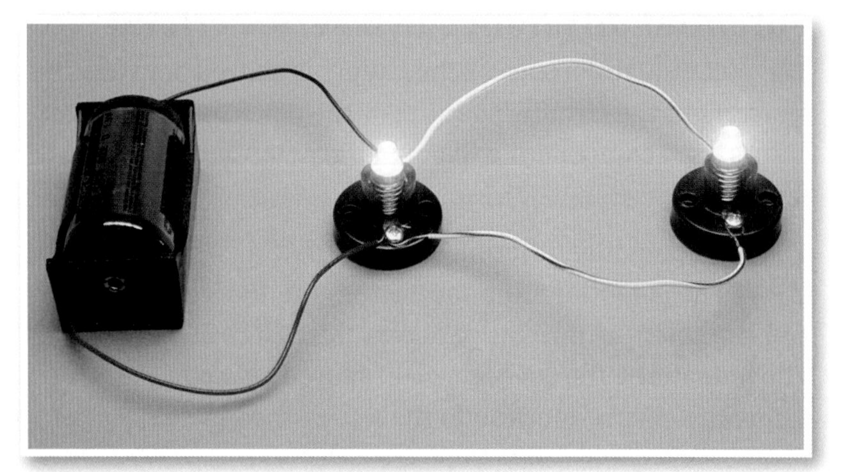

▲ These lamps are connected in parallel

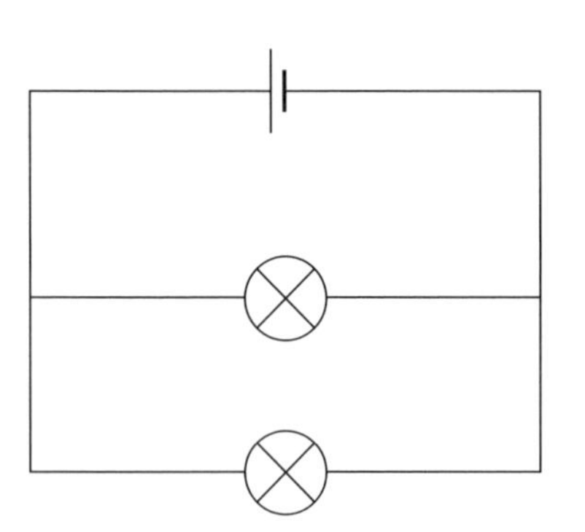

▲ Circuit diagram for the lamps connected in parallel

Resistance in parallel circuits

The resistance in a parallel circuit is lower, because the electrons have alternative ways to flow round the circuit. All the charge does not have to flow through all the components.

A Why is resistance lower in parallel circuits?

Potential difference in parallel circuits

In a parallel circuit, electrons transfer all their energy to the components they pass through. So all components in a parallel circuit have the same potential difference. The potential difference across each branch of the circuit is the same because it does not matter which branch the electrons travel through. Each branch is exposed to the full push of the power supply.

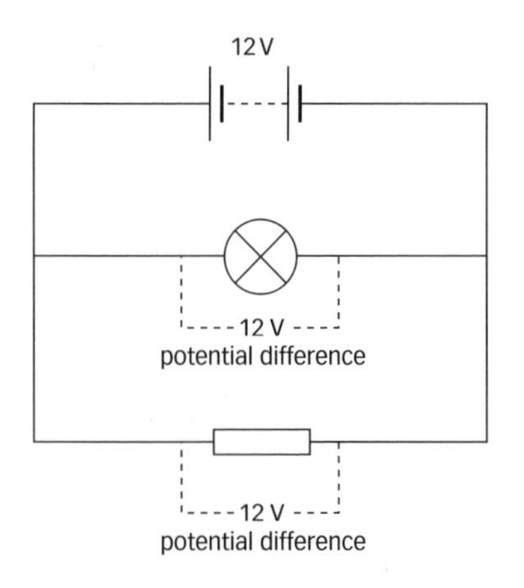

▲ The potential difference is the same across parallel branches in a circuit

B Why is the potential difference the same across the parallel branches in a circuit?

Current in parallel circuits

As there is more than one route for the current to take around a parallel circuit, the current splits between the different branches.

The sum of the currents flowing through all the branches is equal to the total current leaving the battery.

In the circuit shown in the diagram on the right, the current splits into two at **X**. You can see that 0.1 A goes through the lamp, and 0.3 A goes through the resistor. The total current is 0.3 A + 0.1 A = 0.4 A.

At **Y**, the current flowing from the lamp (0.1 A) joins with the current from the resistor (0.3 A) to form a current of 0.4 A again.

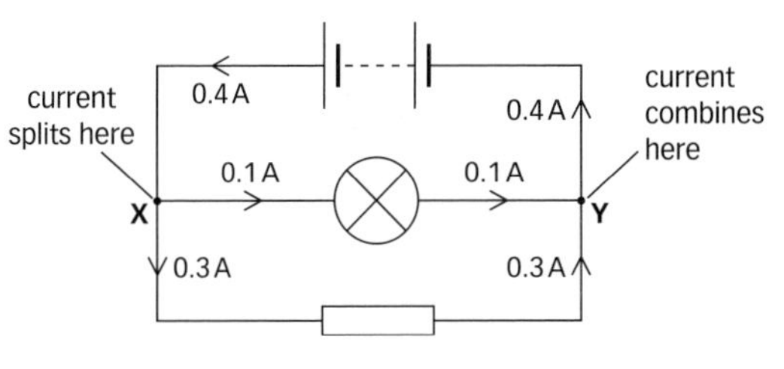

▲ The current from the main part of the circuit is shared between the parallel branches

C What is the sum of the current flowing through the two branches?

Questions

1 What is a parallel circuit?

2 Look at the circuit diagram.

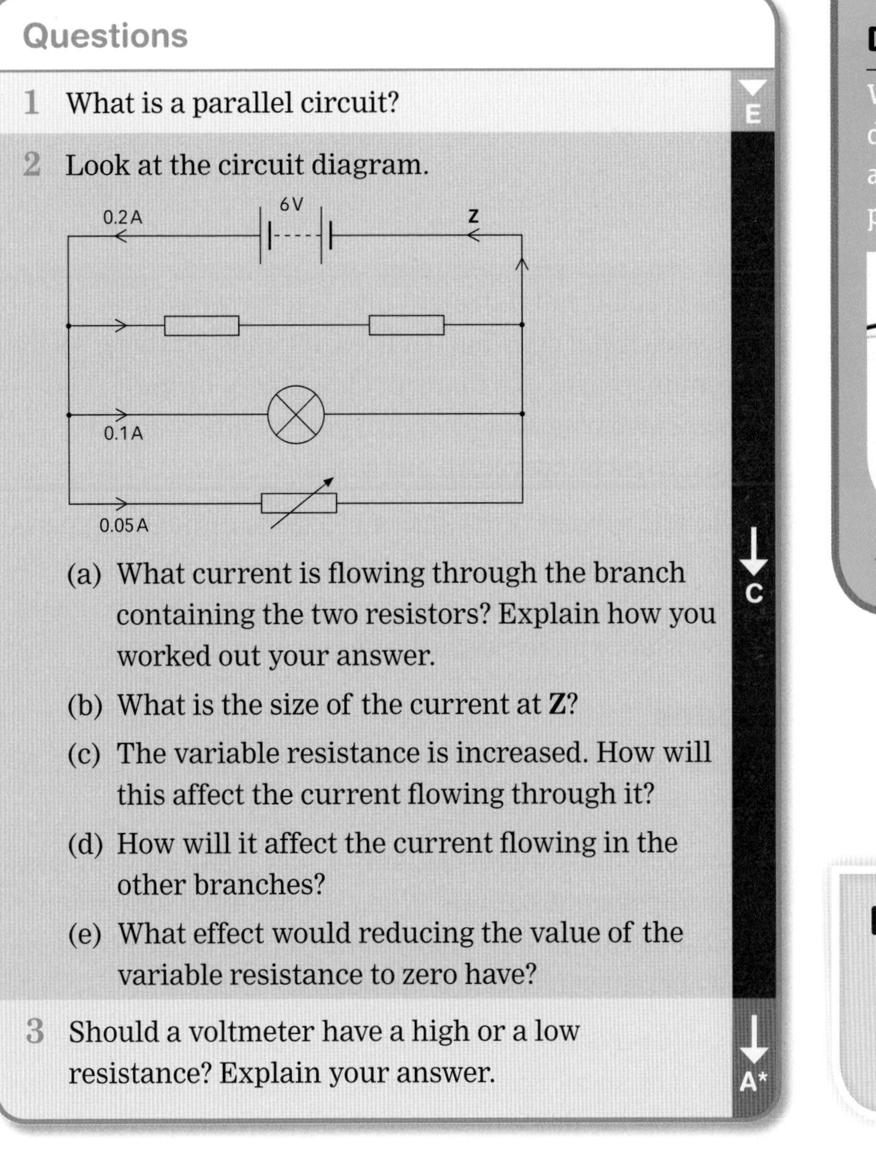

(a) What current is flowing through the branch containing the two resistors? Explain how you worked out your answer.

(b) What is the size of the current at **Z**?

(c) The variable resistance is increased. How will this affect the current flowing through it?

(d) How will it affect the current flowing in the other branches?

(e) What effect would reducing the value of the variable resistance to zero have?

3 Should a voltmeter have a high or a low resistance? Explain your answer.

E ↓ C ↓ A*

Did you know...?

When you measure the potential difference across a component, you always connect the voltmeter in parallel.

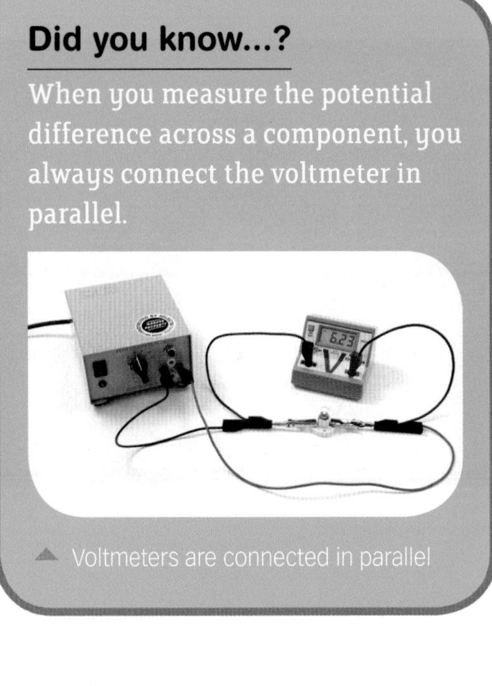

▲ Voltmeters are connected in parallel

Exam tip

✔ Remember that there is more than one path for the current to take in a parallel circuit.

22: Lamps and light-emitting diodes (LEDs)

Learning objectives

After studying this topic, you should be able to:

- describe how the resistance of a filament lamp varies
- describe what a diode does
- understand that a light-emitting diode (LED) emits light when a current flows through it in one direction
- compare the uses of different forms of lighting

Resistance of a lamp

A **filament lamp** consists of a thin coil of wire which is usually made of a metal with a high melting point such as tungsten. When a current flows through the wire, it glows brightly and becomes hot.

When you plot a graph of current against potential difference for a filament lamp, you find that it is not a straight line. It is a curve.

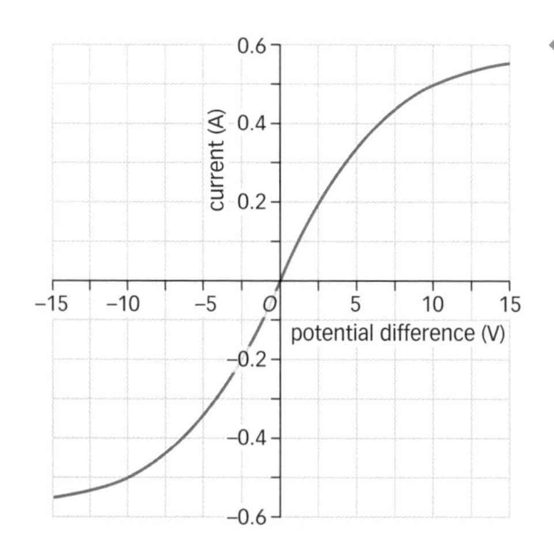

Current–potential difference graph for a filament lamp

A filament lamp

As the graph is not a straight line, it means that the resistance is not constant – it does not stay the same. As the current increases, the filament glows more brightly. The filament also transfers electrical energy into thermal energy, so its temperature increases.

An electric current is the flow of electrons through a material. As the temperature of the coil in the filament lamp increases, it is more difficult for the electrons to flow through the metal. The resistance increases as the temperature increases.

Why does resistance increase with temperature in a filament lamp?

The atoms in a conductor are positively charged ions with electrons that are free to move through the metal. As the temperature increases, the ions vibrate more. This makes it more difficult for the electrons to move through the metal.

Diodes

A diode is a device which lets the current flow through it in one direction only, the 'forward' direction.

At low potential differences, only a very small current flows – typically about 0.001 mA. When the potential difference increases to about 1 V, the resistance of the diode decreases and a current starts to flow through it.

If a diode is connected the other way round, it will allow only a very small current to flow, which is almost zero. This means that the diode has a very high resistance when it is connected in the reverse direction.

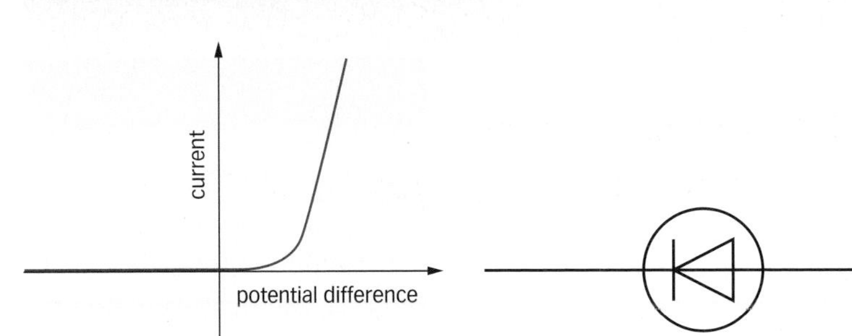

Current–potential difference graph for a diode

Current flows through a diode in the direction of the arrow in the symbol, from positive to negative

LEDs and other types of lighting

An LED, or light-emitting diode, produces light when a current flows through it.

LEDs have a high efficiency and also last a long time. They use a much smaller current than other forms of lighting. They are beginning to be used as replacements for conventional filament light bulbs.

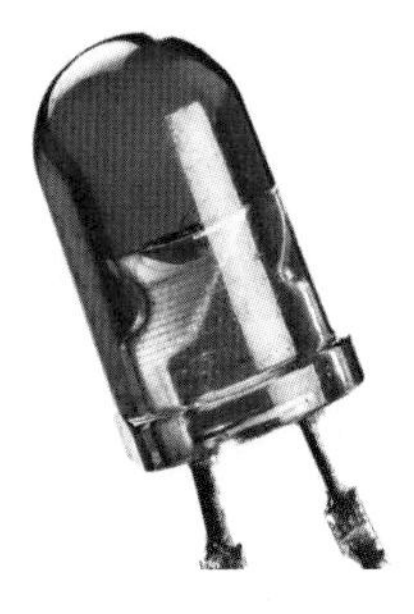

An LED and its circuit symbol

	Type of bulb			
	Filament	Compact fluorescent	Halogen	LED
Power rating	105 W	12 W	20 W	4 W
Efficiency	7.5%	45%	15%	75%
Lifetime	2000 hours	8000 hours	1000 hours	20 000 hours
Cost	£2	£2	£2	£6

A What is an LED?

B Which type of light bulb is the most efficient?

Key words

filament lamp

Exam tip

✔ Remember that when a graph of current against potential difference is a straight line, resistance is constant.

Questions

1. What happens to the resistance of a filament lamp as the temperature increases?

   ↓ E

2. From the graph on the previous page, calculate the resistance of the filament lamp at:
   (a) 5 V (b) 10 V.
   What do you notice about the resistance?

3. (a) How much useful light power is produced by each type of bulb?
   (b) Which type of light bulb has the lowest cost for 20 000 hours of use?
   (c) Which type of light bulb would you recommend using? Explain why.

   ↓ C

4. When a diode is connected in reverse, a very small current of about 0.001 mA flows. Calculate the resistance of the diode if the potential difference is 2 V.

   ↓ A*

23: Light-dependent resistors (LDRs) and thermistors

Learning objectives

After studying this topic, you should be able to:

- understand how a light-dependent resistor (LDR) works
- understand how a thermistor works
- describe applications of LDRs and thermistors

Light-dependent resistors

A light-dependent resistor (LDR) is a special type of resistor. Its resistance changes as the intensity of the light falling on it changes.

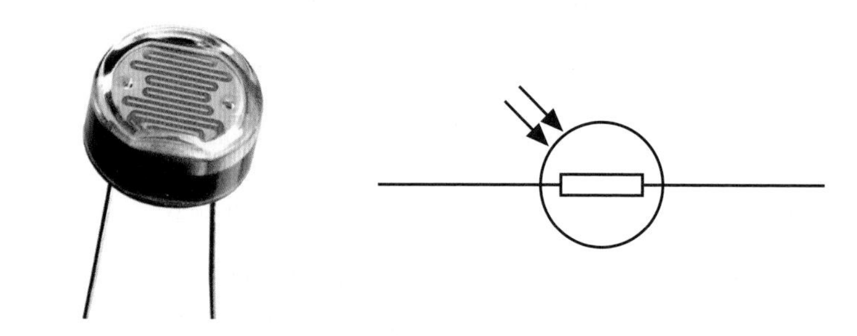

An LDR and its circuit symbol

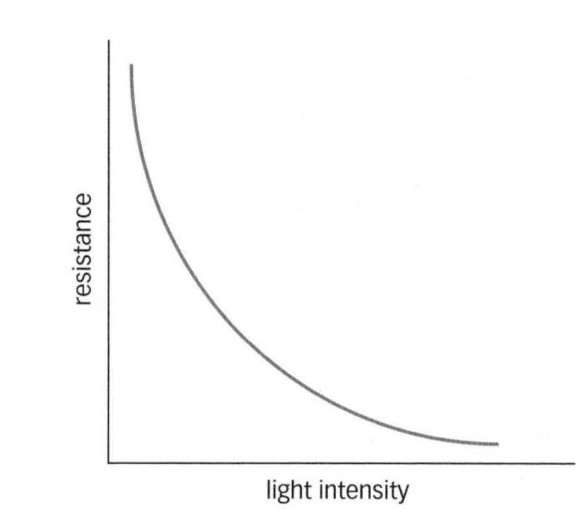

How the resistance of an LDR (light-dependent resistor) varies with intensity of light

A What happens to the resistance of an LDR as the intensity of light increases?

When the light levels are low, the resistance of an LDR is high. This is because an LDR is made of a semiconductor material where the outer electrons are bound weakly to the atoms.

When bright light shines on an LDR, the resistance is much lower. The light energy is transferred to the outer electrons which can then break free from the atoms. They are then free to flow through the LDR.

This change in resistance according to the intensity of light means that LDRs can be used as switches. For example, they can be used to switch on security lights when it gets dark.

Did you know...?

Many street lamps are controlled by light-dependent resistors. If dark clouds reduce the light levels enough, the resistances of the LDRs increase and the street lamps switch on.

Thermistors

A thermistor is another special type of resistor.
Its resistance changes as its temperature changes.

When the temperature of the thermistor is low, its resistance is high. This is because a thermistor is made of a material which does not conduct electricity well at low temperatures. The outer electrons are loosely bound to the atoms, and are not free to flow through the thermistor.

As the temperature increases, more outer electrons gain enough energy to break free from atoms. The electrons are then free to flow through the thermistor. So as the temperature increases, the resistance of a thermistor decreases.

Thermistors can be used as temperature sensors. For example, thermistors are often used in car engines to monitor the temperature of the cooling system. They can be used to switch on a fan if the cooling system goes above a certain temperature. They can also be used to warn the driver if the cooling system is about to overheat.

Thermistors are typically used to measure temperatures in the range –90 °C to 200 °C.

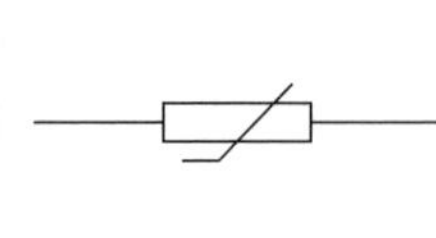

A thermistor and its circuit symbol

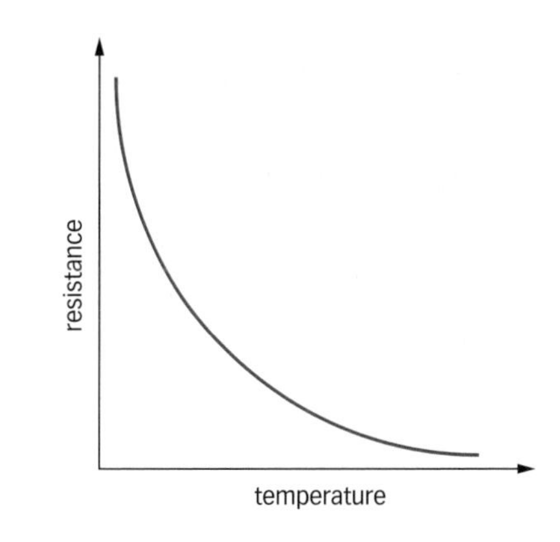

How the resistance of a thermistor varies with temperature

B What happens to the resistance of a thermistor as the temperature increases?

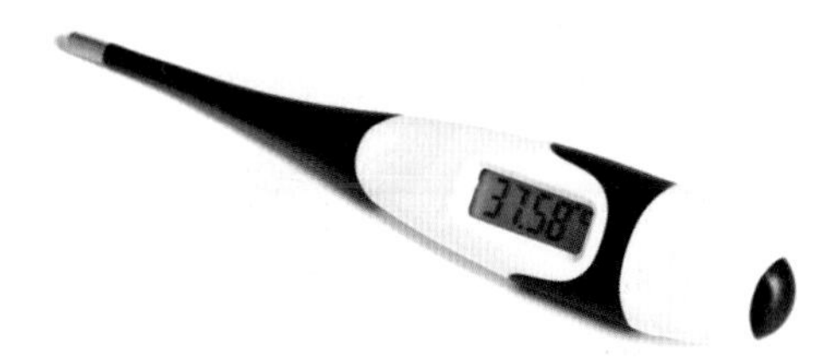

Many digital thermometers use a thermistor for measuring the temperature

Questions

E

1 (a) What is a thermistor?
(b) What is an LDR?

C

2 Describe what will happen to the current in the circuit shown in the diagram as the intensity of light increases.

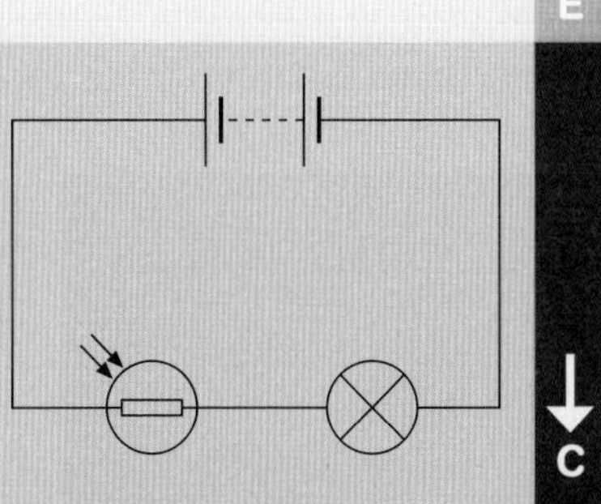

A circuit with an LDR in it

3 Describe how a thermistor could be used to control the temperature in a personal computer.

4 What other applications might LDRs and thermistors be used for?

A*

5 Draw a circuit diagram for a warning system to alert a driver that a car engine is getting too hot.

Exam tip AQA

✔ With a light-dependent resistor, remember that when the light is brighter, the resistance is lower. Similarly, with a thermistor, remember that when the temperature is higher, the resistance is lower.

Learning objectives

After studying this topic, you should be able to:

- ✔ describe electric current as either direct current or alternating current
- ✔ know that mains electricity in the UK is an a.c. supply
- ✔ state that mains electricity in the UK is 230 V a.c. and 50 Hz
- ✔ determine the frequency of an a.c. supply from an oscilloscope

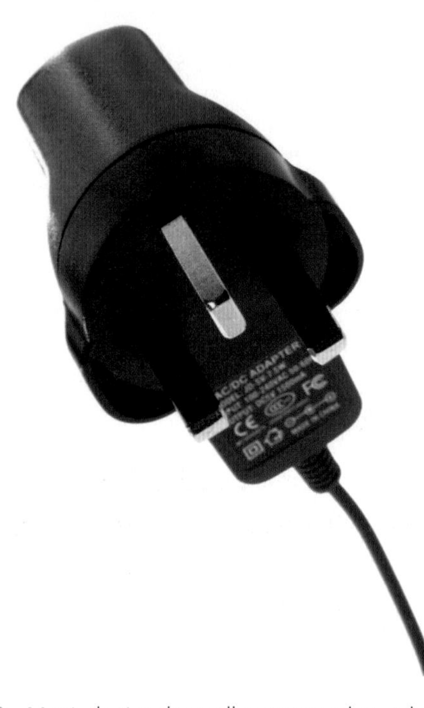

▲ Most electronic appliances need an adaptor to convert a.c. from the mains into the d.c. used by the device

A Give three examples of appliances that use a.c. and three that use d.c.

B What is the frequency of the UK mains supply?

Direct vs alternating current

Think of how many electrical appliances you use. They all require an electric current, either:

- **direct current, d.c.**, or
- **alternating current, a.c.**

You probably use both every day. When you use a mobile phone, laptop or mp3 player you will be using d.c. Direct current is usually produced by batteries or cells. The potential difference from the battery remains the same, and so the current through the appliance is in one direction only.

Electricity from the mains is described as a.c. With alternating current, the potential difference switches smoothly between positive and negative values as part of a repeating cycle. This causes the current to constantly change in size or reverse its direction very rapidly. The number of cycles per second is called the **frequency**. This is measured in hertz (Hz). In the UK the mains supply is 230 V a.c. with a frequency of 50 Hz, that is, 50 cycles per second.

The oscilloscope and a.c.

An **oscilloscope** may be used to produce an image showing how the potential difference of an electrical supply changes with time. With a d.c. supply the potential difference remains constant, and so the screen shows a horizontal line. When an a.c. supply is attached the potential difference changes, and so produces a trace which looks like a sine wave.

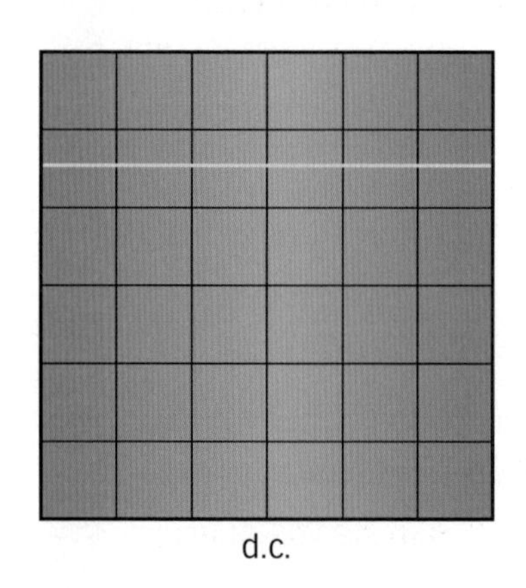

d.c.

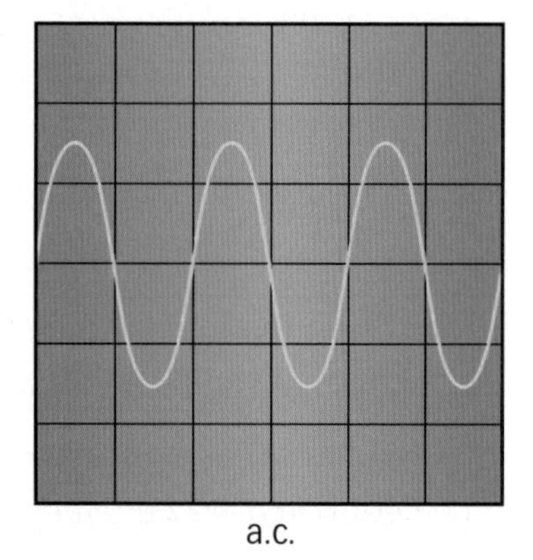

a.c.

▲ Traces on an oscilloscope from d.c. and a.c supplies look very different

The squares on the oscilloscope screen are a bit like the scale on a graph. Both the vertical and horizontal scales can be adjusted on the oscilloscope. The vertical squares measure the potential difference. If the scale is set to 5 volts per division (or 5 V/div) it would mean that each square represents 5 V. A 10 V d.c. supply would cause the green line to jump up two squares.

The horizontal scale is time. Each square can be set to be a number of seconds (or more commonly milliseconds). This is called the **time base**. A time base of 0.4 s/div would mean each square represents 0.4 s.

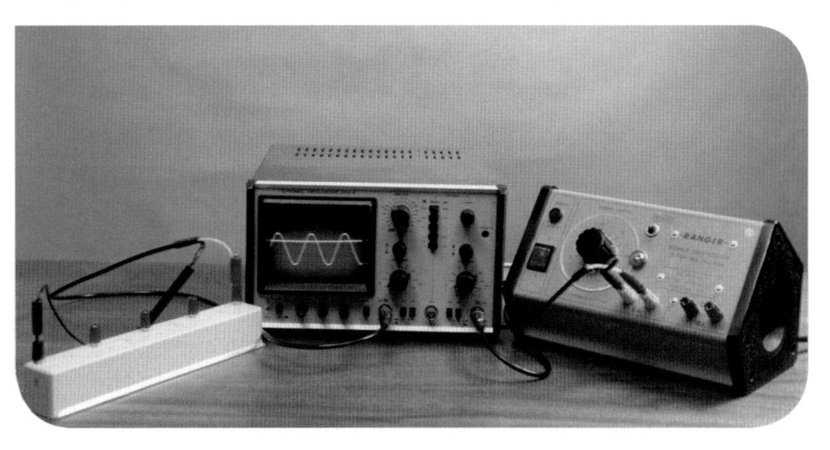

An oscilloscope can be used to see how the potential difference of an a.c. supply changes with time

Finding the frequency of a supply

The frequency of an a.c. supply can be calculated if you know the time for one cycle. This is called the **time period**. Time period and frequency are related as follows in the equation below:

$$\text{frequency (hertz, Hz)} = \frac{1}{\text{time period (seconds, s)}}$$

A time period of 2 s would correspond to a frequency of 0.5 Hz. A time period of 0.01 s would mean that the frequency is 100 Hz.

Using the trace on the screen of an oscilloscope, and knowing the time base, you can determine the time period of an a.c. supply. If the time base is set at 0.05 s/div and the cycle repeats every 2 squares, then the time period is 0.10 s (2 × 0.05 s). From the time period you can calculate the frequency, in this case 10 Hz.

Key words

direct current, d.c., alternating current, a.c., frequency, oscilloscope, time base, time period

Did you know...?

Mains voltage was not standardised in the UK until 1926. Until then different parts of the country had different potential differences and frequencies. If you bought an electrical device from one part of the country it might not have worked where you lived! The supply in the UK used to be 240 V a.c., but in 1993 it was changed to the current value of 230 V a.c. to match the rest of Europe.

Questions

1 What is the potential difference of the UK mains supply? (E)

2 Sketch the trace that would be seen on an oscilloscope for:

(a) a d.c. battery

(b) an a.c. supply. (C)

3 Calculate the frequency of the supply shown in the diagram if the time base is set at 0.02 s/div.

4 Sketch the trace seen on an oscilloscope screen if the time base is set to 0.01 s/div and the V/div is set at 200 V/div and it is connected to the UK mains supply. (A*)

25: Mains electricity in the home

Learning objectives

After studying this topic, you should be able to:

- understand the structure of the UK three-pin plug and its cables
- describe and compare some of the electrical safety features in the home (including circuit breakers and fuses)
- describe what happens when an appliance is earthed

Key words

earth, live, neutral, residual current circuit breaker, double-insulated

A 3-core and a 2-core wire. Both contain a live and a neutral wire, but only the 3-core has an earth wire.

Fuses usually come as replaceable cartridges. The wire inside the fuse heats up and melts if there is too much current.

The UK plug

If you've ever been on holiday abroad, you know that you had to take a plug adapter to use or recharge any of your electrical appliances. Different countries use different plug designs depending on their own electrical systems. The UK plug is unusual as it contain three connections (or pins).

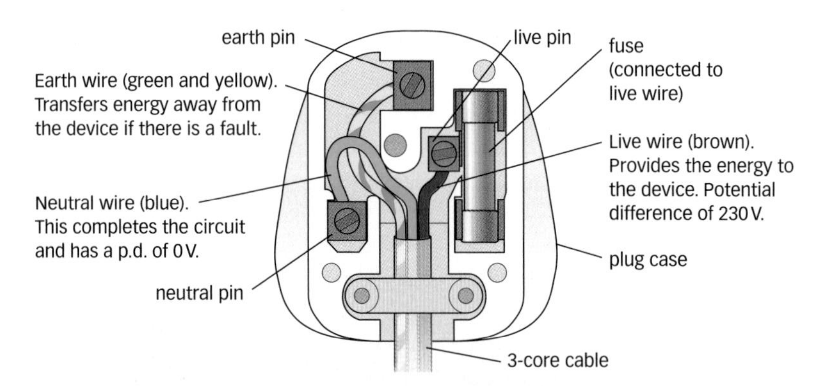

A standard UK three-pin plug

3-core and 2-core cables

The cables connecting an appliance such as a TV to the mains are often described as 3-core wires as they contain three separate wires, each with a copper core. Copper is used as it is an excellent conductor. Some devices do not need an **earth** wire and these use 2-core cables. Every cable must contain both a **live** and **neutral** wire.

A How many pins are there on a standard UK plug?
B Why do most cables have a copper core?

Safety devices

There are several devices designed to improve the safety of the mains electricity supply.

Fuses

In the UK, each plug contains a fuse. This is connected to the live wire and is usually a small cylindrical cartridge.

Each fuse has a rating, and if the current passing through it exceeds this rating, the fuse heats up, melts, and breaks the circuit. This protects your electrical appliances from surges of current – the fuse melts before the wires in the computer.

A fuse with a rating of 13 A would melt if 13 A were to pass through it. A 3 A fuse contains a thinner wire which melts when the current reaches just 3 A. It is important to select the correct fuse rating for each appliance.

Earth wires

The earth wire is another important safety device. One end of this wire is connected to the metal case of the appliance. If the live wire were to come loose and touch the case, the case would become live. If you were then to touch the case, you could receive a very dangerous shock. With the earth wire attached, the case cannot become live as the current passes down the earth wire. This causes a surge in current, and this melts the fuse.

▲ Some appliances are double-insulated, as indicated by the square symbol in the photo. They do not need an earth wire as the case cannot become live.

Circuit breakers

Another safety device is a circuit breaker. Like the fuse, this is connected to the live wire of an appliance, but rather than heat up it detects tiny changes in current.

Residual current circuit breakers (RCCBs) detect a difference in the current between the live and neutral wire. If these values don't match, there must be a fault, and so the circuit breaker shuts off the current.

Despite being more complex than fuses, circuit breakers have a number of advantages. They switch the current off much faster than fuses, and they can be easily reset and used again.

Exam tip

✔ Don't just say a fuse 'blows' when there is too much current. You need to be more precise: it heats up, melts and breaks the circuit.

Did you know...?

Some appliances don't have an earth wire. These are usually made from non-conducting materials (like wood or plastic) and so the case cannot become live. Even some metal appliances are **double-insulated**. The live components are sealed away from the case and so there is no chance of the case becoming live.

Questions

1. Explain why plugs are made of rigid plastic. (E)
2. List the three wires found in a UK plug, state their colour, and explain what they do.
3. Give two advantages of using circuit breakers compared with fuses. (C)
4. Explain how the earth wire protects you if there is a fault with your device.
5. Describe how an RCCB protects a circuit. (A*)

26: Current, charge, and power

Learning objectives

After studying this topic, you should be able to:

- describe what happens when an electrical charge flows through a resistor, and relate this to suitable cable size
- calculate the power of an appliance
- use the equation linking energy transferred, potential difference, and charge

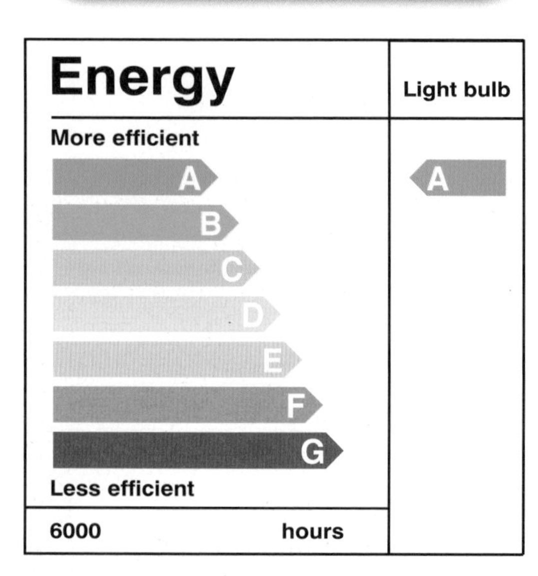

Efficiency label for a light bulb

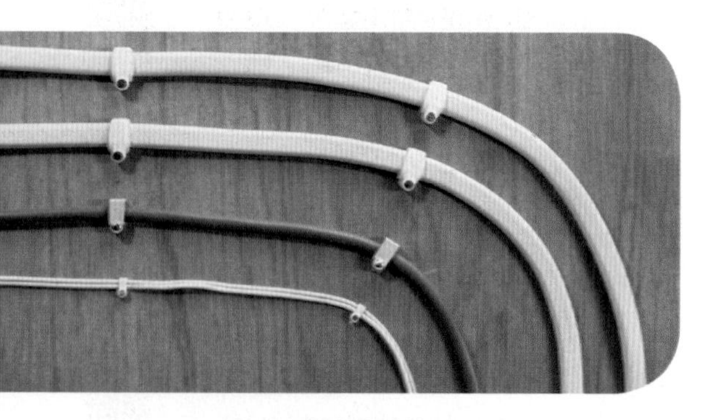

Different sizes of power cable

A A TV transfers 6600 J in 60 s. What is its power?

Resistance and efficiency

When an electrical charge flows through a resistor, the resistor gets hot. Energy has been transferred from the charge to the resistor.

Many electrical appliances waste energy by transferring electrical energy to heat energy. For example, filament lamps transfer a much higher percentage of electrical energy to heat energy than compact fluorescent lamps do.

When you are buying new appliances, you can check how much energy they waste by looking at the efficiency rating label on the packaging. An appliance which wastes less energy will have a higher rating. The most efficient appliances are rated A++.

Power cables

The size of a power cable for an appliance is related to the amount of energy it transfers. The cable needs to be large enough not to heat up when the current flows through it. An appliance that needs a higher power will have a larger cable.

A kettle with a power rating of 3 kW has a thick cable. The cables connecting a laptop to the mains have smaller thicknesses: the one leading to the mains socket supplies a power of 250 W, so it is thinner than the cable for the kettle, and the cable from the adapter to the laptop supplies only about 65 W, so it is thinner still.

Power

Different appliances transfer energy at different rates. The rate is given by the equation:

$$\text{power (watts, W)} = \frac{\text{energy transferred (joules, J)}}{\text{time taken (seconds, s)}}$$

If power is P, the energy transferred is E, and the time taken for the energy transfer is t, then:

$$P = \frac{E}{t}$$

Power, potential difference, and current

The power of an appliance can also be calculated using the equation:

$$\text{power (watts, W)} = \text{current (amperes, A)} \times \text{potential difference (volts, V)}$$

When power is P, current is I, and the potential difference provided by the supply is V, then:

$$P = I \times V$$

You can use this equation to calculate the current through an appliance and decide what size of fuse you should use for the appliance. The two most common sizes for fuses are 3 A and 13 A.

Worked example 1

A computer has a power rating of 200 W. What fuse should be fitted to the plug? Assume that the mains supply is 230 V.

power = current × potential difference or $P = I \times V$

$$\text{current} = \frac{\text{power}}{\text{potential difference}}$$

$$= \frac{200\ \text{W}}{230\ \text{V}} = 0.87\ \text{A}$$

The fuse with the smallest current rating that can be fitted to the plug is 3 A, so this should be used.

How much energy is transferred?

You might remember that the potential difference is the energy transferred (or work done) for each coulomb of charge passing through an appliance. So we can calculate how much energy is transferred.

energy transferred (joules, J) = potential difference (volts, V) × charge (coulombs, C)

If energy transferred is E, potential difference is V, and charge is Q, then:

$$E = V \times Q$$

Worked example 2

A charge of 50 C flows through a TV that is plugged into the mains. How much energy is transferred?

energy transferred = potential difference × charge

= 230 V × 50 C

= 11 500 J

B A toaster has a power rating of 1100 W. What fuse should be fitted to the plug?

C A 24 V battery transfers 75 C when starting an engine. How much energy is transferred to the engine?

Exam tip

✔ Remember that you should use watts rather than kilowatts in this type of calculation.

Questions

1. What are the two most common types of fuses?
2. A food processor transforms 50 000 J of energy in 50 seconds. What is its power?
3. What size of fuse should be fitted to the following appliances (assume 230 V mains supply):
   (a) a hair dryer with a power rating of 1500 W
   (b) a plasma TV with a power rating of 450 W?
4. (a) A charge of 5000 C flows through the hair dryer. How much energy is transferred?
   (b) A charge of 3000 C flows through the plasma TV. How much energy is transferred?
5. What links can you see between the equations for power, energy transfer, and charge?

27: The atom and the nucleus

Learning objectives

After studying this topic, you should be able to:

- describe the structure of an atom and the relative charges and masses of its components
- explain how the charge of an atom changes when it loses or gains electrons
- understand the meaning of the term isotope
- describe how new evidence caused scientists to change their model for the atom

The atom

Atoms make up all matter. They are the smallest part of every substance. We now know that the atom has a very small central **nucleus**, with **electrons** in orbit.

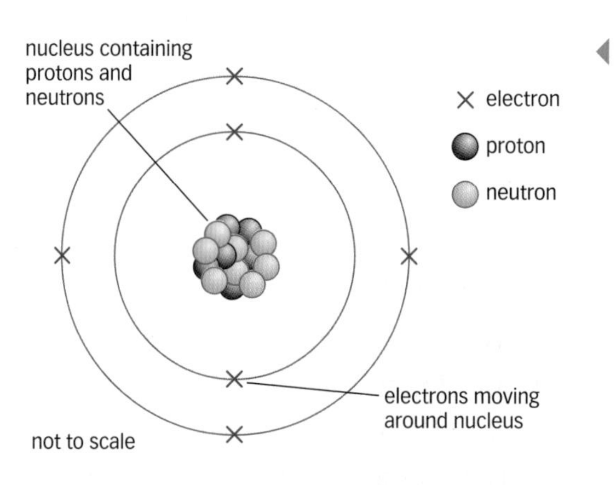

All atoms have a central nucleus with electrons in orbit around it. The nucleus is around 10 000 times smaller than the atom. If the nucleus was 1 mm across, the atom would have a diameter of around 10 m!

	Proton	Neutron	Electron
relative mass	1	1	~0 (1/2000)
relative charge	+1	0	−1

A Which particle has a positive charge?

Protons and **neutrons** are about the same size, but protons are positively charged and neutrons are neutral. Electrons are negatively charged, and much smaller than the particles in the nucleus.

All atoms have an **atomic number** and a **mass number**. The atomic number is the number of protons in the atom. All atoms of the same element have the same number of protons, and so the same atomic number. For example, all carbon atoms contain six protons.

The mass number refers to the number of protons plus the number of neutrons. A carbon 12 nucleus contains six protons and six neutrons (12 particles in total).

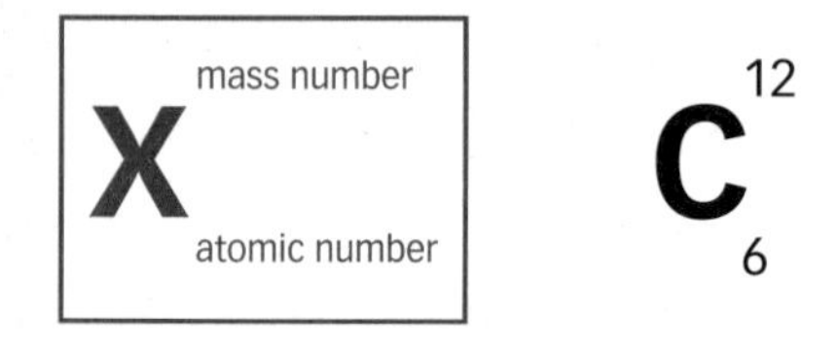

Each atom has a mass number and an atomic number

Isotopes and ions

Isotopes are atoms with the same number of protons, but a different number of neutrons. Isotopes of carbon include carbon 12 (six neutrons), carbon 13 (seven neutrons), and carbon 14 (eight neutrons). Each atom contain six protons – remember that all carbon atoms contain six protons.

Atoms are usually neutral. They have the same number of protons as electrons, so there is no overall charge. However, it is possible to add or remove electrons from atoms, creating charged **ions**. When ions have the same type of charge they repel each other, whereas oppositely charged ions attract.

B Explain the meaning of the term isotope.

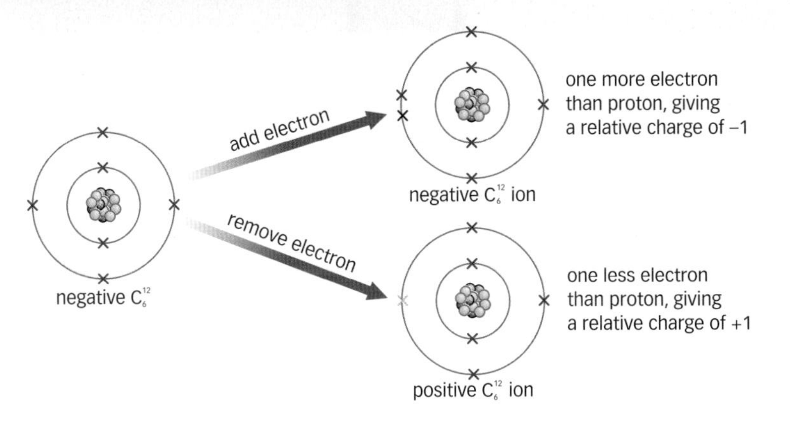

▲ Adding or removing electrons creates ions

Key words

nucleus, electron, proton, neutron, atomic number, mass number, isotope, ion

The discovery of the nucleus

Before the nucleus was discovered, scientists thought the atom was like a plum pudding. They thought the atom was a sphere with electrons scattered throughout it, rather like blueberries in a muffin. This changed when students of the famous physicist Ernest Rutherford collected some evidence from a scattering experiment.

They fired tiny positive particles, called alpha particles, at a very thin piece of gold foil (or gold leaf). They were extremely surprised to find that some of these particles were scattered back from the foil. Rutherford concluded that the atom must be very different from the 'plum pudding model'. Instead, he suggested that instead there must be a small, dense, positive centre. He called this the 'nucleus'. He knew that the nucleus must be small because most of the alpha particles went straight through the gold atoms. Scattering occurred only with the alpha particles that happened to be lined up with the nucleus and were repelled back the way they came.

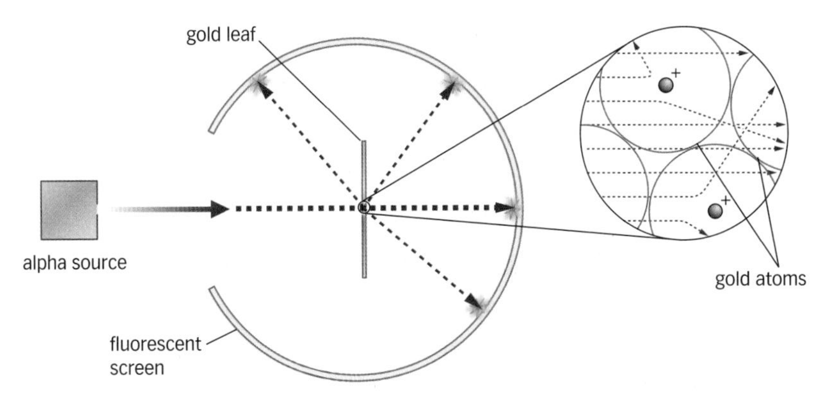

▲ Although most alpha particles went straight through the gold leaf, some were scattered back

Did you know...?

Rutherford was right. The nucleus is really tiny; most of the atom is just empty space. The nucleus is so small that if it were the size of a pea and it was placed on the centre spot at Wembley stadium, the electrons would be orbiting in the surrounding car parks.

Questions

1. Draw a diagram of an atom, labelling all the important parts. (E)
2. State the number of protons and the number of neutrons in an atom of U^{238}_{92}.
3. Explain what an ion is, and how one might be formed.
4. Describe how Rutherford discovered the nucleus. (C)
5. Explain how Rutherford was able to conclude that the nucleus was small, dense, and positive. (A*)

28: Radioactive decay and half-life

Learning objectives

After studying this topic, you should be able to:

- ✔ describe radioactive decay
- ✔ list some of the sources of background radiation
- ✔ understand the meaning of the term half-life, and how to determine half-life from a graph of activity against time

Key words

radioactive decay, ionising radiation, background radiation, activity, random, half-life

Radioactive decay and background radiation

Nuclear radiation is naturally around us all of the time; it is not only man-made. The nuclei of some atoms are unstable, and **radioactive decay** occurs when a nucleus from one of these atoms breaks down and emits one of the three types of **ionising radiation**:

- alpha particles (α)
- beta particles (β)
- gamma rays (γ).

The radiation around us is called **background radiation**. It comes from a variety of sources. Less than 15% comes from man-made activities:

- use of radiation in hospitals
- nuclear weapons testing
- nuclear power, including accidents.

Most of it occurs naturally due to the breakdown of radioactive atoms found within rocks and from cosmic rays arriving from space.

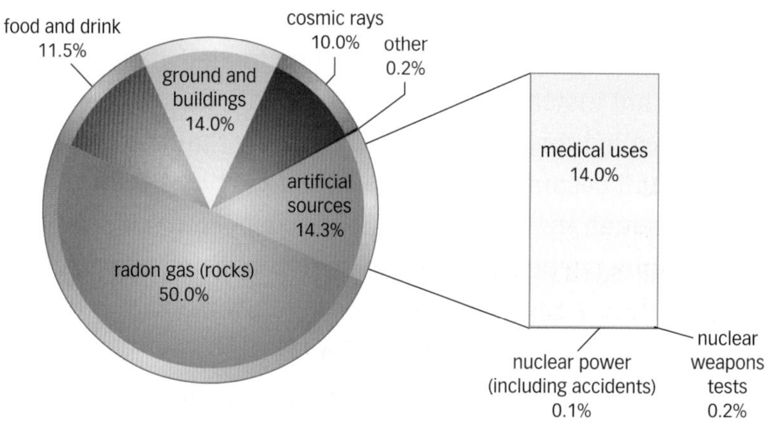

▲ Sources of background radiation found in the UK

Your exposure to radiation depends on where you live and, for example, on the nature of your job. Different parts of the country have different levels of background radiation, and an airline pilot or someone working in a hospital can receive a higher dose than other people.

A What are the three most common sources of background radiation in the UK?

Detecting radiation

Radiation may be detected using a Geiger–Muller (GM) tube. When the radiation enters the tube, it ionises the atoms of the gas inside the tube. It removes electrons from the atoms in the gas and this causes a small electric current which is detected by a counter.

If atoms within human cells are ionised this can damage the cell. The DNA within the cell can be affected. This might ultimately kill the cell or cause it to mutate, potentially leading to cancer.

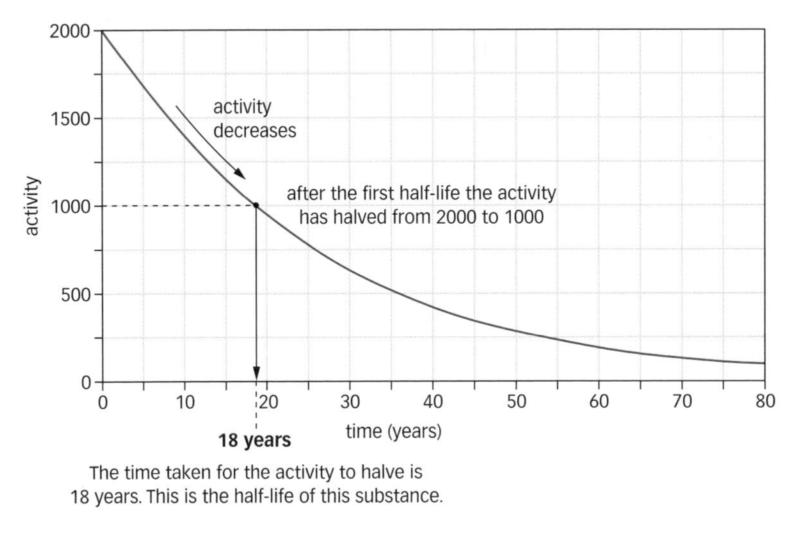

▲ Half-life can be determined from a graph of activity against time

Half-life

A substance which contains radioactive atoms gives out radiation all of the time. The nuclei within it decay, and the number of these decays per second is called the **activity**. You cannot stop a radioactive substance decaying, no matter what you do to it. This radioactive decay is also a truly **random** process. There is no way of knowing which nuclei will decay next, nor when any particular atom will decay.

As the radioactive atoms within a substance decay, fewer and fewer radioactive nuclei remain. This leads to a drop in activity as time passes. The average time taken for half of the radioactive nuclei in a sample to decay is called the **half-life**. This is the time it takes for the activity to halve.

Different isotopes have different half-lives. These range from milliseconds to billions of years. The table below shows a few examples.

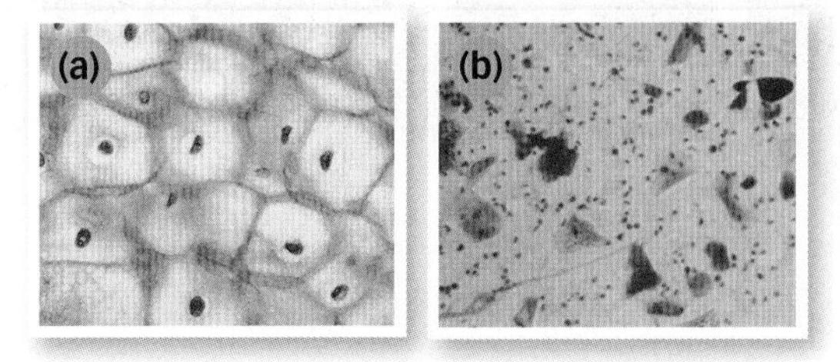

▲ (a) Healthy cells, and (b) cells that have been exposed to a high dose of ionising radiation

B What effect does ionising radiation have on human cells?

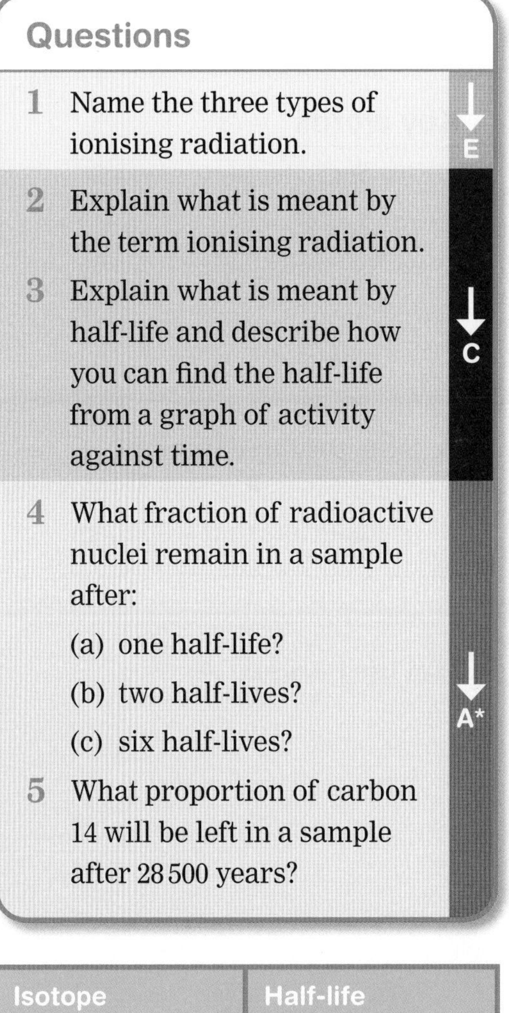

Questions

1 Name the three types of ionising radiation. (E)

2 Explain what is meant by the term ionising radiation.

3 Explain what is meant by half-life and describe how you can find the half-life from a graph of activity against time. (C)

4 What fraction of radioactive nuclei remain in a sample after:

(a) one half-life?

(b) two half-lives?

(c) six half-lives?

5 What proportion of carbon 14 will be left in a sample after 28 500 years? (A*)

Isotope	Half-life
nitrogen 17	4 seconds
radon 220	3.8 days
carbon 14	5700 years
uranium 238	4.5 billion years

29: Alpha, beta, and gamma radiation

Learning objectives

After studying this topic, you should be able to:

- ✔ describe some properties of alpha, beta, and gamma radiation in terms of the changes in the nucleus and their penetrating power
- ✔ describe how to handle radioactive sources safely

Key words

alpha particle, beta particle, gamma ray

A State the three types of radiation and give examples of materials that stop them.

Did you know...?

In November 2006 Alexander Litvinenko, a former officer in the Russian Federal Security Service, died in University College Hospital London. It is believed he was murdered. The cause of death was radiation poisoning from polonium 210. He consumed just 10 millionths of a gram. It was not possible to detect the radiation as the alpha particles did not leave his body. However, they caused fatal damage to his internal organs.

Types of decay and penetrating power

The three types of ionising radiation all come from the nucleus of unstable atoms, but they are each very different.

alpha particles α^4_2	These are very ionising particles made up of two protons and two neutrons. This is the same as a helium nucleus. Alpha particles have a large mass and a large positive charge. This makes them the most ionising type of radiation.
beta particles β^0_1	A beta particle is a fast electron from the nucleus. Beta particles have a negative charge. They have less mass and a lower charge than alpha particles, making them less ionising.
gamma rays γ^0_0	A gamma ray is a high frequency electromagnetic wave. This kind of radiation is not very ionising, but travels a long way.

How far the radiation travels depends on how ionising it is. The more ionising it is, the more quickly it slows down, and stops. Alpha radiation travels only a few centimetres through air, beta particles a few metres and gamma radiation several kilometres.

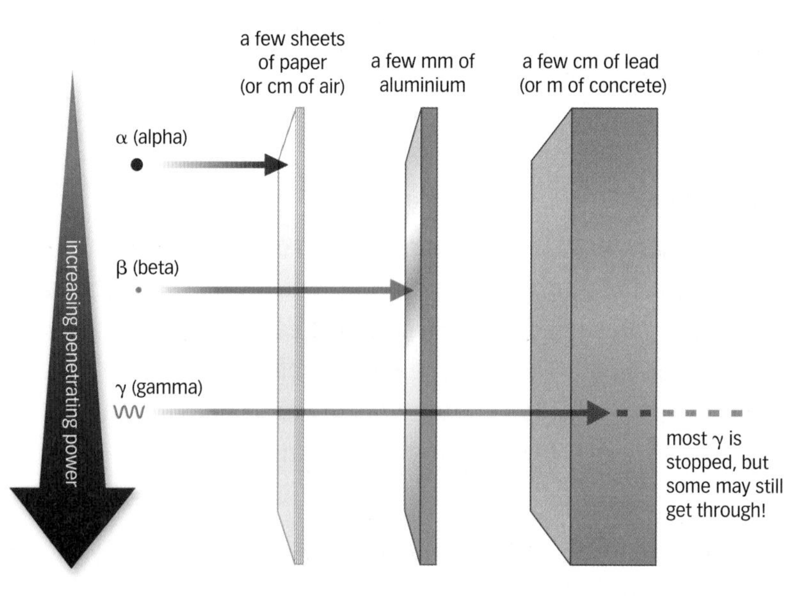

▲ The different penetrating powers of the three types of radiation

Deflection of radiation

Alpha and beta radiation are both deflected by electric and by magnetic fields. Alpha particles are deflected less than beta particles, and they are deflected in the opposite direction from beta particles. Gamma radiation is not deflected.

This is explained by the charge and mass of the radiation. Alpha and beta radiation have opposite charges, so they are deflected in opposite directions. Gamma radiation has no charge, so it is not deflected. Alpha particles have a greater mass than beta particles, so alpha radiation is deflected less.

Nuclear equations

When a radioactive nucleus emits an alpha particle, the mass number and atomic number of the nucleus changes. The mass number drops by 4 and the atomic number drops by 2. This means that the element changes. For example, uranium 238 becomes thorium 234 after emitting an alpha particle.

$$U^{238}_{92} \rightarrow Th^{234}_{90} + \alpha^{4}_{2}$$

In beta decay a neutron breaks up into a proton and a beta particle. As a result the mass number of the atom stays the same (it has lost one neutron but gained one proton), but the atomic number goes up by 1. For example, carbon 14 forms nitrogen 14 after beta decay.

$$C^{14}_{6} \rightarrow N^{14}_{7} + \beta^{0}_{-1}$$

These decay equations must always balance, with the atomic numbers and mass numbers adding up to the same value on both sides.

Handling radioactive sources safely

Radioactive sources must be handled very carefully. Depending on the type of radiation and the activity of the radioactive source, different precautions are needed. In each case you need to think about the shielding, the length of time of exposure and the distance from the source.

Shielding can be as simple as wearing gloves, but if the source emits more penetrating radiation then denser shielding such as lead, lead crystal glass, or special radiation suits might be needed.

To reduce the time you are exposed to radiation, sources are only taken out when they are being used.

You should always keep your distance from radiation sources. The further away you are from the source the lower your exposure. Using tongs keeps them at a safer distance.

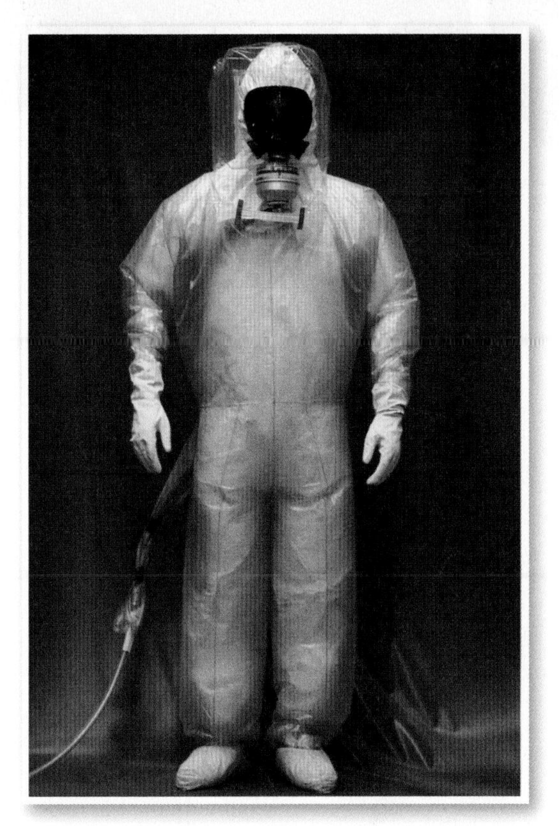

▲ An example of the clothing worn to protect against strong radioactive sources

Questions

1 List the three types of radiation in order of their penetrating power, from most penetrating to least penetrating. (E)

2 Describe some of the safety precautions needed when handling radioactive sources. (C)

3 Using a GM (Geiger–Müller) tube and an assortment of different absorbers, describe an experiment you could do to determine the types of radiation emitted from a radioactive source.

4 Complete the following decay equations: (A*)

(a) $Pu^{239}_{94} \rightarrow U^{?}_{?} + \alpha^{?}_{?}$

(b) $Pb^{210}_{82} \rightarrow Bi^{?}_{?} + \beta^{?}_{?}$

30: Uses of radiation

Learning objectives

After studying this topic, you should be able to:

- ✔ describe some of the uses of radiation

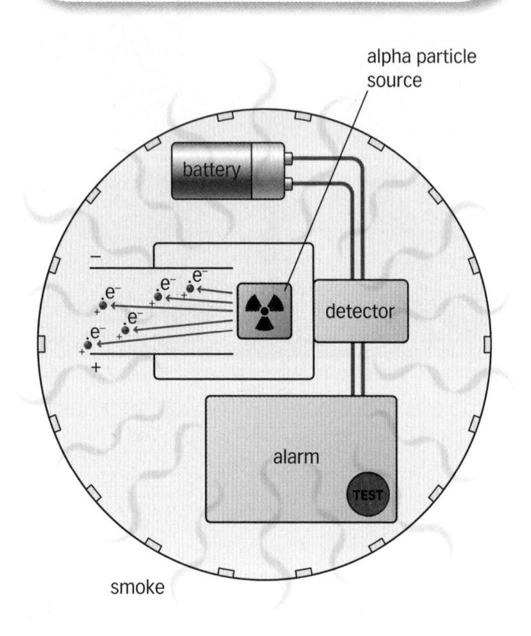

▲ When smoke absorbs the particles, the alarm goes off

Smoke alarms and alpha particles

Alpha particles don't travel very far, but are very ionising. They are ideal for use in smoke alarms. Most smoke alarms contain a weak source of alpha radiation. This ionises air inside the alarm and creates a very small electric current (just like in the GM tube). When smoke enters the alarm this current drops, setting off the alarm. If there is no smoke, there is no change in current, and so the alarm stays quiet.

The source in the alarm must have a long half-life – the activity must remain fairly constant, so that you will not need to replace it regularly.

A Describe how radiation is used to detect smoke in a smoke alarm and why alpha sources are used.

Beta particles and paper mills

Compared with alpha radiation, beta particles can pass through thicker materials. Most beta particles are able to travel through several sheets of paper. In paper mills, the sheets of paper pass between a beta source and a detector.

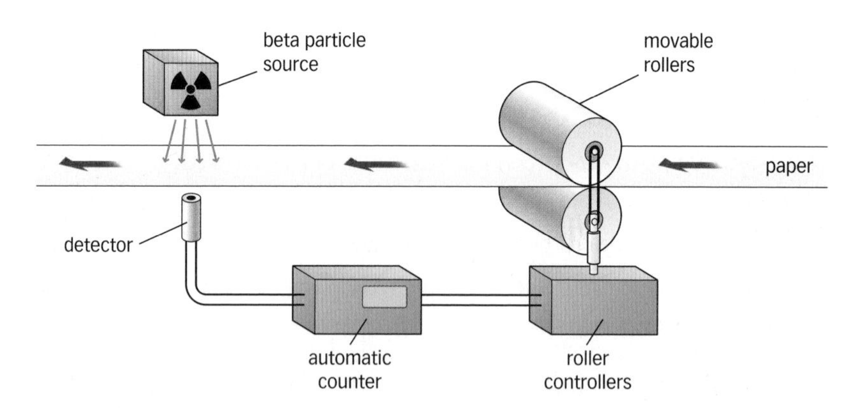

▲ Beta radiation may be used in paper mills to automatically monitor the thickness of the paper

The thicker the paper, the fewer beta particles get through. This makes the counts recorded by the counter go down. If the counts fall too low, this means that the paper is too thick and so the rollers squeeze together.

As with the smoke alarm, the source selected must have a long half-life to keep the activity fairly constant.

Key words

gamma knife, sterilising

B Describe what would happen to the recorded counts if the paper was too thin.

Medical and other uses of gamma rays

An important use of gamma rays is in killing cancerous cells within the body as part of radiotherapy. Doctors may use a special machine called a **gamma knife.** This contains a movable source of gamma rays that are focussed on the tumour and are fired into the body. The gamma source is moved around in order to reduce the exposure of the healthy tissue to the radiation while still providing a high enough dose to kill the cells inside the tumour.

Gamma radiation is used for **sterilising** medical equipment. It kills microorganisms such as bacteria

Gamma rays are also used in medical tracers. A weak source of gamma rays is either ingested by the patient or injected. Gamma rays travel out of the body and special cameras are used to monitor the movement of the source around the body. Doctors can then identify problems such as blockages or leaks within internal organs.

A similar technique can also be used to detect leaks in underground pipes.

In both cases, gamma sources with fairly short half-lives are used, usually just a few hours. This means the radioactivity falls to a low level very quickly.

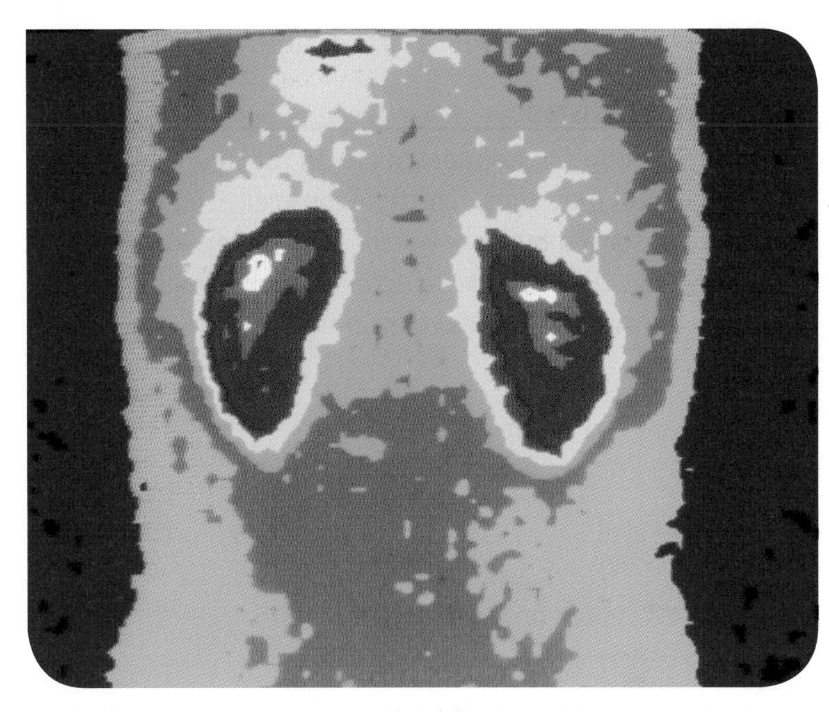

▲ An image of the kidneys obtained by using radioactive tracers injected into the body

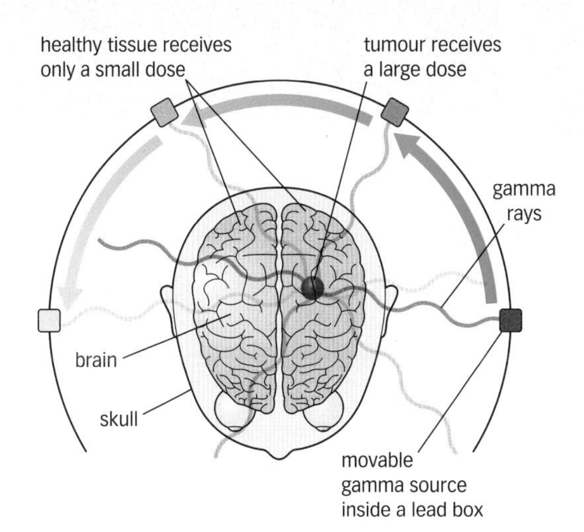

▲ A gamma knife is used to kill cancerous cells in a brain tumour

Questions

1 Give three different examples of uses of radiation.

E

2 A similar arrangement to the one used in the paper mill is used to ensure steel sheets are kept at a constant thickness. Explain how this process might work, describing the type of radiation you would use.

C

3 Give two reasons why gamma sources are used as medical tracers rather than alpha or beta sources, and explain why the source must have a short half-life.

4 Explain why a beta source with a long half-life is used in paper mills. What would happen if a source with a short half-life were to be used instead?

A*

5 Describe the advantages and disadvantages of using a radioactive medical tracer to diagnose a serious medical condition.

31: Fission and fusion

Learning objectives

After studying this topic, you should be able to:

- describe the process of nuclear fission
- describe the process of nuclear fusion
- compare the advantages and disadvantages of each method for producing energy

Key words

nuclear fission, uranium 235, plutonium 239, chain reaction, nuclear fusion

A Give two examples of fissionable materials.

B Describe how a nuclei of uranium 235 may be split and give the name of this process.

Did you know...?

In a nuclear reactor the chain reaction is carefully controlled. In a nuclear bomb, the number of fissions increases dramatically, releasing vast amounts of energy in a very short time. The Russian Tsar bomb released the energy equivalent of exploding 50 million tonnes of TNT. That is an incredible 2.1×10^{17} J, enough to power over 55 million TVs for a year.

Nuclear fission

Nuclear fission is the splitting of atoms. It is the reaction which takes place inside all nuclear reactors. Fission releases energy in the form of heat. This is used to turn water into steam which turns turbines, which then turn generators to produce electricity.

Either **uranium 235** or **plutonium 239** atoms are split into two smaller nuclei. Most reactors use uranium 235. These smaller nuclei are often very radioactive, and make up the radioactive waste produced by nuclear reactors.

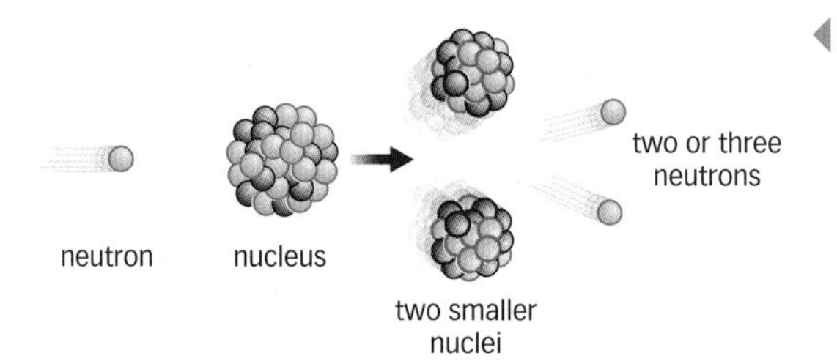

When a fissionable nucleus absorbs a neutron, it splits into two smaller nuclei, and fires out two or three neutrons in its turn

In nuclear fission, a nucleus first absorbs an extra neutron. This makes the nucleus spin and distort. After a few billionths of a second it splits into two smaller nuclei and releases two or three further neutrons.

Uranium 235 and plutonium 239 are described as fissionable substances as they can both be split easily.

Chain reactions

If there are enough fissionable nuclei in the material, then a **chain reaction** may start. The neutrons released in the first fission are absorbed by more nuclei, which then also split. These fissions release more neutrons, which lead to even more fissions and the process continues.

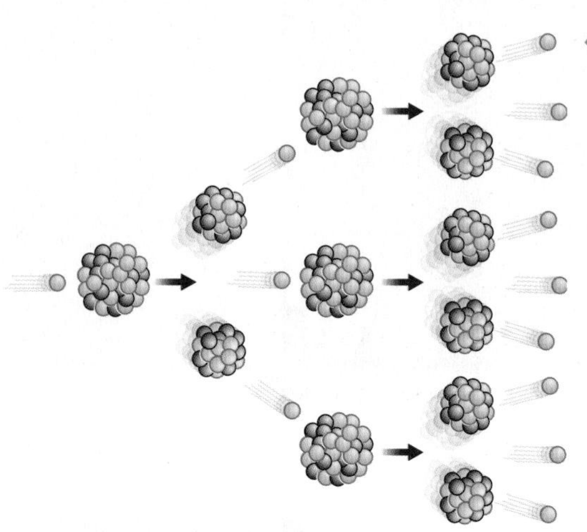

In a chain reaction the neutrons from one fission go on to create further fissions

Nuclear fusion

In **nuclear fusion** atomic nuclei are stuck (or fused) together. This forms heavier nuclei. Like nuclear fission, this process releases heat energy. Nuclear fusion is the process by which energy is released in all stars, including our Sun.

Use of nuclear fusion to generate electricity in the future would have the same advantages as fission nuclear power. No carbon dioxide is produced and very large amounts of electricity can be generated. Unlike nuclear fission, nuclear fusion does not produce radioactive waste.

Fusion on Earth

So far it has proved to be very difficult to sustain a nuclear fusion reaction on Earth. Atomic nuclei are positively charged because of their protons. This means they repel one another other when they get close together. The nuclei have to move very, very fast to get close enough to fuse. This happens in the core of stars like our Sun because the core is so hot. The nuclei are moving around at very high speeds and smash together.

Several experimental fusion reactors are being built. Some use superstrong magnetic fields to try to squeeze the nuclei together. Others use incredibly powerful lasers to heat up a tiny volume of gas to enormous temperatures.

▲ The interior of the experimental JET Tokamak fusion reactor in Oxfordshire

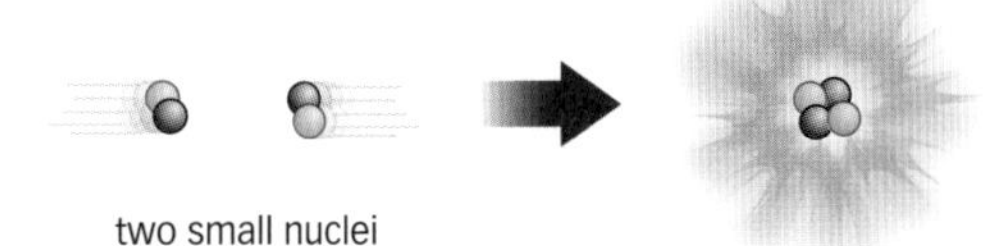

▲ In nuclear fusion two smaller nuclei fuse together to make a larger nucleus

Exam tip

- ✔ Do not confuse fission and fusion with radioactive decay (alpha, beta, or gamma). Nuclear reactors or nuclear bombs are not an example of a use of radioactivity.
- ✔ Be careful not to mix up fission and fusion. In nuclear *fu*sion nuclei are *fu*sed together.

Questions

1. What is the name of the particle which is absorbed by a nucleus to trigger nuclear fission?
2. What is the fissionable material that is most often used in nuclear reactors?

E

3. Describe the process of nuclear fusion.
4. Give an advantage of using nuclear fusion rather than nuclear fission to generate electricity.

C

5. Describe how a nuclear chain reaction might take place. Draw a diagram to illustrate your answer.

A*

32: Star life cycles

Learning objectives

After studying this topic, you should be able to:

- explain how stars are formed
- describe what happens when a star approaches the end of its life
- describe the complete life cycle of a star

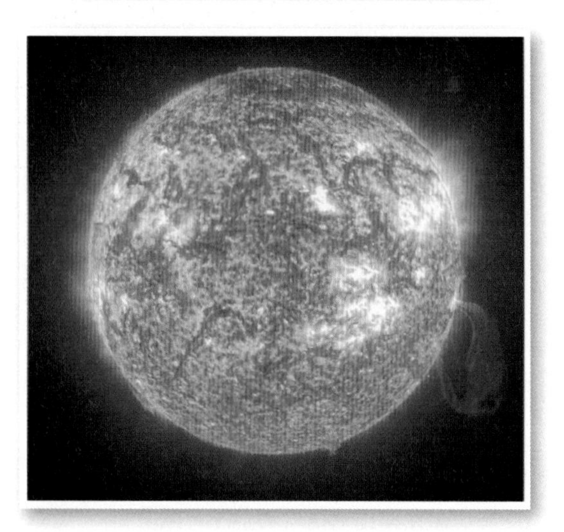

The Sun is in the main phase of its life. This will last for billions of years.

A What is the name given to a large cloud of gas in space?

B What force pulls the gas together to form a star?

Did you know...?

It is amazing to think you are made of star dust. All the elements in your body above hydrogen were made in the core of large stars. The atoms of elements heavier than iron were created when these stars went supernova, blowing these atoms into space to eventually form you!

Star birth and life

The Universe contains billions of stars. All of them are formed in the same way. A huge cloud of gas (mainly hydrogen) called a **nebula** begins to be pulled together by gravity. A large ball of gas forms in the centre of this cloud.

As it gets denser, more gas is pulled in. The ball gets hotter and hotter, forming a **protostar** (a similar process forms planets around the star). Eventually the protostar gets so big, and its centre gets so hot, that nuclear fusion happens in its core. A star is now born. Hydrogen nuclei are fused together to make helium.

The energy released in nuclear fusion pushes out against gravity and keeps the star stable. These forces are balanced. This stage is described as the main sequence of the star, and it can last for billions of years. Our Sun has been in its main sequence for around 5 billion years. It has another 5 billion years to go before it runs low on hydrogen in its core, and begins to die.

Dying stars

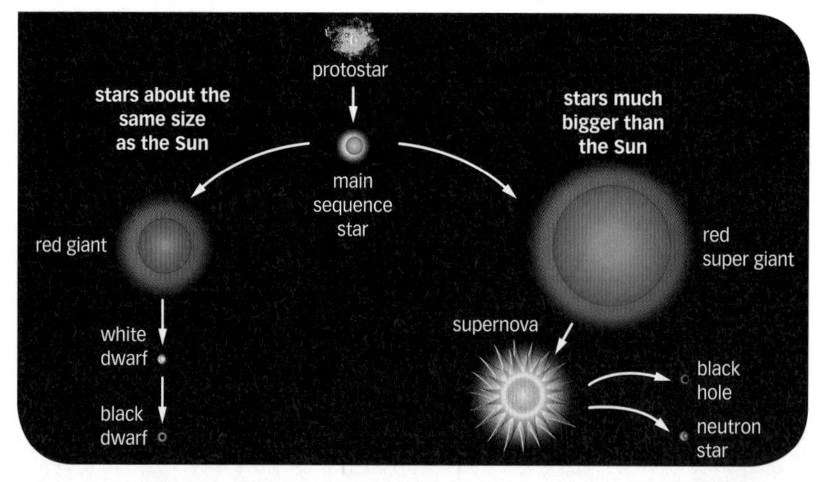

The life cycle of stars

Smaller stars

All stars eventually die, but what happens to them depends on their mass. Small stars, such as our Sun, gradually cool and expand into a **red giant**. The outer layers of the star break away and form a planetary nebula. All that remains is the white hot core of the star. This **white dwarf** gradually cools over time.

The Crab Nebula is left over from a star that went supernova in 1054

Larger stars and new elements

Stars that are much larger than our Sun have much more dramatic lives. They too eventually expand and cool, but they grow much bigger, turning into **red super giants**. They are so big that they are able to fuse heavier nuclei together to make other elements. When the Universe was formed in the Big Bang it was mainly made of hydrogen – the heavier elements, up to iron, have all been made in the core of dying stars.

The star then explodes in a gigantic explosion called a **supernova**. When a star goes supernova it often outshines all the other stars in the galaxy put together. All the elements heavier than iron are created here. This gigantic explosion spreads these elements throughout the Universe.

C What is the name given to the explosion of a large star at the end of its life?

Neutron stars and black holes

During a supernova, the core of the star is crushed down by large gravitational forces. This can form a very dense kind of star that is made up only of neutrons. This **neutron star** spins very fast and sends pulses of radio waves through space that can be detected on Earth.

If the star is even bigger, the core is crushed down into a tiny space – it forms a **black hole**. These mysterious objects have infinite densities. All the mass of the core is crushed down into a space smaller than an atom. The gravitational force is so large that nothing can escape, not even light.

Questions

1. What is the name given to the hot remains of a red giant star after the outer layers have broken away, leaving just the core? (E)
2. Describe why our Sun will remain stable for billions of years.
3. Explain how a large star first produces heavier elements, and then spreads them throughout the Universe. (C)
4. Describe the life cycle of a star:
   (a) The size of our Sun
   (b) A star much bigger than our Sun. (A*)

Key words

nebula, protostar, red giant, white dwarf, red super giant, supernova, neutron star, black hole

Exam tip

✔ What happens to a star depends on how big it is (its mass). Our Sun is too small to supernova and form a black hole. Only stars with much more mass explode at the end of their lives.

Course catch-up

Revision checklist

- Some materials gain an electrostatic charge when rubbed together. Opposite charges attract, like repel.
- Electric current is the flow of charge around a circuit.
- The potential difference between two points in an electronic circuit is the work done (energy transferred) per coulomb of charge that passes through the points.
- Circuit symbols represent electrical components.
- Resistance is a measure of how difficult it is for electrons to pass through a component.
- Current must pass through all components in a series circuit one after the other. Current in a parallel circuit can take more than one route.
- The resistance of a filament lamp increases as its temperature increases.
- An LED (light-emitting diode) produces light when a current flows through it, and is efficient and long-lasting.
- The resistance of an LDR (light-dependent resistor) varies with intensity of light. LDRs are used in switches.
- The resistance of a thermistor varies as its temperature changes. Thermistors are used in temperature sensors.
- Electric current has two forms: direct current (d.c.), usually produced by batteries or cells; and alternating current (a.c.), such as UK mains electricity (230 V a.c. with frequency 50 Hz).
- UK three-pin plugs contain a live wire (brown), a neutral wire (blue), an earth wire (green and yellow), and a fuse. Fuses, earth wires, and circuit breakers are safety devices.
- When an electrical charge flows through a resistor, energy is transferred. Larger appliances need larger power cables. The power of an appliance is its rate of energy transfer.
- Atoms have a small central nucleus of protons (positively charged) and neutrons (no charge) surrounded by negatively charged electrons in orbit. Positively or negatively charged ions are created when atoms gain or lose electrons.
- Isotopes are atoms with the same number of protons (same atomic number) but a different number of neutrons.
- Radioactive decay is the breakdown of the nucleus of an atom to form ionising radiation. Half-life is the average time taken for activity (decays per second) to halve.
- Nuclear fission is used in nuclear reactors to produce electricity.
- Nuclear fusion produces energy in stars.

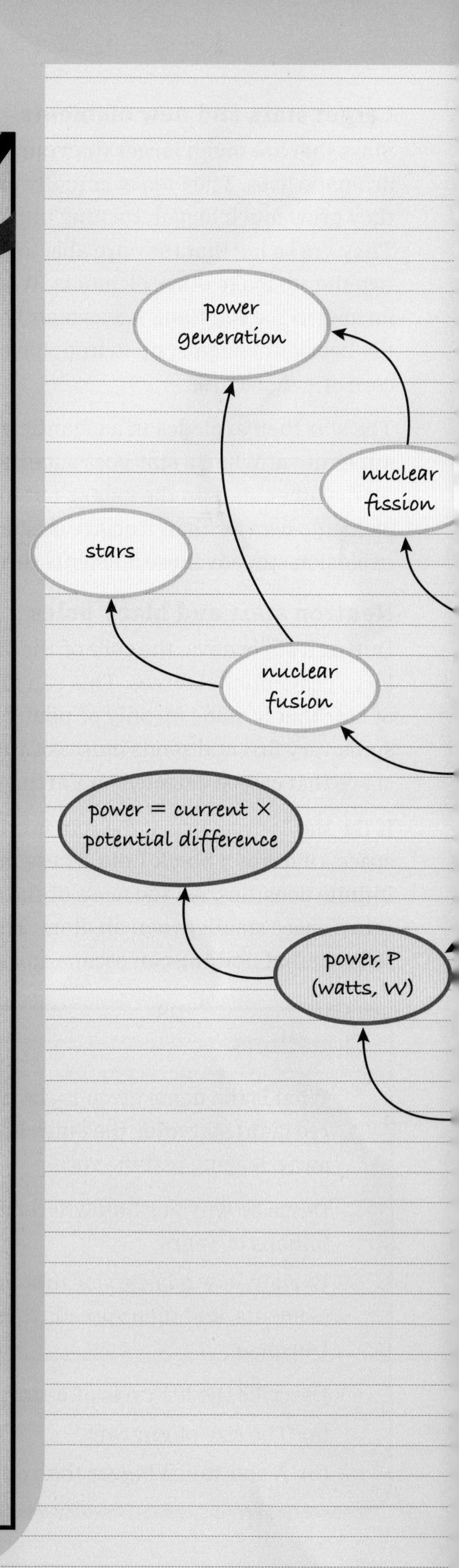

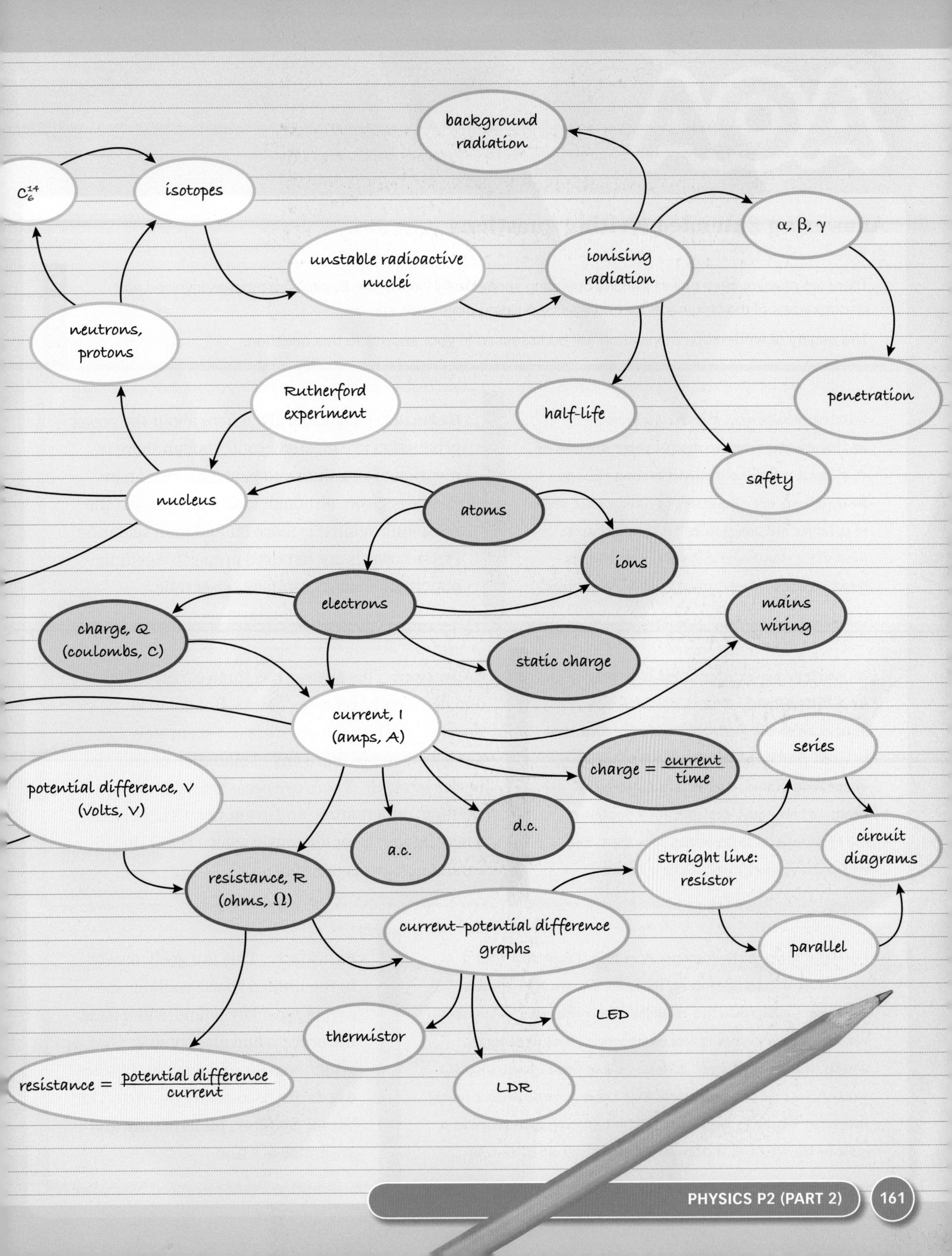
background radiation
C_6^{14}
isotopes
unstable radioactive nuclei
ionising radiation
α, β, γ
neutrons, protons
Rutherford experiment
half-life
penetration
safety
nucleus
atoms
ions
electrons
charge, Q (coulombs, C)
mains wiring
static charge
current, I (amps, A)
series
charge = $\frac{\text{current}}{\text{time}}$
potential difference, V (volts, V)
d.c.
a.c.
circuit diagrams
straight line: resistor
resistance, R (ohms, Ω)
current–potential difference graphs
parallel
LED
thermistor
LDR
resistance = $\frac{\text{potential difference}}{\text{current}}$

Answering Extended Writing questions

QUESTION

There are many nuclear power stations operating successfully worldwide. Explain the process involved and discuss some of the arguments for and against building more such stations.

The quality of written communication will be assessed in your answer to this question.

G–E

Nuclear power gets electricity from radioactivity from uranum there are lots of α, β, and γ flying around hiting things and geting hot. The stuff is dangerous becos if it leaks out an you eat it you get sick and die. I don't want it near me! But its good becose it doesn't make grenehous gas.

Examiner: The facts given here about the power production are mostly wrong, though there is mention of uranium and heat. One reason in favour is given, although lacking detail; but the reasoning against is more tabloid than scientific. Physics words are not used properly. Spelling, punctuation, and grammar are erratic.

D–C

In nuclear power, atoms hit each other and brake apart. This is called fision. This makes the fuel get very hot, and that is used for the power. The electricity is chepe and clean. But It is dangerous because they can explode, and poison a lot of people, also terorists can steal the radioactive fuel and make a bomb.

Examiner: This answer has some correct ideas, and deals with some key points. But detail is missing: the fission process is not accurately described; nor is there any explanation of why the power is 'clean'. There are occasional errors in spelling, punctuation, and grammar. Crucially, the word 'fission' is ambiguously spelt – could the candidate have meant 'fusion'?

B–A*

Nuclear power uses fission. A moving neutron hits a uranium nucleus; this splits into two smaller atoms, and more neutrons are released. They hit other uranium nuclei, and a chain reaction happens. It causes heat, which makes the power. It doesn't use fossil fuel, and no greenhouse gas is produced. But there is radioactive waste material which is difficult to store safely; and there is a risk of dangerous material leaking like at Chernobyl.

Examiner: This is a good answer. In the limited space available it covers most key points, addressing the process and also some arguments for and against. Physics words are used correctly, except for 'atom'. Spelling, punctuation, and grammar are fine.

Exam-style questions

1 Match these quantities with their units.

A01

current	watts
potential difference	ohms
charge	joules
resistance	coulombs
power	amps
energy	volts

2 A circuit is shown below.

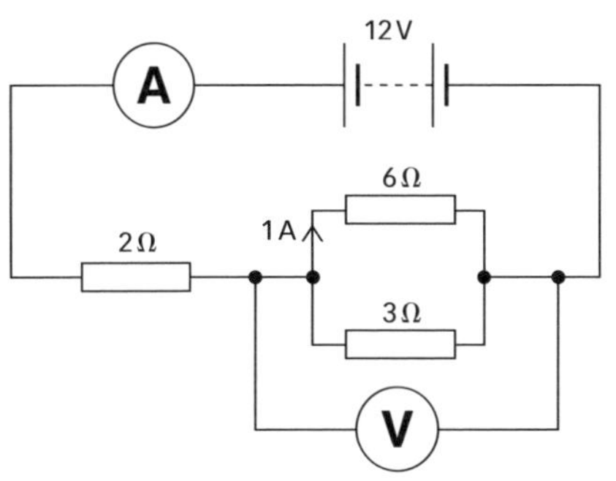

A01 **a** How are the 3Ω and 6Ω resistors connected together?

A01 **b** How are the 2Ω resistor, the ammeter, and the battery connected together?

A02 **c** Calculate:

i the reading on the voltmeter

ii the current through the 3Ω resistor

iii the current through the ammeter.

3 A cathode-ray oscilloscope (CRO) trace is shown below.

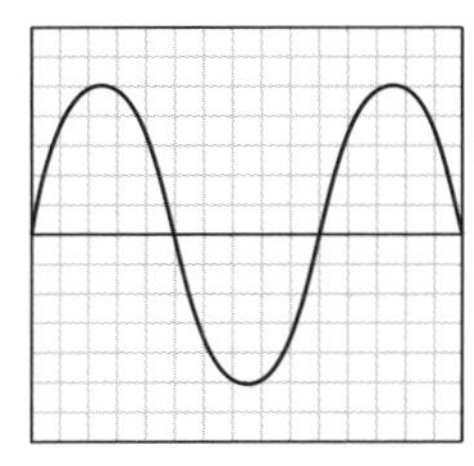

A01 **a** Explain why this trace illustrates an alternating current.

A01 **b** Describe how a d.c. trace would appear.

A02 **c** The time base of the CRO is set at 0.01 ms/division. Calculate:

i the time period of the a.c.

ii the frequency of the a.c.

A02 **d** If the frequency of the trace was doubled, how would the trace change?

G–E D–C B–A*

Extended Writing

4 Explain how fuses, circuit breakers, and the earth wire help to keep mains electricity safe. A01 G–E

5 You have three radioactive isotope sources available, emitting α, β, and γ radiations. Explain which you would use in a smoke detector, a paper-thickness detector, and as a tracer to find where an underground pipe is leaking. A02 D–C

6 Describe the life cycle of a star about the same size as our Sun, and a star much bigger than our Sun. A01 B–A*

A01 Recall the science

A02 Apply your knowledge

A03 Evaluate and analyse the evidence

P3 Part 1

Physics in medicine

AQA specification match

P3.1	3.1	X-rays
P3.1	3.2	Using X-rays
P3.1	3.3	Ultrasound
P3.1	3.4	Medical uses of ultrasound
P3.1	3.5	More on refraction
P3.1	3.6	Introduction to lenses
P3.1	3.7	Describing images
P3.1	3.8	Converging lenses
P3.1	3.9	Diverging lenses
P3.1	3.10	Magnification
P3.1	3.11	The human eye
P3.1	3.12	Problems with eyesight
P3.1	3.13	The camera
P3.1	3.14	Lens power
P3.1	3.15	Total internal reflection
P3.1	3.16	Optical fibres and lasers
Course catch-up		
AQA Upgrade		

Why study this unit?

Physics is extremely important in a wide range of industries, including medicine. You've probably already had at least one X-ray, and maybe even an ultrasound. The technologies used by doctors in today's modern hospitals rely heavily on ideas found in physics. As part of this unit you will study how X-rays are used, learn about their benefits and dangers, and understand why sometimes it's better to use ultrasound.

You will learn more about light, how it is refracted by different materials, and how simple lenses can focus it into a point. In this unit you will also find out how the human eye works, learning how images are focussed, why some people need to wear glasses, and how these glasses help them see crystal clear images.

Finally, you will learn about how light can be sent down optical fibres, not only providing high-speed broadband, but also allowing doctors to look inside you without the need to cut you open and take a peek!

You should remember

1. The properties of electromagnetic waves.
2. The nature of sound waves and the meaning of ultrasound.
3. How waves are refracted when they travel from one medium to another.
4. The dangers of ionising radiation and some of the properties of gamma rays.

Medical imaging has come a long way since Röntgen discovered X-rays at the end of the nineteenth century. This machine is a CT scanner. It uses X-rays to produce hundreds of images of slices through the body. These are stitched together by powerful computers to make a 3D image of a patient's insides. These machines are not cheap, each one costing around a quarter of a million pounds!

1: X-rays

Learning objectives

After studying this topic, you should be able to:

- state that X-rays are electromagnetic waves with a short wavelength
- understand some properties of X-rays

Key words

X-rays, electromagnetic waves, frequency, wavelength, transmitted, ionisation

Did you know...?

The Diamond synchrotron, located just outside Oxford, sends electrons around a high-speed particle accelerator ring. This makes the electrons emit very high energy X-rays. These are used to study different things, from the structures of chemical compounds and viruses to learning about the best conditions for manufacturing chocolate!

The Diamond Light Source synchrotron can use intense high-energy X-rays to probe matter at a molecular level

What are X-rays?

If you've ever been unlucky enough to break a bone, you will have gone for an X-ray scan. Scans allow the doctor to treat your injury without needing to cut you open to take a look.

increasing frequency and energy

decreasing wavelength

radio waves	micro-waves	infrared (IR)	visible light	ultraviolet (UV)	X-rays	gamma rays

X-rays are part of the electromagnetic spectrum

An X-ray scan uses **X-rays** to produce an image of the inside of your body. X-rays are similar to the gamma rays in radioactive decay. They are both examples of **electromagnetic waves**. X-rays are produced by very fast-moving electrons, while gamma rays come from the nucleus of radioactive atoms. In hospitals, electron beams are accelerated by high voltages and then smashed into a metal plate, causing the emission of X-rays.

A What kind of wave are X-rays an example of?

All electromagnetic waves travel very fast: they all travel at the speed of light in a vacuum. This includes X-rays. However, compared with visible light, X-rays have a much higher **frequency** and so a much shorter **wavelength** (only gamma rays have shorter wavelengths). Some X-rays are just 0.000 000 000 01 m long (0.01 nm). This means you could fit around one billion of them across your fingernail! The wavelength is about the same size as the diameter of the atom.

B What part of the electromagnetic spectrum has an even shorter wavelength than X-rays?

As X-rays have such a tiny wavelength, they are easily **transmitted** through healthy body tissue. They are able to pass through the body with only some being absorbed by denser materials such as the bones. The denser the material, the more X-rays are absorbed. They are able to travel through most materials and are only fully absorbed by thick slabs of very dense material such as lead or steel.

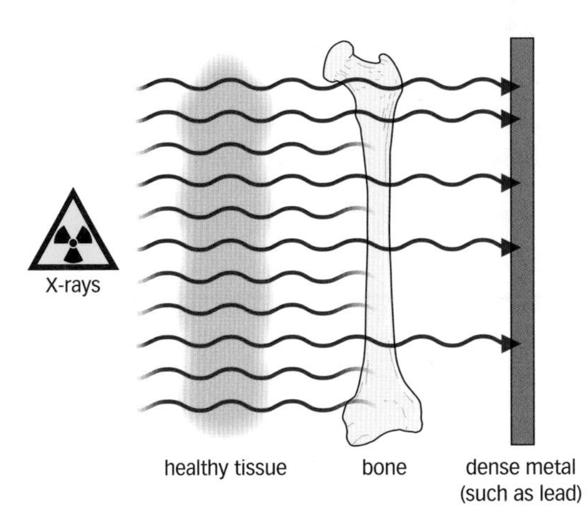

X-rays are easily transmitted through healthy tissue; bone absorbs some X-rays, and lead absorbs most X-rays

X-rays and ionisation

Like gamma rays, high-energy X-rays can cause **ionisation**. When an X-ray strikes an atom it transfers all of its energy to the atom. This gives the electrons in the atom so much energy that they break free and fly away at high speed. This creates an ion and this ionisation can be dangerous. Ionisation from high-energy X-rays has exactly the same effect as alpha, beta, and gamma radiation. The ionisation can kill body cells, or potentially lead to cancer.

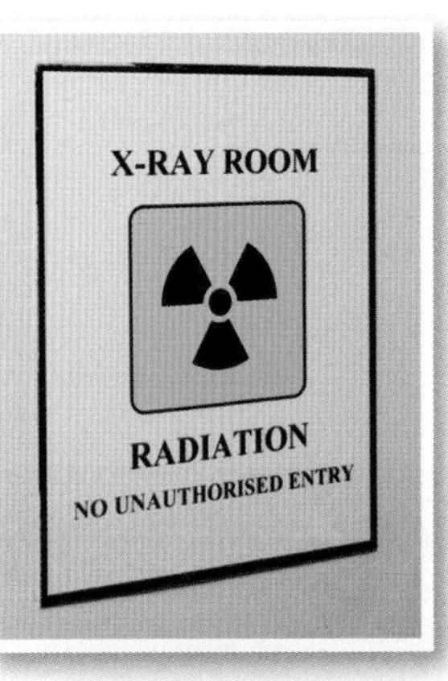

Like other forms of ionising radiation, X-rays are potentially hazardous

When photographic film is exposed to X-rays, the effect is similar to exposing it to light or to radiation from a nucleus. Where the film is ionised by the X-rays there is a chemical change, and a picture can be developed. Exposing an area of photographic film to X-rays darkens the film, leaving a cloudy patch. The longer the exposure, or the greater the intensity of the X-rays, the darker the area of the film becomes.

Questions

1. Describe the similarities and differences between X-rays and visible light. **E**
2. Describe and explain the effect X-rays have on photographic film.
3. Explain why exposure to X-rays is potentially dangerous to living tissue. **C**
4. Describe what happens to X-rays when they are transmitted through the body.
5. Suggest a technique that could be used to reduce a hospital worker's exposure to X-rays. **A***

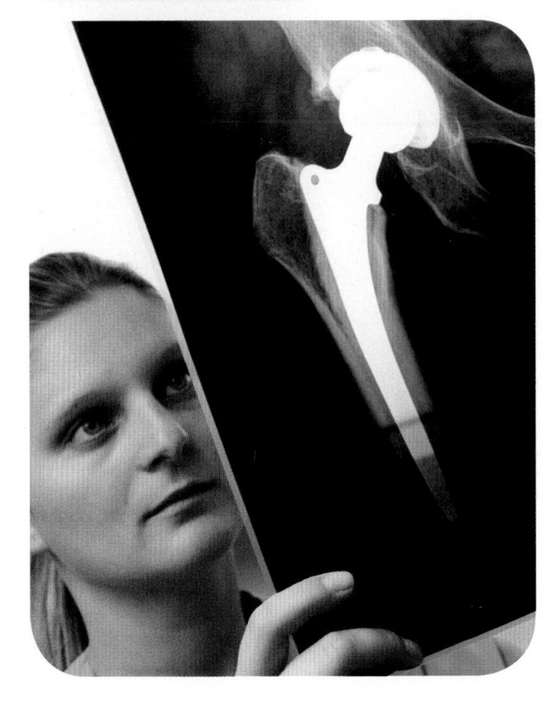

Photographic film turns cloudy when exposed to X-rays

Learning objectives

After studying this topic, you should be able to:

- give some examples of how X-rays are used
- describe how X-rays can be used to diagnose and treat some medical conditions
- describe the dangers of X-rays and some of the precautions taken when using them

Key words

CCD, CT scan, radiotherapy, radiographer

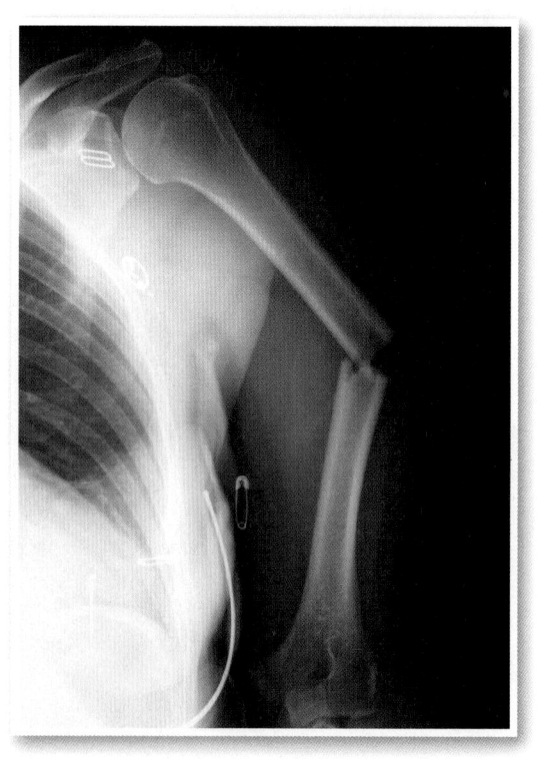

X-ray scans are used to look inside the body without the need for surgery

X-rays in use

In spite of their dangers, X-rays have some very valuable uses. The most common is X-ray photography. This has an important role in security. Luggage is scanned by X-ray machines at airports, and customs officers can even use high-energy X-rays to scan entire vehicles when they enter the UK.

Most modern X-ray machines make use of **CCDs** (charge-coupled devices), rather than photographic film, to detect the X-rays. This is the same kind of sensor as that found in digital cameras. It produces an image that can be processed by a computer.

X-ray scans are widely used in hospitals to diagnose many different medical conditions. The X-rays are used to produce an image, usually of part of a patient's skeleton.

X-rays are fired into the patient. Most pass through the patient's body and they then strike a photographic film or a CCD. But the denser material inside the patient (such as their bones) absorbs some of the X-rays. This causes a variation in the intensity of the X-rays received by the detector, and so an image is produced.

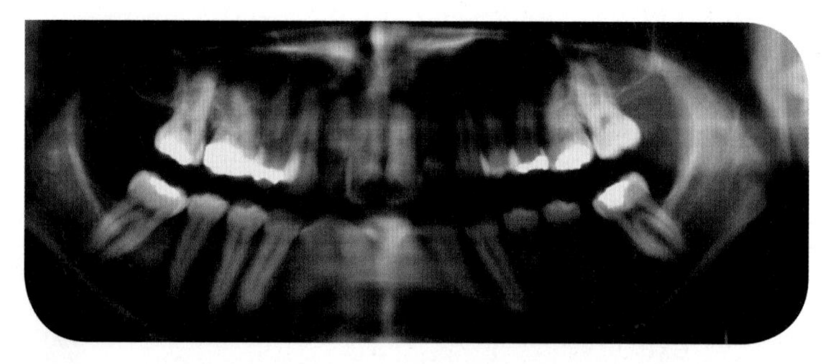

Dentists can use dental X-ray images to check fillings (shown here in bright white) and look for any tooth decay

The same principle is used in dental scans. Denser materials, such as fillings, absorb more X-rays, whereas a decaying tooth inside the gum will absorb slightly less than a healthy tooth. This leads to differences in the intensity of the X-rays detected by the photographic film or CCD, allowing the dentist to make an accurate diagnosis.

A Give two examples of uses of X-ray photography.

A computed tomography scan (**CT scan** for short) is an advanced technique that uses X-rays and CCDs. X-rays are fired through the patient and a series of images, like slices through the patient, are produced. These are processed by a computer to create a 3D image of the inside of the body.

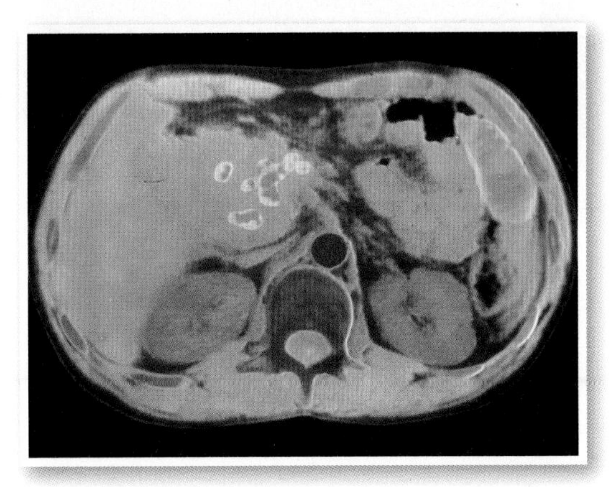

A CT scan produces an image of a slice through the body, allowing the internal organs to be seen

X-ray treatment

X-rays are not only used for imaging inside the body. They can be used as part of **radiotherapy** treatment. They can be used to kill cancerous cells within a tumour.

Working with X-rays

An X-ray machine in a hospital is operated by a **radiographer**, who is responsible for ensuring that X-rays are used safely. X-rays can be dangerous, particularly if you are exposed to them on a regular basis.

B What is the 'CT' in CT scan short for?

C What is the name given to the person who takes X-ray images in a hospital?

To reduce their own exposure, the radiographer leaves the room or stands behind a large lead screen whenever the machine is used. X-rays are ionising and prolonged exposure can be very dangerous. People working with X-rays every day have to monitor their own exposure carefully.

Exam tip AQA

✔ When describing a use of X-rays don't just say 'an X-ray'. Instead say 'producing an X-ray image of a broken bone'. Make sure you give a specific example.

Questions

1. Give two examples of X-ray detectors. (E)
2. Describe how X-rays may be used to produce an image of a broken bone.
3. Describe what happens during a CT scan. (C)
4. Explain why a radiographer leaves the room whenever X-ray photographs are being taken.
5. Suggest some advantages of using X-rays as part of radiotherapy compared with using gamma rays. (A*)

Did you know...?

X-rays were discovered and given their name by the German physicist, Wilhelm Röntgen. He used his new discovery to produce an X-ray image of his wife's hand. She was so disturbed by what she saw that she proclaimed 'I have seen my death'. In fact, X-ray photography rapidly spread to hospitals and has been used to save hundreds of thousands of lives.

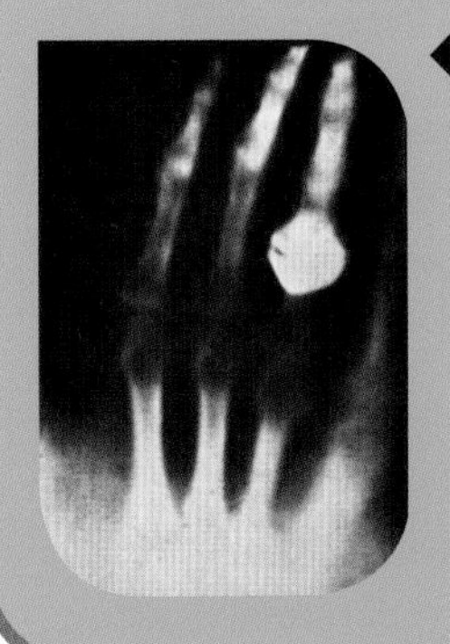

X-ray technology has come a long way since Röntgen's first photograph of his wife's hand

3: Ultrasound

Learning objectives

After studying this topic, you should be able to:

- state that ultrasound waves are sound waves above 20 000 Hz
- describe how ultrasound may be produced
- describe how ultrasound interacts with materials

A What does 'frequency' mean?
B State the range of human hearing.

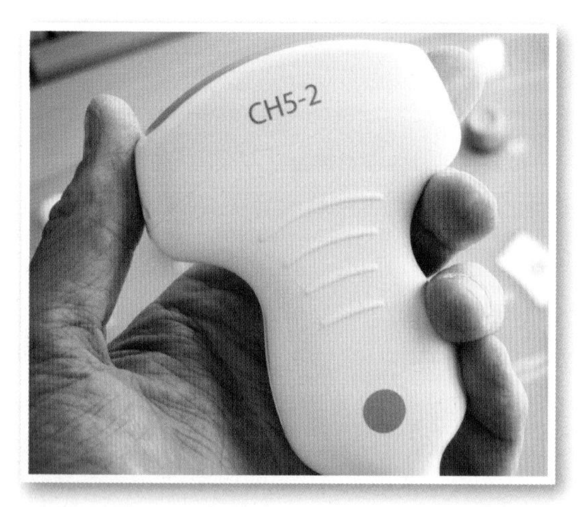

▲ Electronic systems are often used to produce ultrasound

Exam tip AQA

- Just like 20:20 vision, our hearing might be described as 20:20 hearing. We can hear sounds from 20 Hz to 20 kHz.
- When describing ultrasound waves don't forget to say that they are sound waves, with a frequency above the range of human hearing, that is above 20 000 Hz.

Can you hear that?

Your ears are amazing sensory organs. They can detect a wide variety of sound waves over a massive range of volumes, from the quietest of whispers to the loudest bass at a concert. You can hear sound waves with a range of wavelengths from just over 1 cm to over 16 m. However, some sounds are too low-pitched or too high-pitched for us to hear.

You should remember that a sound wave is a series of oscillations, or vibrations, that travel through air or another medium. The number of vibrations each second is called the frequency. A frequency of 50 Hz would mean 50 vibrations every second.

When these vibrations reach your ear they are channelled down the ear canal, causing your ear drum to vibrate. These vibrations pass through a series of tiny bones and eventually an electric signal is sent to the brain. The brain interprets the signal as a sound.

The human ear can detect a very wide range of frequencies. Human hearing ranges from just 20 Hz to up to 20 000 Hz. Sound waves above this frequency are called **ultrasound**. They are too high-pitched for the ear to detect.

Some animals, such as bats, dolphins. and orcas, produce ultrasound naturally. Humans use a variety of electronic methods to produce ultrasound. For example, one way is to use a small electronic speaker, called an ultrasonic transducer. Inside the speaker a small crystal is made to vibrate at a very high frequency, up to 20 million times per second. This produces ultrasound.

Reflections of ultrasound

Ultrasound has a very short wavelength, so it does not spread out (diffract) as much as sound as it travels through the air. This allows the ultrasound to be focussed into a narrow beam.

Ultrasound is able to pass through most materials. When a beam of ultrasound passes from one medium to another, the change in density causes some waves to be **reflected** back the way they came.

The greater the difference in density, the stronger the reflection. The change in density might be at the boundary between two layers of different materials, or it might even happen because there is a fine crack inside the object.

By timing how long it takes for the reflection to come back, we can calculate the distance to the boundary or crack.

distance travelled (metres, m)	=	speed of sound through material (metres per second, m/s)	×	time taken (seconds, s)

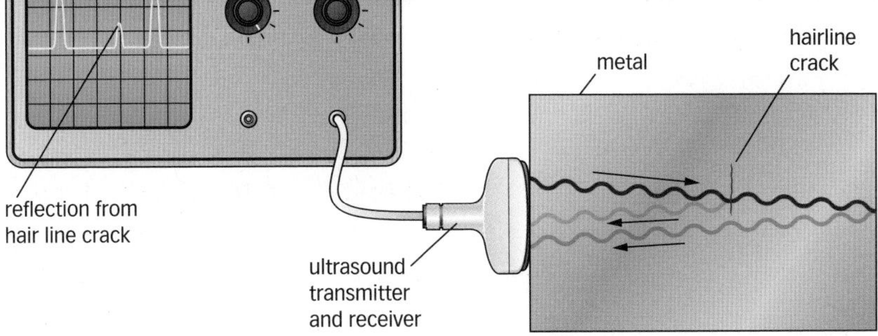

Reflections from tiny cracks inside a piece of metal can been seen on an oscilloscope trace

An oscilloscope can be used to calculate the position of a crack inside a material, as shown in the diagram above. The ultrasound transmitter also acts as a receiver and detects any reflected ultrasound pulses. These are displayed on the screen. The large peak is a reflection from the end of the object; the smaller peak is a reflection from the crack. Using the time base of the oscilloscope, the time taken for the sound pulse to travel to and back from the crack can be calculated, and so the position of the crack can be determined.

If there are several different layers or cracks inside the object, the reflected waves return at different times, depending on the depth of the layers. A computer can process this information to build up a picture of the inside of the object that is being scanned.

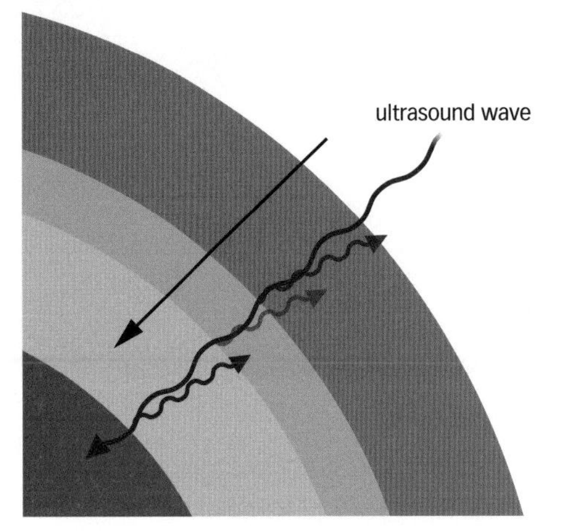

Ultrasound is reflected whenever there is a change in density in the medium it is passing through

Key words

ultrasound, reflection

Questions

1. Describe what happens to ultrasound when it passes from one medium to another. (E)
2. Define what is meant by ultrasound.
3. A beam of ultrasound is fired into a long metal bar that is known to have a crack inside. The sound travels at 3000 m/s through the metal. After 0.004 s a refection is received back at the transmitter. How far along the bar is the crack? (C)
4. Describe how the traces on an oscilloscope can be used to determine the distance to a crack. Include a diagram of a sample trace in your answer. (A*)

4: Medical uses of ultrasound

Learning objectives

After studying this topic, you should be able to:

- describe some medical uses of ultrasound
- compare some advantages and disadvantages of X-rays and ultrasound scans

Key words

ultrasound scan, kidney stone

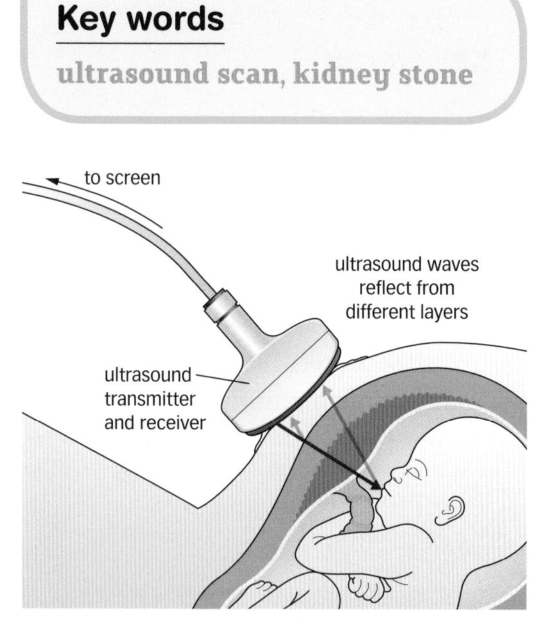

Ultrasound reflects from different layers in the mother and baby

- When describing a use of ultrasound don't just say 'an ultrasound'. Be more precise, say 'an ultrasound body scan', for example, or 'an antenatal scan'.

Ultrasound and medicine

You've probably already had an **ultrasound scan**, but you would have been too small to remember. In fact, you were not even born.

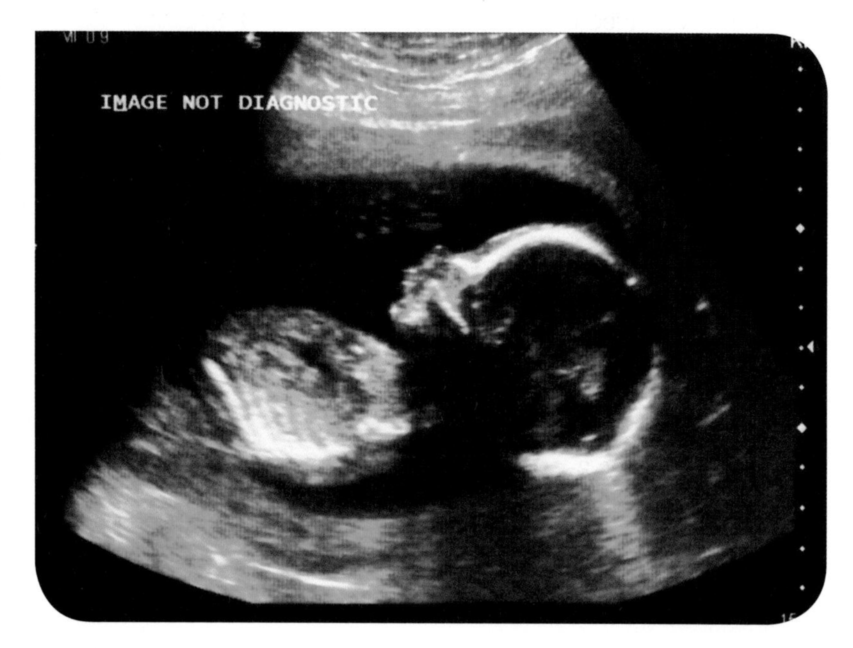

Ultrasound waves are used to safely produce an image of an unborn baby

As you may remember, ultrasound waves have a high frequency and a short wavelength. This means they are able to travel inside the body and produce images useful for doctors. Ultrasound scans are commonly used to check that babies are developing without any problems, whilst still inside their mother.

An ultrasound beam passes into the mother and then reflects off different layers inside her. There is a strong reflection back from the unborn baby's skeleton. The reflections are then processed by a computer to produce an image.

Ultrasound scans are not only used for scans of unborn babies; they are also used to check on a patient's heart, kidneys or liver.

A Give three examples of things that might be scanned as part of an ultrasound body scan.

Ultrasound has other important medical uses. It can even be used to monitor the blood flow inside a patient's veins.

Another common medical use of ultrasound is to break up **kidney stones**. These sometimes form in a patient's kidney. They can block important ducts and can be very painful.

Ultrasound is used to break up the kidney stones inside the body, preventing the need for surgery. Ultrasound is directed at the kidney stone. This makes the stone vibrate at a very high frequency, causing it to break up into small enough pieces to pass out of the body in the patient's urine.

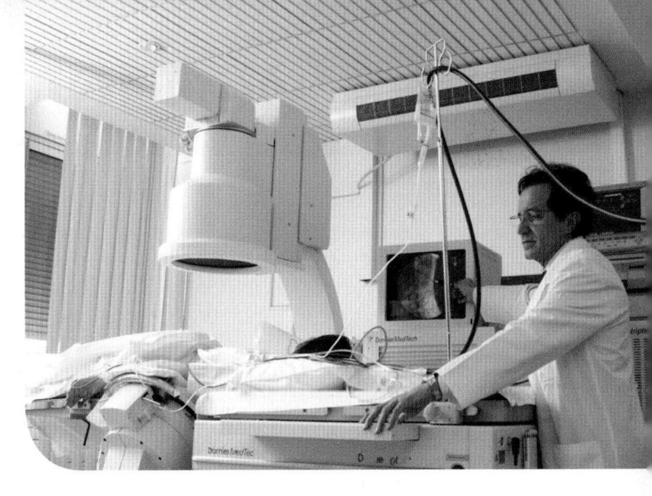

▲ Ultrasound is used to break up painful kidney stones

B What is a kidney stone and why is it a problem?

Comparing ultrasound and X-ray scans

Both ultrasound and X-rays are used to obtain images of the inside of the human body. Ultrasound body scans have advantages over X-ray scans. An ultrasound scan does not damage living cells as some higher energy X-rays can. The ultrasound waves are non-ionising, unlike X-rays. This makes them suitable for scanning unborn babies. A pregnant woman should never have an X-ray as the baby's cells are particularly vulnerable to the effects of ionising radiation.

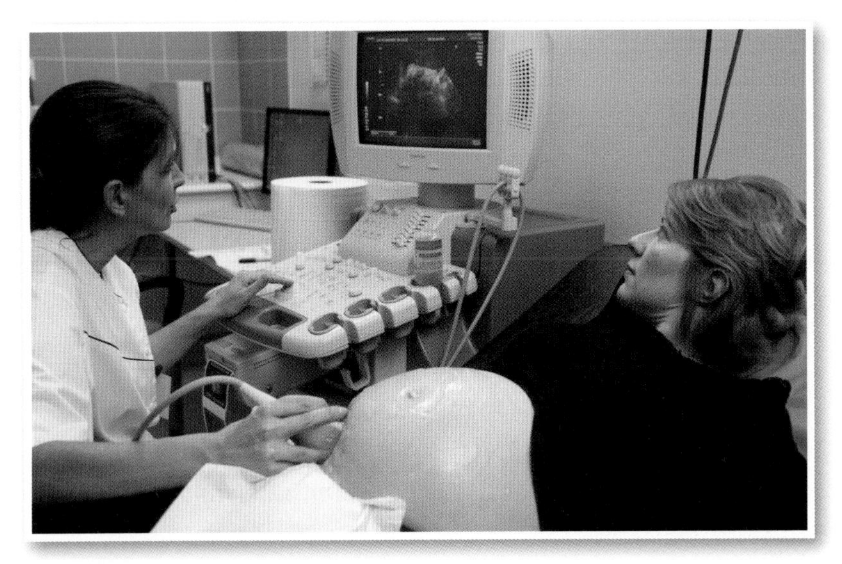

▲ Unlike X-rays, ultrasound waves are non-ionising, and so they can be used to scan women who are pregnant

In spite of their dangers, X-rays have one significant advantage over ultrasound. X-rays have much shorter wavelengths. The images produced are therefore often much clearer, giving more detail than an ultrasound scan. This allows tiny, hairline cracks or small anomalies to be spotted by doctors.

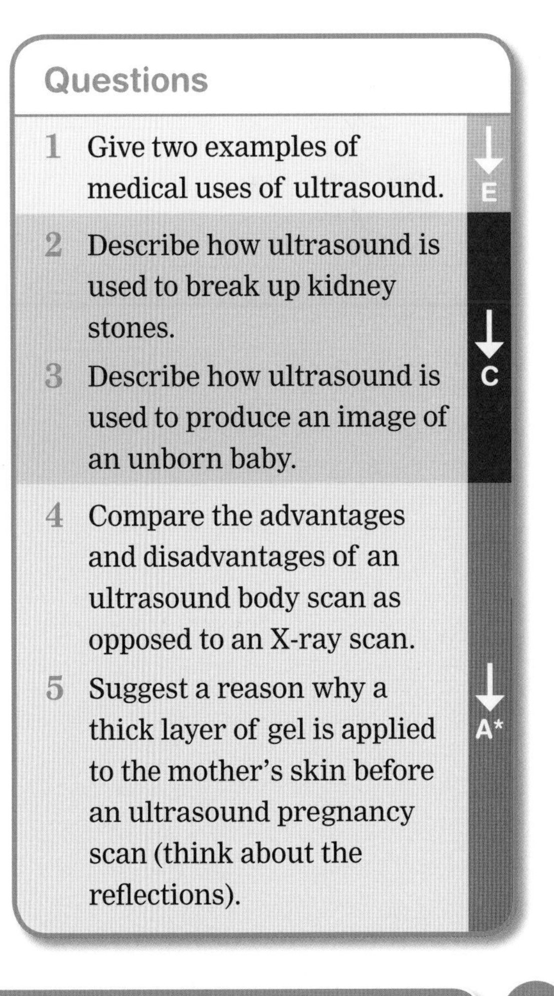

Questions

1 Give two examples of medical uses of ultrasound. (E)

2 Describe how ultrasound is used to break up kidney stones.

3 Describe how ultrasound is used to produce an image of an unborn baby. (C)

4 Compare the advantages and disadvantages of an ultrasound body scan as opposed to an X-ray scan.

5 Suggest a reason why a thick layer of gel is applied to the mother's skin before an ultrasound pregnancy scan (think about the reflections). (A*)

5: More on refraction

Learning objectives

After studying this topic, you should be able to:

- describe the process of refraction
- explain Snell's law
- understand what is meant by refractive index

Key words

refraction, refractive index

The reflection of light leads to some strange optical effects

Did you know...?

Back in 1621, the Dutch physicist Willebrord Snellius rediscovered the mathematical relationship between the refractive index and the angles of incidence and refraction. It is sometimes called Snell's law in his honour. In France it is called 'Snell–Descartes' law' as Descartes derived it independently in 1637. But maybe the honour should go to Ibn Sahl of Baghdad, who accurately described it in a manuscript called *On Burning Mirrors and Lenses* – in 984!

Bending rays of light

The **refraction** of light leads to some unusual optical effects. Mirages in deserts are caused by refraction, swimming pools look shallower than they actually are, and fish that are seen from above the surface of the water are not where they appear to be.

A Give an example of an unusual optical effect caused by refraction.

As you will remember, refraction is the bending of light (or any wave) when it travels from one medium to another. When light moves from one medium to another its speed changes, depending on the density of the material. This speed change causes a change in the direction of the light (unless the light wave hits the new medium head-on, in which case the direction of the ray is along the normal).

The direction that the light bends in depends on the relative density of the two media. If the light enters a denser material, such as when travelling from air to glass, the light slows down. This makes the light bend towards the normal.

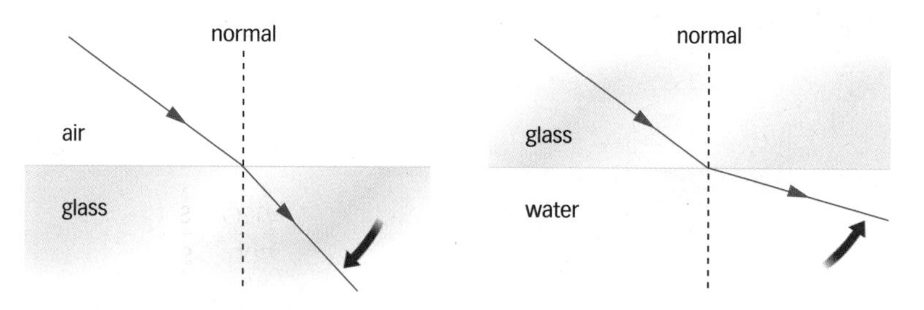

Light either refracts towards or away from the normal depending on the relative density of the two media

If the light goes from a denser medium to one that is less dense, such as when travelling from glass to water, the light speeds up. This makes the light bend away from the normal.

B What happens to the direction of light when it passes from a medium into one that has a lower density?

Refractive index

The **refractive index** of a material is a measure of the speed of light through the material compared with the speed of light in a vacuum. The more slowly the light travels through the material, the higher its refractive index. The denser the material, the higher its refractive index. Water has a refractive index of 1.3, and the refractive index of glass is 1.4.

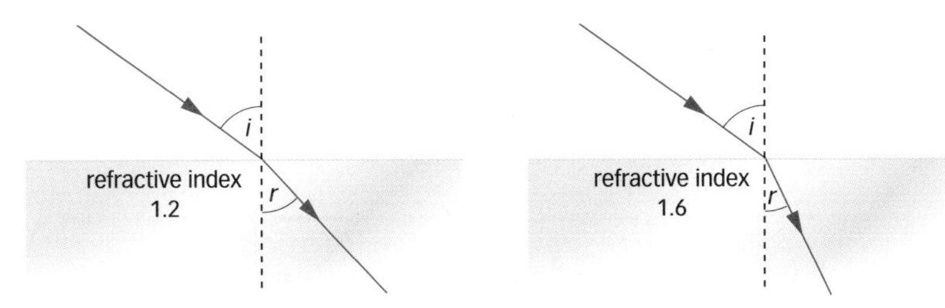

▲ The greater the refractive index, the more the light ray bends when it enters the material

Materials with a higher refractive index cause the light to slow down more and bend more towards the normal.

The refractive index of a material can be calculated using this equation:

$$\text{refractive index} = \frac{\text{sine of the angle of incidence}}{\text{sine of the angle of reflection}} = \frac{\sin i}{\sin r}$$

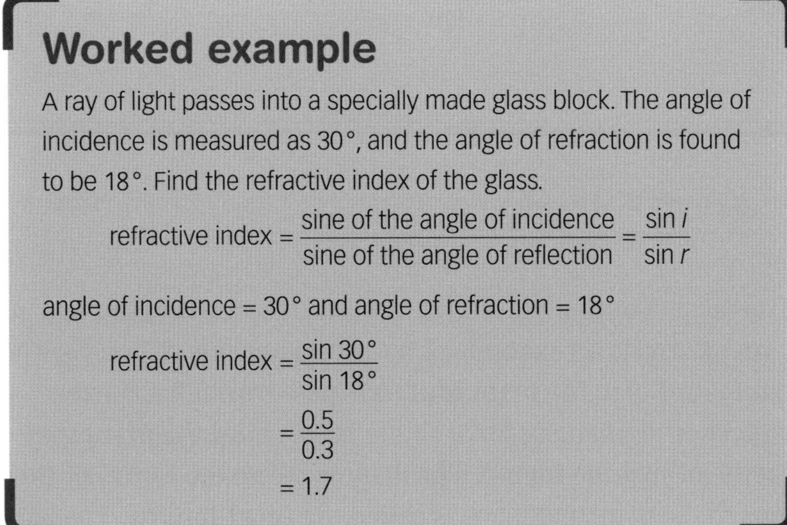

Worked example

A ray of light passes into a specially made glass block. The angle of incidence is measured as 30°, and the angle of refraction is found to be 18°. Find the refractive index of the glass.

$$\text{refractive index} = \frac{\text{sine of the angle of incidence}}{\text{sine of the angle of reflection}} = \frac{\sin i}{\sin r}$$

angle of incidence = 30° and angle of refraction = 18°

$$\text{refractive index} = \frac{\sin 30°}{\sin 18°}$$

$$= \frac{0.5}{0.3}$$

$$= 1.7$$

Plotting a graph of sin *i* against sin *r* gives a straight line through the origin. Sin *i* is directly proportional to sin *r*. The refractive index can be found by measuring the gradient of the graph. Materials with a higher refractive index show a steeper line.

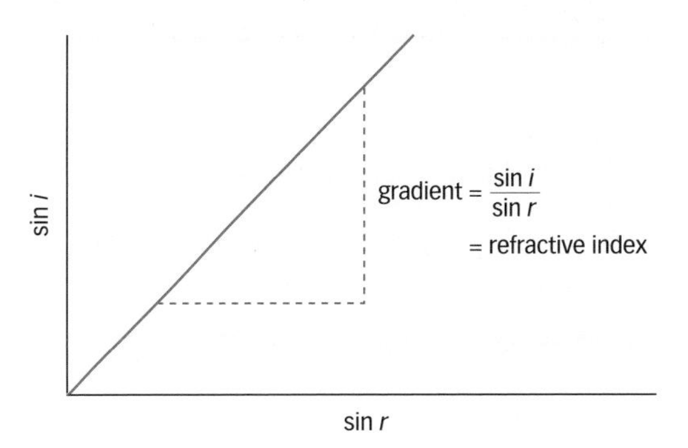

▲ A graph of sin *i* (the sine of the angle of incidence) against sin *r* (the sine of the angle of reflection) shows a straight line through the origin. This is a directly proportional relationship.

Questions

1. Describe what happens to the direction of light when it passes from a low density medium into one with a higher density.
2. Explain what causes a ray of light to change direction when it passes from one material to another.
3. Explain what is meant by refractive index and describe how it might be calculated.
4. A ray of light passes into a glass block. The angle of incidence is 49°, and the angle of refraction is found to be 30°. Find the refractive index of the glass.
5. Describe an experiment you could do, including the measurements you would take, to determine the refractive index of a transparent block.

E C A*

6: Introduction to lenses

Learning objectives

After studying this topic, you should be able to:

- ✔ understand that a lens forms an image by refracting light
- ✔ describe the difference between a converging and a diverging lens

Key words

converging, **convex**, **principal axis**, **principal focus**, **focus**, **focal length**, **concave**, **diverging**, **virtual focus**

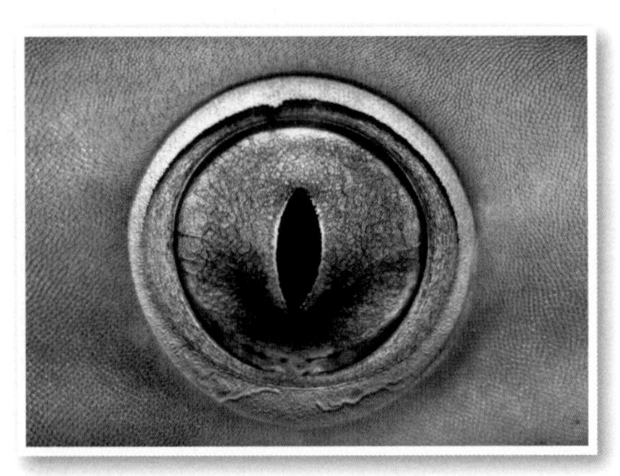

▲ Like most eyes, the eyes of a shark contain a lens to focus the light

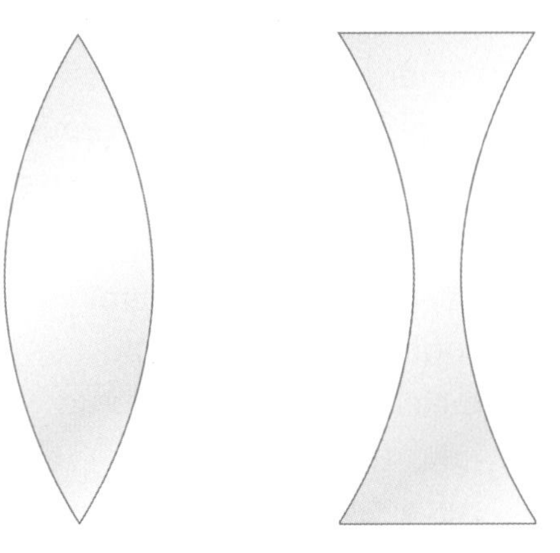

▲ A simple convex, converging lens

▲ A simple concave, diverging lens

Life through a lens

Lenses are found in cameras, telescopes, glasses, and in the eyes of most animals. There are lots of different types of lens, but they all work in exactly the same way.

A lens is used to refract the light that passes through it. The light then forms an image. The lens is usually denser than the medium around it, so when light enters the lens it bends towards the normal. When it leaves the lens it bends away from the normal.

A Give two examples of uses of lenses.

Although the light refracts both when it enters and when it leaves the lens, we often simplify our drawings by showing the light refracting at the centre of the lens.

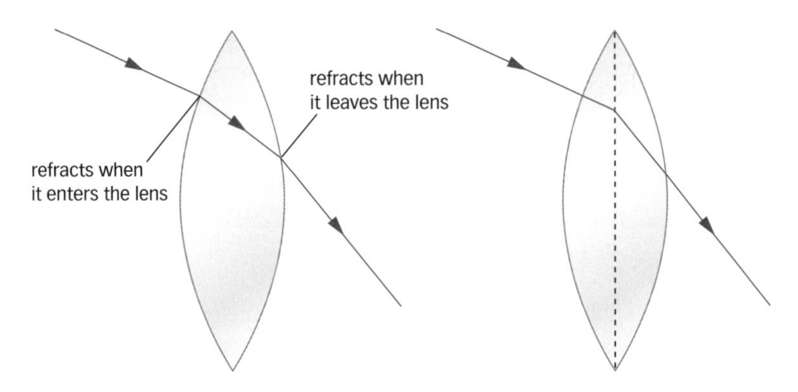

▲ A lens refracts light when it enters and when it leaves the lens. To simplify this we just draw the light refracting in the middle of the lens.

The most commonly used type of lens is a **converging** one. A **convex** lens is an example of a converging lens. When rays of light parallel to the **principal axis** pass through a convex lens, they are focussed to a single point called the **principal focus** (or just the **focus**). The distance from the centre of the lens to the principal focus is called the **focal length**. The lens has two principal foci, one on each side of the lens.

B What type of lens can be used to focus light at a principal focus?

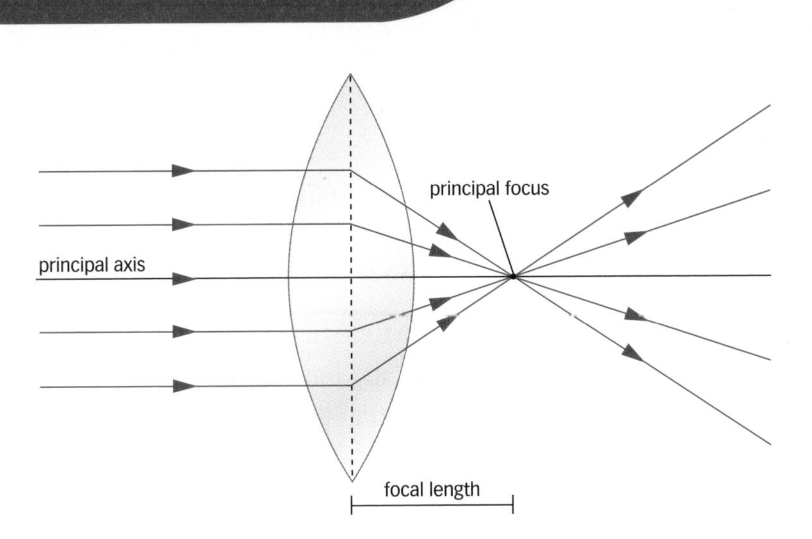

▲ Parallel rays of light through a convex lens are focussed at the principal focus

▲ The converging lens used in a magnifying glass focuses the sunlight. This can create a hot spot.

Other types of lens

A **concave** lens is an example of a **diverging** lens. Here the parallel rays of light spread out when they pass through the lens.

On a diagram such as that shown on the right you can extrapolate the rays back to a **virtual focus**. There is no actual focus here; it is just where the rays of light appear to come from, hence the use of the term 'virtual'.

Diverging lenses also have a focal length. This is the distance from the centre of the lens to the virtual focus.

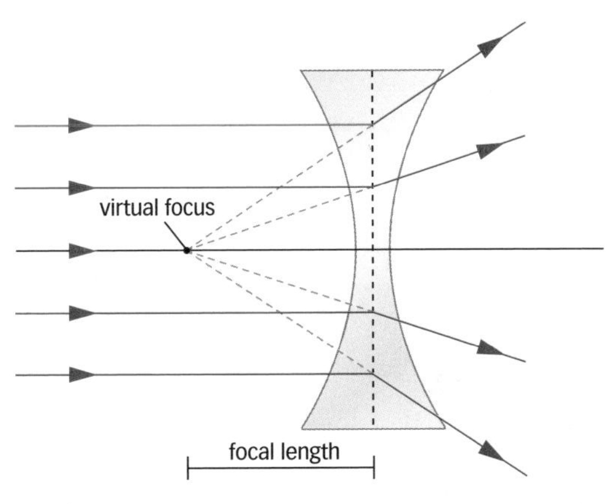

▲ A diverging lens spreads out the rays of light. The light then appears to come from a virtual focus.

Questions

1. What are the most common types of lens? (E)
2. Draw a diagram to show how a converging lens can focus rays of light.
3. Draw a diagram to show how a diverging lens spreads out rays of light, and how the virtual focus can be found. (C)
4. Two convex lenses are 10 cm apart along the same principal axis. They both have a focal length of 5 cm. Draw a diagram to show the path followed by parallel rays of light that enter the lens on the left, before passing through the lens on the right. (A*)

Did you know...?

The oldest known artificial lens is the Nimrud lens. This is a 3000-year-old piece of rock crystal discovered in modern Iraq. Scientists are not sure what this lens was used for. Some think it may have been used as a magnifying glass, others believe it was used to start fires by focussing sunlight.

Exam tip

✔ Don't mix up convex and concave lenses. A concave lens curves inwards – like a cave.

Learning objectives

After studying this topic, you should be able to:

- ✔ describe the nature of an image formed by a lens

Key words

diminished, **magnified**, **upright**, **inverted**, **real**, **virtual**

▲ A projector contains a converging lens to focus an image on a screen

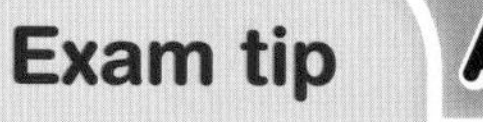

Exam tip AQA

- ✔ When describing an image, make sure you use all four of the characteristics in the table.

Forming an image

We use lenses to form images of a wide variety of objects. From distant planets right down to microscopic cells, lenses are used to help us make sense of the world around us.

All lenses are designed to use refraction to bend the light in a specific way, depending on the lens. Inside most optical instruments, a lens is used to refract the light to form an image.

> **A** What is the name of the bending of light that happens when lenses form an image?

Lenses are used to make very different types of image. A digital camera uses a lens to produce a tiny image on film or a light-sensitive CCD. A digital projector produces an image on a screen much larger than the tiny screen inside the projector.

The images formed depend on the lens used and the distance of the object from the lens.

> **B** What two factors affect the nature of the image formed by a lens?

Image characteristics

The image formed by a lens is described according to four features (characteristics).

Feature	Description
Position	Is the image closer to or further from the lens than the object?
Size	Is the image smaller (**diminished**) or larger (**magnified**) than the object?
Orientation	Is the image the same way up (**upright**) as the object or has it been flipped upside down (**inverted**)?
Type of image	Is the image a **real** image or a **virtual** image?

A real image is one that can be formed on a screen or surface. Rays of light from the same part of an object take different paths through the lens to cross at a focus on the other side of the lens. If you place a screen at that focus, where the rays intersect, it will show an image of the object. A virtual image is when the rays of light appear to the observer to have come from a focus. But the rays have not intersected: there is nowhere you could place a screen to form an image.

▲ You can use a convex lens to produce a real image on a screen

A convex lens can be used to produce an image of an object far away from the lens. In this case the image is closer to the lens than the object, diminished, inverted, and (as it can be formed on a screen) it must also be real.

Here is another example of using a simple converging lens to form an image. This time the object is much closer to the lens and so the characteristics of the image are different.

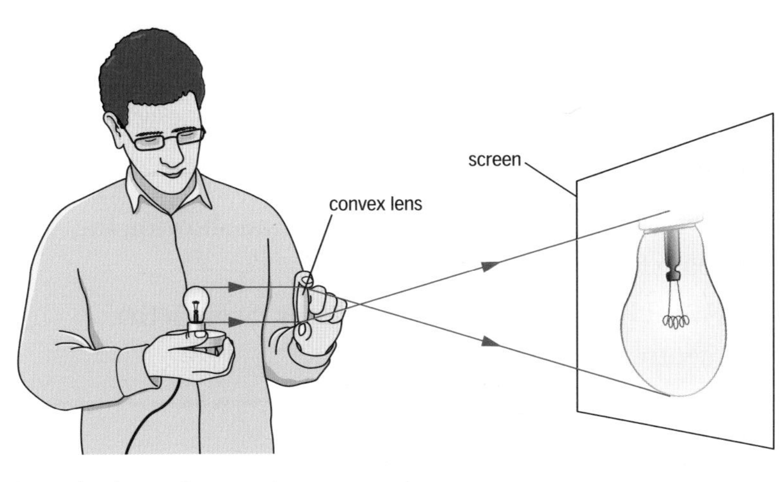

▲ Projecting an image using a convex lens

Questions

1. Give two examples of optical instruments that use lenses to form images. (E)
2. State the four characteristics that must be used when describing the image formed by a lens.
3. Describe the difference between a real image and a virtual image. (C)
4. Describe the characteristics of the image formed by a convex lens when the object is close to the lens (as seen in the diagram on the left).
5. Using the four characteristics in the table, describe the image formed inside a digital camera. (A*)

8: Converging lenses

Learning objectives

After studying this topic, you should be able to:

- construct ray diagrams to show how images are formed by a converging lens
- describe the images formed by a converging lens

Key words

ray diagram

This symbol represents a converging (convex) lens. It is often used when drawing ray diagrams as an alternative to drawing a full lens.

Some telescopes use a pair of lenses to produce an enlarged image of objects in the night sky. In the seventeenth century, Galileo used one to provide evidence that the Earth was not at the centre of the Universe.

Constructing ray diagrams

We often draw (or construct) **ray diagrams** showing the path of light through a lens to help us determine the nature of the image formed.

In these examples an arrow is used to represent the object. Light is reflected off all parts of the object, but we just consider the light that is being reflected from the tip.

Light is being reflected from the tip in many directions, but we can find out where the image will be by considering light that follows one of three paths.

1. *Light passing through the centre of the lens.* The light that passes through the centre of a lens continues through in a straight line.
2. *Light travelling parallel to the principal axis.* The light that is travelling parallel to the principal axis is refracted by the lens so that it passes through the principal focus on the other side.
3. *Light passing through the focus.* The light that passes through the principal focus is refracted by the lens so that it comes out travelling parallel to the principal axis.

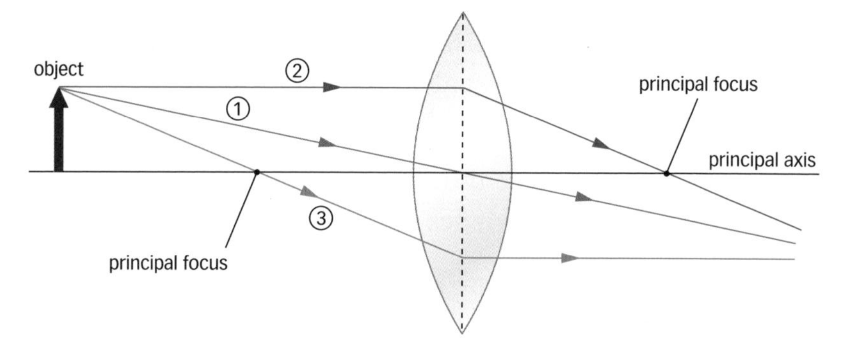

The three rules for rays of light passing through a converging (convex) lens

A What is the name given to a diagram showing the path followed by a ray of light?

B What happens to light that passes through the centre of a converging lens?

Forming different images

Using these three paths, we can construct ray diagrams to show how images are formed.

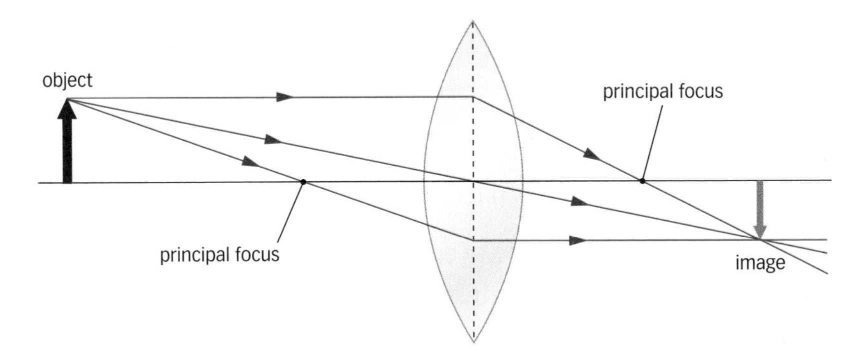

▲ Using a converging lens to form an image of an object that is far away from the lens

The image forms where the rays cross. If an object is far away, the image formed is closer to the lens, smaller, inverted and real (the rays do cross over).

If the object is much closer to the lens, then the image is very different.

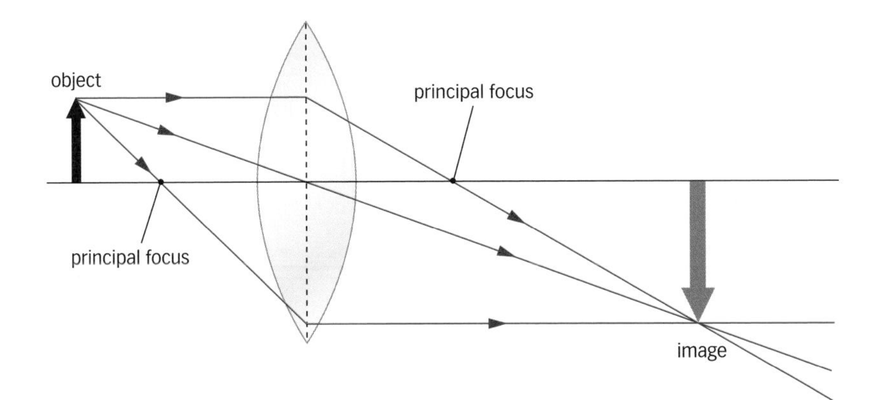

▲ If the object is much closer to the lens, the image formed is very different

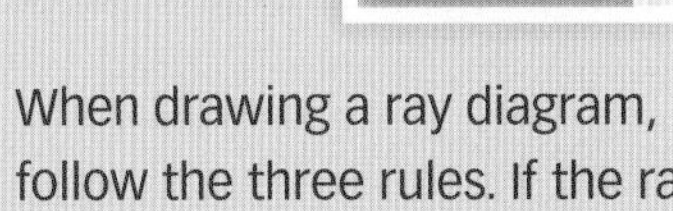

✔ When drawing a ray diagram, follow the three rules. If the rays cross, the lens forms a real image that can be projected onto a screen.

Questions

1 What happens to light that enters a converging lens parallel to the principal axis? (E)

2 Carefully draw a diagram showing the three rules for the path followed by rays of light through a converging lens.

3 Using the four image characteristics in the table on spread P3.7, describe the image formed in the second ray diagram on this page. (C)

4 Use graph paper to carefully produce your own ray diagrams for the two examples on this page. (A*)

9: Diverging lenses

Learning objectives

After studying this topic, you should be able to:

- ✔ construct ray diagrams to show how images are formed by a diverging lens
- ✔ describe the images formed by a diverging lens

▲ This symbol represents a diverging (concave) lens. It is often used when drawing ray diagrams as an alternative to drawing a full lens.

▲ Some types of glasses contain diverging lenses. They can help the wearer focus on objects that are far away.

Ray diagrams for diverging lenses

A concave lens is an example of a diverging lens. In a similar way to how we used a converging lens, we can use a diverging lens to form an image of an object.

When the rays of light pass through the lens they are refracted and, just as with a converging lens, we look at the paths of three rays to help us construct a ray diagram.

1. *Light passing through the centre of the lens.* Just like a converging lens, when light passes through the centre of a diverging lens it continues through in a straight line.
2. *Light travelling parallel to the principal axis.* When light travelling parallel to the principal axis passes through the lens, it is refracted away from the principal axis. When we trace back along the path of the refracted ray we find that it passes through the virtual focus.
3. *Light heading for the virtual focus.* When light that is heading for the virtual focus on the other side of the lens passes through the lens, it is refracted so that it comes out travelling parallel to the principal axis.

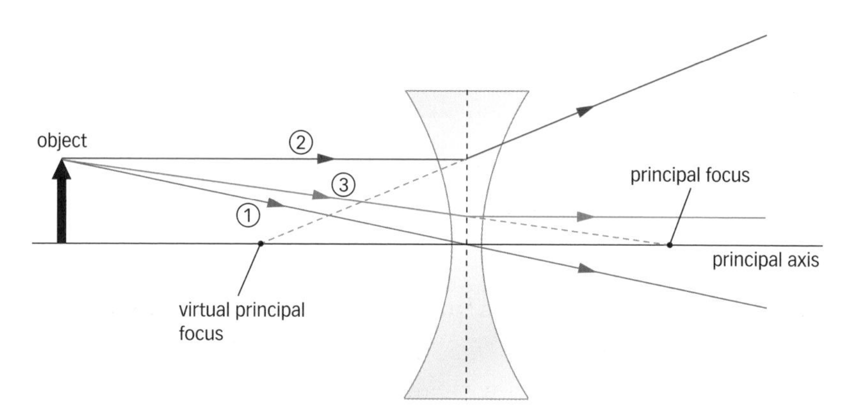

▲ The three rules for rays of light passing through a diverging (concave) lens

A What happens to the path of a ray of light that passes through the centre of a diverging lens?

Describing the images formed by a diverging lens

Unlike a converging lens, the image from a diverging lens always forms on the same side of the lens as the object. In order to see the image, you must look through the lens.

The image from a diverging lens is always closer to the lens than the object, smaller than the object (diminished), upright and virtual.

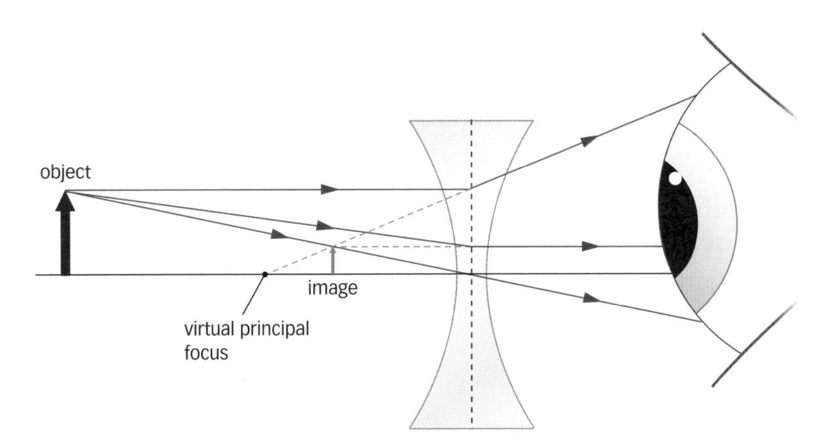

Forming a virtual image with a diverging lens

The image formed by a diverging lens is always a virtual image. It cannot be formed onto a screen or camera film.

B Why is it not possible to form an image from a diverging lens on a screen?

When the object moves further away from a diverging lens, the virtual image gets smaller and smaller. This allows you to see a wider field of view through the lens. This makes this kind of lens useful as a security peep hole in front doors. If there was no lens in the hole you would not be able to see much at all.

A diverging lens is used in a peephole to provide a wider field of view

Questions

1. What happens to light that enters a diverging lens parallel to the principal axis? (E)
2. Carefully draw a diagram showing the three rules for the path followed by a ray of light through a diverging lens. (C)
3. Use graph paper to carefully produce your own ray diagram showing the image formed by a diverging lens.
4. Describe what happens to the image formed by a diverging lens as the object moves further from the lens. Construct two ray diagrams to help illustrate your answer. (A*)

10: Magnification

Learning objectives

After studying this topic, you should be able to:

- calculate the magnification of an object
- describe how a converging lens is used in a magnifying glass

Key words

magnification

Calculating magnification

Lenses are often used to magnify tiny objects. As the light passes through the lens, it is refracted in such a way as to produce a larger image. This allows the observer to see minute details usually invisible to the naked eye.

The **magnification** of an object can be calculated using the equation:

$$\text{magnification} = \frac{\text{image height}}{\text{object height}}$$

If the magnification is 1, the image is the same height as the object. If the magnification is less than 1, the image is smaller than the object. For example, a magnification of 0.5 would mean the image is half the size of the object. A 40 cm object would produce a 20 cm image.

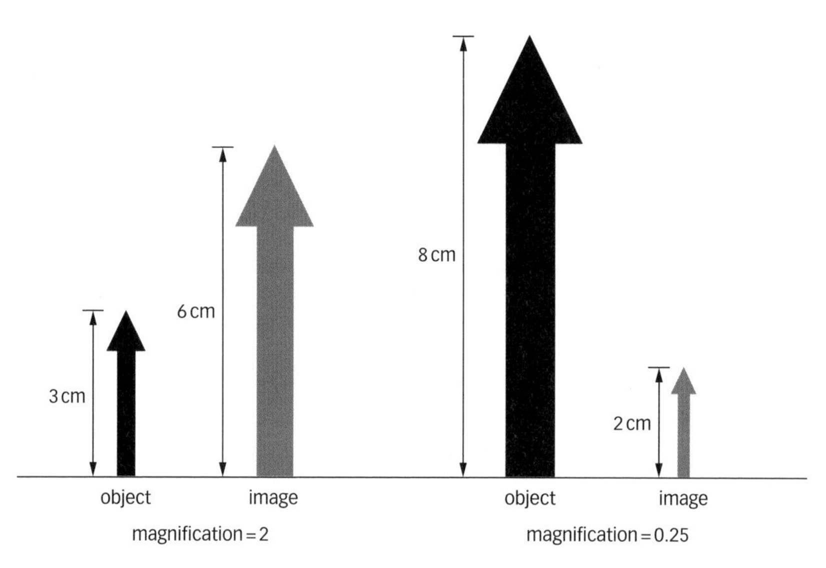

▲ Examples of different magnifications

Any magnification greater than 1 means the image is larger than the object. A magnification of 5 would mean the image is five times larger than the object: a 2 cm object would produce a 10 cm image.

A What is meant by a magnification of 3?

B An image, 2 cm high, is focussed by a lens onto a photographic film. The object being photographed is 10 cm high. Calculate the magnification.

Worked example

Light from a 4 cm high object passes through a lens onto a screen. The image is 36 cm high. Calculate the magnification.

$$\text{magnification} = \frac{\text{image height}}{\text{object height}}$$

image height = 36 cm and object height = 4 cm

$$\text{so, magnification} = \frac{36\text{ cm}}{4\text{ cm}} = 9$$

The magnifying glass

Telescopes and microscopes use combinations of lenses to magnify an object. A magnifying glass uses a single converging (convex) lens.

For the lens to act as a magnifying glass, the object being magnified must be close to the lens. It needs to be nearer to the lens than the principal focus (its distance from the lens must be less than the focal length). The magnifying glass produces an image on the same side of the lens as the object, so the user must look through the lens in order to see the magnified image.

Using a convex lens as a magnifying glass produces a magnified, upright, virtual image, which is further away from the lens.

A magnifying glass is a single convex lens

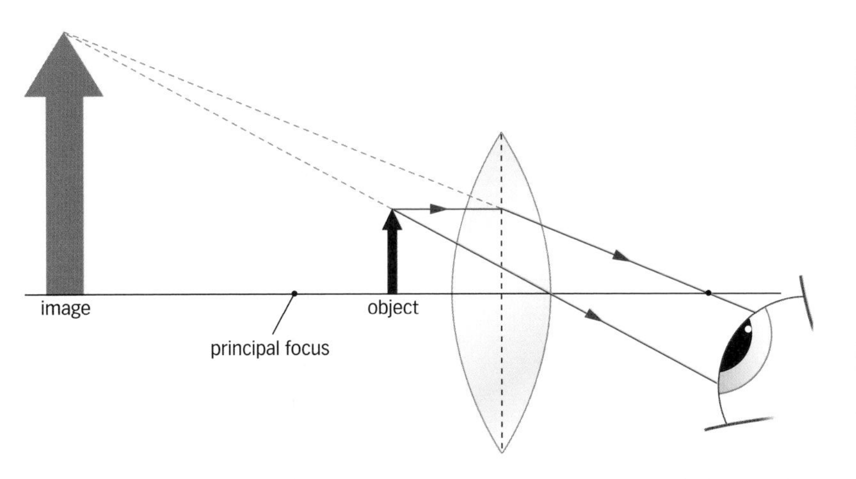

Using a convex lens as a magnifying glass produces a magnified, virtual image

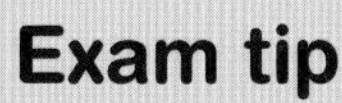

Exam tip AQA

- When constructing a ray diagram for a magnifying glass, only two rays can be drawn.
- A converging lens can produce both real and virtual images. A virtual image is only produced when the distance from the object to the lens is less than the focal length.

Questions

1. What type of lens is used in a magnifying glass?
2. Light from a 2 cm high object is projected through a lens onto a screen, producing an image 56 cm high. Calculate the magnification.
3. An image, 4.8 cm, high is focussed on a CCD by a lens. The object being photographed is 19.2 cm tall. Calculate the magnification.
4. Use graph paper to carefully construct a ray diagram for a magnifying glass and calculate the magnification.
5. Compare the similarities and differences between images formed by a magnifying glass and a diverging lens.

E ↓ C ↓ A*

11: The human eye

Learning objectives

After studying this topic, you should be able to:

- ✔ describe the structure of the human eye, including the functions of the key parts
- ✔ state the range of vision of the human eye

Key words

suspensory ligaments, retina, cornea, iris, pupil, ciliary muscles, range of vision

▲ The human eye is an amazing optical instrument

Did you know...?

It is not just light which causes the pupil to open wider. When you see someone you fancy, your pupil dilates. You can't stop it: it is a biological reaction aimed at making you look more attractive to a potential mate!

Eye, eye

Imagine what the world be like if the human eye had not evolved. The eye is an amazing object. It allows us to see the world around us, to make sense of the beauty and complexity of the natural world.

The human eye, like the eyes of most animals, contains a convex lens. This is held in place by a series of **suspensory ligaments**. Unlike the glass lenses we have been discussing, the lens in your eye is made of a jelly-like substance which allows it to change its shape. Light passes through the lens, is refracted, and is focussed onto the back of the eye (called the **retina**).

The retina contains special cells that detect light intensity and colour. They send a signal, via the optic nerve, to your brain.

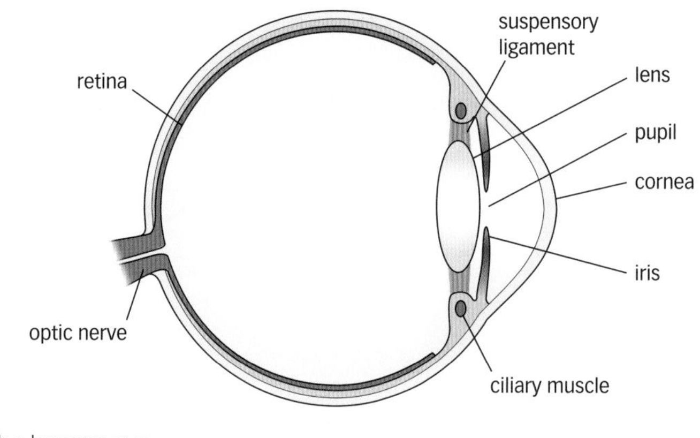

▲ The human eye

A What is the name of the part of your eye that has light-sensing cells?

Other important parts of the eye help to produce clear images. The **cornea** is the transparent front part of the eye that covers the **iris** and **pupil**. Like the lens, the cornea also refracts light passing through it, helping to form a focussed image. Refraction at the cornea plays a very important role in forming an image, but unlike the lens it is a fixed shape and cannot change its focus.

The iris is the coloured part of the eye. It can expand or contract, allowing differing amounts of light to pass through the gap in the centre (the pupil). The larger the pupil, the more light can enter the eye. At night your iris slowly relaxes, so that the pupil becomes larger. This dilated pupil helps you to see more clearly in low light conditions. It takes around 20 minutes to reach full 'night vision'. A quick glance at a bright light causes the pupil to shrink back, losing the night vision

> **B** What is the name given to the coloured part of the eye?

▲ The eye can focus on objects that are very far away, as well as on objects that are much closer, for instance when reading

Changing focus

The shape of the lens inside the eye can be changed by the **ciliary muscles**. When these contract, they pull on the suspensory ligaments and change the shape of the lens. This allows the eye to focus on nearby objects, and then rapidly change the focus to look at objects far away.

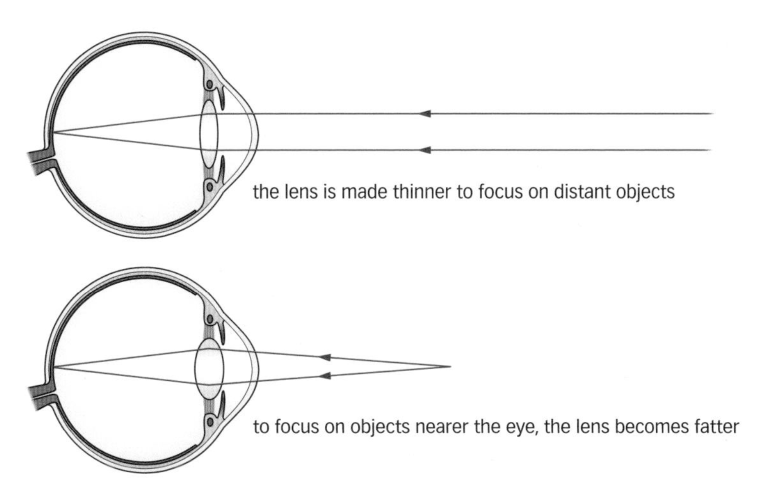

▲ The shape of the lens changes so that light from objects that are far away as well from those that are closer to the eye can be focussed on the retina

The lens is stretched and made thinner for looking at distant objects. It springs back to a much fatter shape to focus the light from nearby objects.

The **range of vision** describes the distance from the nearest point to the furthest point that the eye can focus on. If you move your fingers too close to your eye, it can't focus and you see a blurry image. This near point is around 25 cm; objects much closer than this cannot be focussed by the eye alone.

The far point is much further away. The eye can focus on objects at infinity; these objects are so far away that the light entering the eye from all parts of these objects is effectively parallel to the principal axis.

Questions

1. Describe what happens to light when it enters the eye.
2. Produce a table summarising the function of the following structures in the eye: iris and pupil; retina; lens; cornea; ciliary muscle; suspensory ligaments.

↓ E

3. What is the range of human vision?
4. Describe how the lens changes shape to focus on objects at different distances.

↓ C

5. Suggest a reason why it becomes more difficult to focus on nearby objects as you age.

↓ A*

12: Problems with eyesight

Learning objectives

After studying this topic, you should be able to:

- understand the reasons for long- and short-sightedness
- describe how corrective lenses can be used

Key words

short-sightedness, long-sightedness, corrective lenses

Sometimes people have difficulty focusing on some objects, and see a blurry image

B Which problem with vision becomes more common as people age?

Seeing near and far

The eye is such a complex organ that it is not surprising that we experience problems with our vision. This becomes more frequent as you age, but often people are born with slight vision defects. The lens, the cornea, or even the shape of the eye itself can lead to problems when focussing on objects. This produces a blurry image, with some loss of detail.

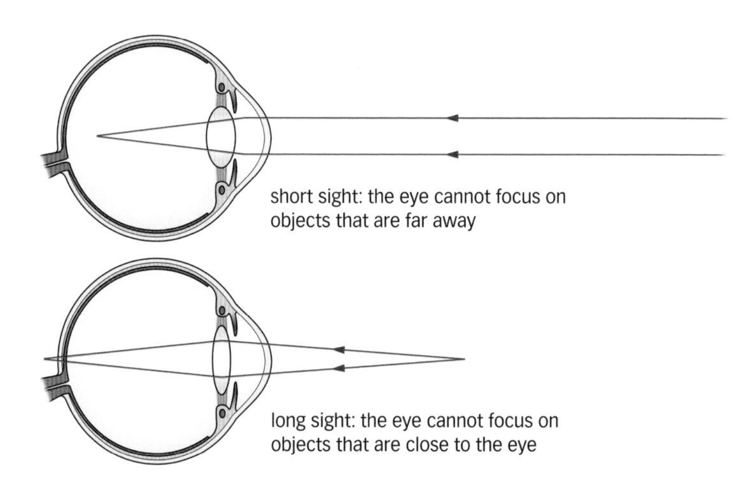

If the eye cannot focus the light onto the retina, a blurry image is seen

Two common problems with vision are **short-sightedness** and **long-sightedness**, often just called short sight and long sight.

A What are the two most common problems with vision?

People with short sight cannot focus on objects that are far away. This can happen because the eye is slightly too long, or cannot make the lens thin enough to focus the light on the retina.

People with long sight cannot focus on objects that are close to the eye. This may be because the eye is slightly too short, or cannot make the lens fat enough to focus the light. This often happens as people get older. The lens loses some of its elasticity and so does not spring back into a fat enough shape to focus the light.

Using glasses

To help form a clear image, **corrective lenses** are used. These are used in glasses and refract the light passing through them in such a way as to produce a sharp image on the retina.

To correct short-sightedness a diverging (concave) lens is used. This allows the light from distant objects to be focussed on the retina instead of in front of it, producing a clear image.

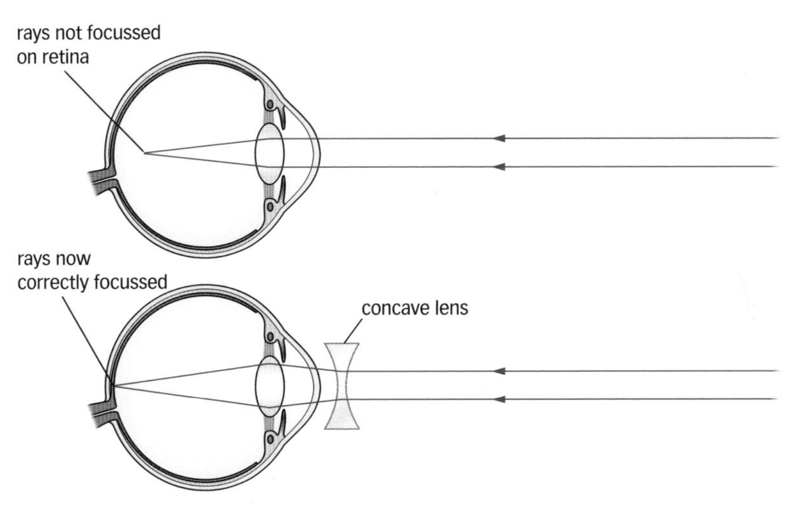

▲ A concave lens is used to correct short-sightedness

To correct long-sightedness a converging (convex) lens is used. The lens helps the eye produce a clear image by converging the light from nearby objects on to the retina.

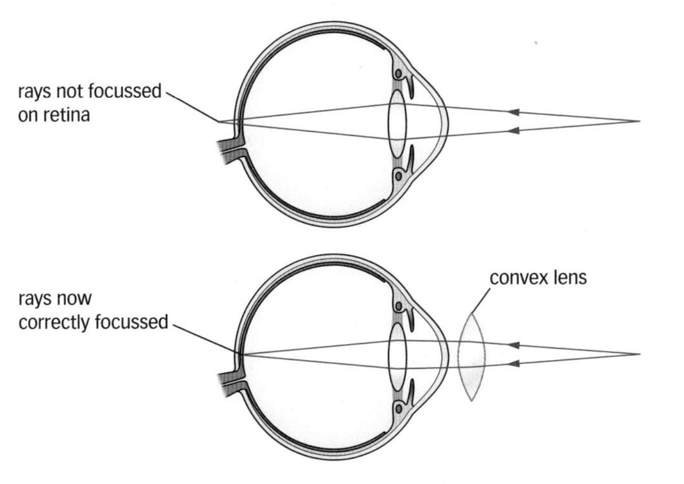

▲ A convex lens is used to correct long-sightedness

▲ Wearing glasses allows an image to be clearly focussed on the retina

Did you know...?

Lenses were first used to help with eyesight in ancient Egypt. Hieroglyphs from the fifth century BC show the use of simple lenses. It is not clear exactly when modern glasses (that you wear on your face all the time) were invented, or who invented them. It is likely that glasses were invented between 1280 and 1300, in Italy.

Questions

1 What is a corrective lens? (E)

2 Explain what is meant by short sight and long sight, and suggest a reason for each.

3 Draw diagrams to show how in long sight and short sight a clear image cannot be focussed on the retina. (C)

4 Draw ray diagrams showing how glasses can be used to correct for long sight and short sight. (A*)

13: The camera

Learning objectives

After studying this topic, you should be able to:

- describe how a camera forms an image
- compare the structure of the eye with a camera

Key words

camera

Cameras are everywhere! There is probably a camera on your mobile phone, and there are thousands of CCTV cameras in towns and cities.

Despite recent advances with digital images, the mechanical parts of cameras, such as the lens, haven't really changed much in principle since the earliest cameras.

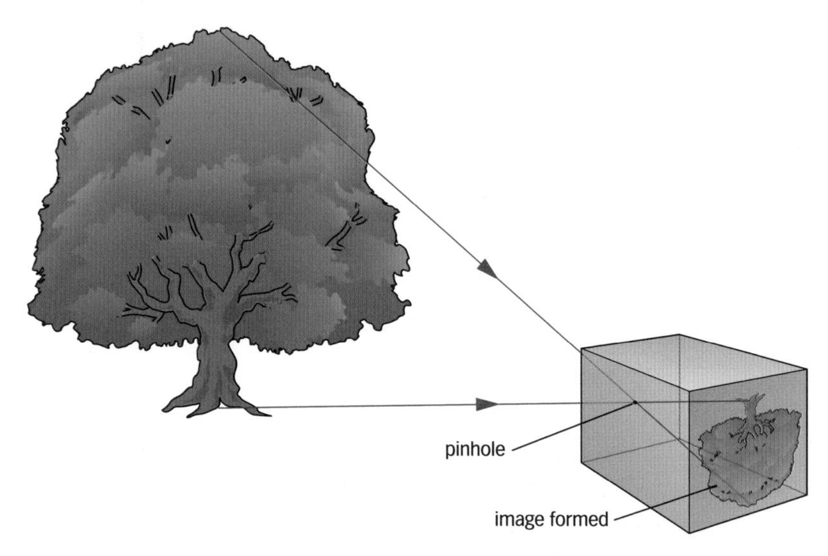

An image can be produced using a simple pinhole camera

A pinhole camera is the simplest form of camera. It does not even have a lens. Light passes through a tiny hole into a light-proof box. This produces an inverted image on the back of the box.

It was not until around 1600 that a simple converging (convex) lens was added to the pinhole camera. However, the camera was still not able to take a photo – it could only project an image, not record it.

A camera shares some features of the human eye. Both have a lens to focus the light. However, to get clear images of objects at different distances, lenses in cameras are often moved forwards or backwards while the lens in the human eye changes its shape. The shutter in a camera performs a similar role to the iris and pupil. It allows more or less light in, depending on the conditions.

A What was the first type of lens to be used in a pinhole camera?

Over 200 years later, the first photographic image was taken. Light was focussed onto a plate that contained chemicals that reacted differently depending on the brightness of the light shone onto them. Over the years scientists developed better chemical compounds. These were more sensitive to light and produced clearer images.

A modern camera uses a lens to focus the light onto either a very fine-grain chemical film or onto a light-sensitive component called a CCD (charge-coupled device).

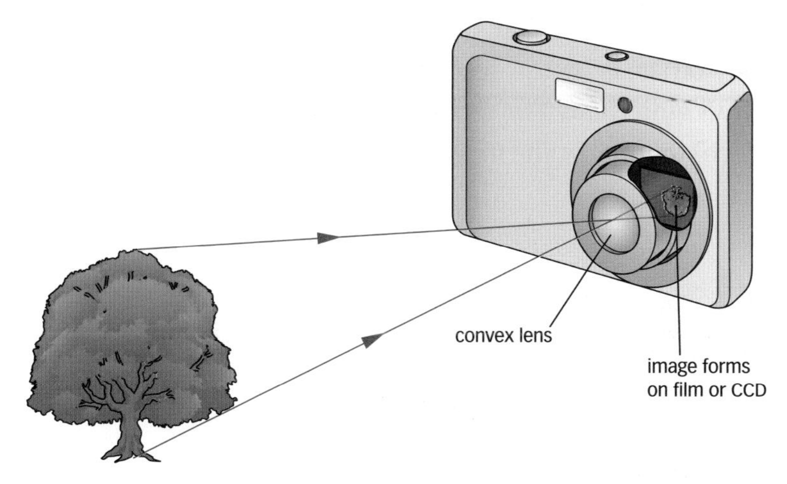

▲ A camera focuses an image onto film or a CCD

Detecting the light

The light-sensitive parts of a camera are very similar to the retina in the human eye. Camera film and CCD chips react when light shines on them. In camera film there is a permanent chemical change. Each photograph needs a fresh piece of film.

A CCD has millions of tiny light-sensitive pixels that each produce an electric current when light strikes them. This signal is then processed and stored as a digital picture. This only takes a fraction of a second. The camera is then ready to take another picture.

Questions

1 What part of a digital camera is equivalent to the retina in the human eye? (E)

2 Describe how a modern camera focuses an image onto a CCD or a piece of film.

3 Describe the differences between the use of camera film and the use of CCDs found in digital cameras. (C)

4 Give some advantages and disadvantages of using a CCD to take photos. (A*)

Did you know...?

Nearly all commercial cameras now use CCDs, which contain millions of megapixels. A 7 MP camera takes pictures made up of 7 million pixels. Each pixel forms a tiny part of the image. However the number of megapixels is not always the best indicator of quality. Other factors such as the size and type of lens are very important.

The cells within your retina are much more sensitive than most CCDs. Your eye produces a much clearer image than most digital cameras.

B Give two examples of a light-sensitive part of a camera.

▲ The CCD used in digital cameras behaves in a similar way to the retina in the eye

14: Lens power

Learning objectives

After studying this topic, you should be able to:

- describe the factors affecting the power of converging and diverging lenses
- calculate the power of a lens
- explain the relation of lens shape and refractive index for a given focal length

Key words

lens power, dioptres

Worked example 1

A convex lens used in a pair of glasses has a focal length of 50 cm. Find its optical power in dioptres.

$$\text{optical power} = \frac{1}{\text{focal length}} \quad \text{or} \quad P = \frac{1}{f}$$

focal length, $f = 50\text{ cm} = 0.50\text{ m}$

$$\text{optical power} = \frac{1}{0.5\text{ m}} = +2\text{ D}$$

Worked example 2

The focal length of a diverging lens is negative (because the principal focus is virtual). A diverging concave lens has a focal length of –0.4 m. Find its optical power in dioptres.

$$P = \frac{1}{f}$$

focal length, $f = -0.4\text{ m}$

$$\text{optical power} = \frac{1}{-0.4\text{ m}} = -2.5\text{ D}$$

The power of a lens

The lenses we use are of many different shapes and are made of a range of transparent materials. The **lens power** of any of these lenses describes how good it is at converging or diverging light. The more powerful the lens, the shorter its focal length. In the diagram, the lenses on the left are much ‘stronger’.

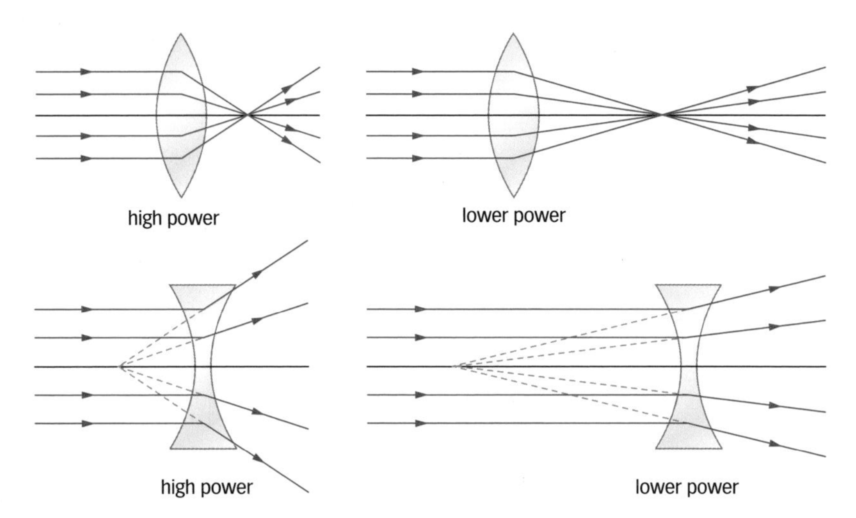

▲ The higher power lenses have shorter focal lengths

A What is meant by the power of a lens?

Lens power is measured in **dioptres**. A typical lens found in a pair of glasses may have a power of +3 D. The more powerful a lens, the higher this number.

The power of a converging lens is always a positive number. The power of a diverging lens is always a negative number.

B A lens has a power of –2 dioptres. Is it a converging or diverging lens?

The power of a lens is given by:

$$\text{optical power (dioptres, D)} = \frac{1}{\text{focal length (metres, m)}}$$

If P is power in dioptres, and f is focal length in metres, then:

$$P = \frac{1}{f}$$

What affects the focal length of a lens?

Two things determine the power of a lens and therefore the focal length.

- Refractive index of the material of the lens. The higher the refractive index, the more the lens will bend the rays of light. The lens has a higher power and a shorter focal length.
- Curvature of the two surfaces of the lens. The greater the curvature (the fatter the lens is), the more the lens bends the rays of light. The lens power is higher and the focal length is shorter.

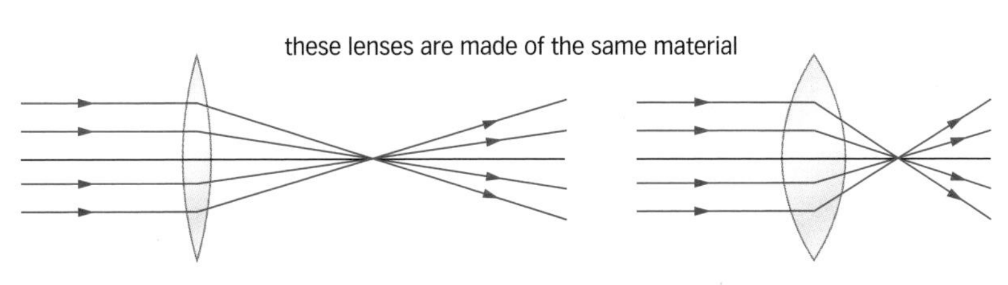

▲ The refractive index of the material is the same for these two lenses. The fatter lens is more powerful and has a shorter focal length.

The lenses in modern glasses can be very thin, yet they still have a high power. This is done by making the lenses from a special glass that has a much higher refractive index (it bends the light more) than normal glass. The lens is still just as powerful, with the same focal length as the older type made with normal glass, but does not need to be so curved. Because the lens can be flatter (lower curvature), it can also be manufactured thinner.

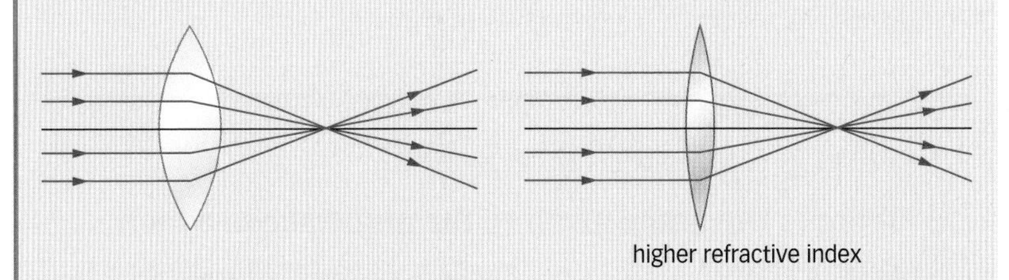

▲ A lens with a higher refractive index can have a lower curvature and still have the same focal length as a fatter lens

Exam tip

✔ When you calculate optical power in dioptres, the focal length must be in metres.

◀ Lenses manufactured for modern glasses can be made thinner. The higher refractive index means that a less curved lens can provide the same power and focal length. Flatter lenses can be made thinner.

Questions

1. What two factors affect the focal length of a typical lens? E
2. Two convex lenses are used in a telescope. The first has a focal length of 20 cm, the second a focal length of 60 cm. Find the optical power of each lens in dioptres. C
3. A diverging lens is used in a peep hole. It has a focal length of –10 cm. Find its optical power in dioptres.
4. Describe the effect on its power of changing the curvature of a lens. Use diagrams to help illustrate your answer.
5. Explain why a lens made from a material with a higher refractive index will have a higher power than a similarly shaped lens made from glass with a lower refractive index. A*

Learning objectives

After studying this topic, you should be able to:

- describe the conditions needed for total internal reflection
- calculate the critical angle

Key words

total internal reflection (TIR), critical angle

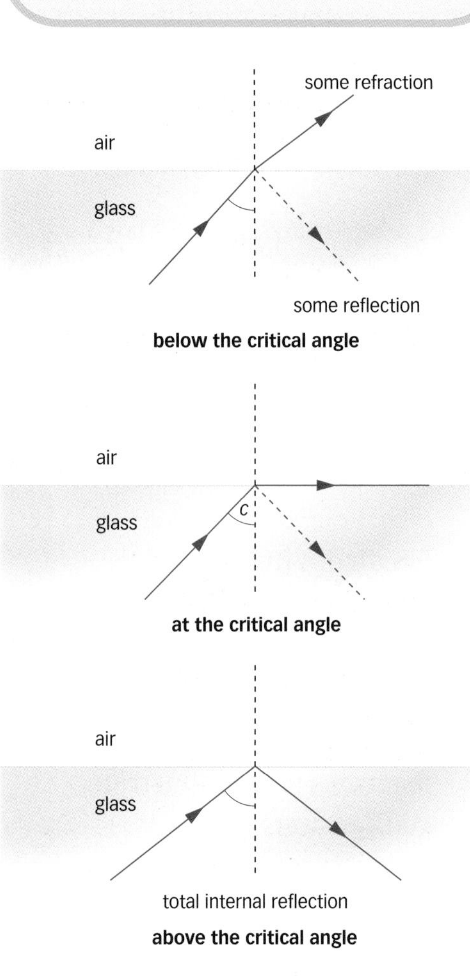

Total internal reflection

Total internal reflection

When light travels from one medium to another it refracts (bends), but there is also a small amount of internal reflection. For example, if light travels from glass to air it bends away as it leaves the glass, but a small amount is reflected back into the glass.

If the light hits the boundary between the glass and air at a large enough angle from the normal, all of the light is reflected back inside the glass. This is called **total internal reflection** (sometimes TIR).

This happens if the angle at which the light hits the boundary is above the **critical angle** for the glass.

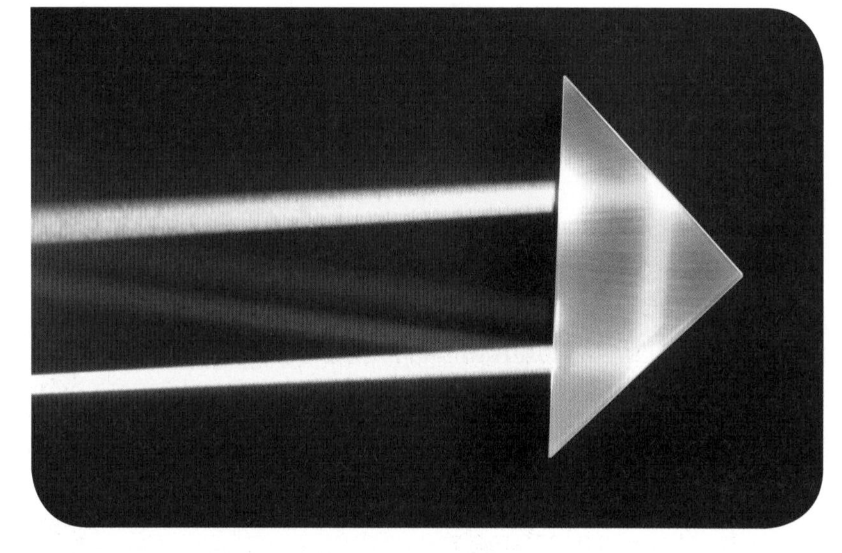

Total internal reflection inside a glass prism

A What is total internal reflection?

B What is the name given to the angle above which this occurs?

The effect can be seen whenever light travels from an optically dense material to an optically less dense one (for example, from water to air, glass to air or Perspex to air). Total internal reflection only happens when:

- the angle of the light is above the critical angle, *and*
- the light is travelling in the more optically dense of the two materials.

Refractive index and the critical angle

The critical angle of a material depends on its refractive index. The higher the refractive index, the lower the critical angle.

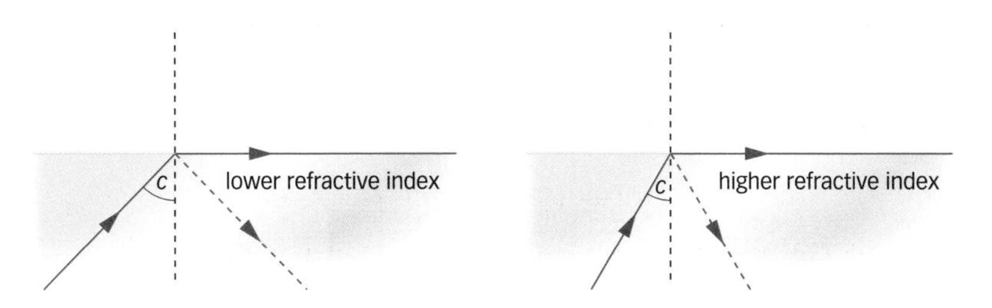

▲ Materials with a higher refractive index have a smaller critical angle

The refractive index of a material and its critical angle are related in the equation below:

$$\text{refractive index} = \frac{1}{\sin c}$$

c is the critical angle of the material.

Questions

1. State the two requirements for total internal reflection.
2. Draw a diagram to show what happens to the ray of light inside a glass block when it hits the edge of the block:
   (a) below the critical angle
   (b) at the critical angle
   (c) above the critical angle.
3. A block has a critical angle of 30°. Find the refractive index of the block.
4. A different block has a refractive index of 1.7. Calculate the critical angle of the block.
5. Describe an experiment that could be carried out to determine the critical angle of a glass block.

E ↓ C ↓ A* ↓

Worked example 1

The critical angle of a specially made glass block is found to be 42°. Find the refractive index of the block.

$$\text{refractive index} - \frac{1}{\sin c}$$

critical angle, $c = 42°$

$$\text{refractive index} = \frac{1}{\sin 42°}$$

Worked example 2

A Perspex block has a refractive index of 1.2. Calculate the critical angle of the block.

$$\text{refractive index} = \frac{1}{\sin c}$$

$$1.2 = \frac{1}{\sin c}$$

$$\sin c = \frac{1}{1.2}$$

$$= 0.83$$

critical angle, c = the angle whose sine is 0.83 (or $\sin^{-1} 0.83$)

critical angle, $c = 56°$

Exam tip

✔ Remember that there are two conditions for total internal reflection: the light must be travelling in the denser of the two materials, and the angle of the light from the normal must be above the critical angle of the material.

16: Optical fibres and lasers

Learning objectives

After studying this topic, you should be able to:

- describe how light is sent along optical fibres by total internal reflection (TIR)
- give examples of uses of total internal reflection
- describe how lasers are used as an energy source

A What material are most optical fibres made from?

Optical fibres

If you have cable TV or broadband then you already have an **optical fibre** running into your house. Optical fibres are very fine glass cables that can be used to transmit large amounts of information very quickly. Fibre optic broadband is generally much faster than using normal phone lines.

Information is transmitted very quickly along fibre optic cables

Pulses of visible light or infrared are sent down optical fibres. They travel along the fibre by total internal reflection, reflecting off the inside of the glass fibre until they reach the other end.

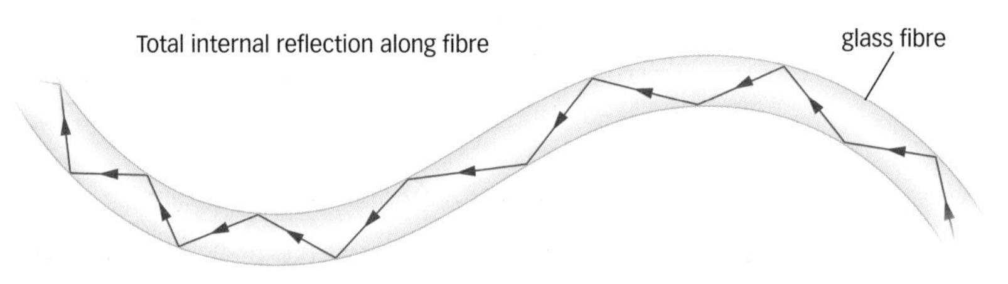

Light is totally internally reflected along an optical fibre

An endoscope uses fibre optics to produce images from inside the body

Optical fibres are not only used for communication. A laparoscope is a medical instrument that is inserted through a tiny keyhole incision to get an image of the inside of the body without having to cut the patient open. An **endoscope** also uses optical fibres, but in this case there is no incision; for example the long tube is often passed through the patient's mouth down to the stomach to obtain images.

Inside the endoscope there are several bundles of optic fibres. The endoscope is inserted into the body and visible light is sent down one of the bundles of cables. The light illuminates the inside of the body and is then reflected back down a separate bundle of fibres. The doctor either looks along the fibre, or the image is sent to a TV screen providing a clear picture of the patient's insides.

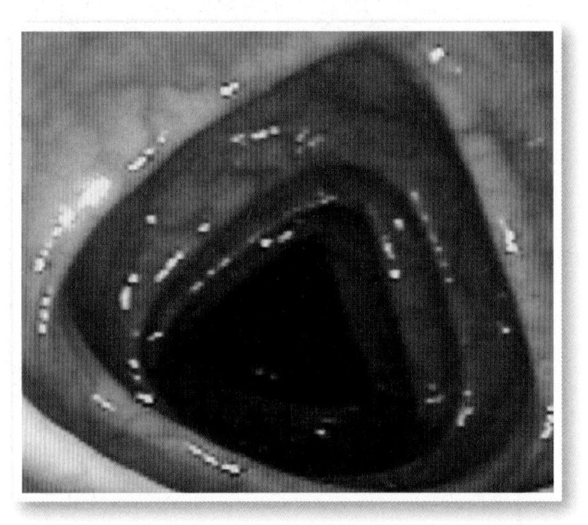

Light travels along an endoscope, allowing the doctor to see inside the patient's body

Laser light

Lasers produce a narrow beam of light that does not spread out very much as it travels through the air. You may have seen the effects at a laser light show or from a laser pointer. The beams remain very narrow, even over long distances.

▲ Lasers have many uses

As the light from a laser is very intense, it is often used as an energy source. The light from powerful lasers is used to cut metal sheets, burn through or etch materials, and cauterise objects.

Specially designed lasers are even used for corrective eye surgery. A laser is used to reshape the cornea so that it will focus a clear image on the retina.

Questions

1 Give two examples of uses for fibre optic cables. (E)

2 Draw a diagram to show how light is totally internally reflected along an optical fibre.

3 Explain why light must first be sent down an endoscope before an image of the inside of the body can be seen.

4 Describe the differences between laser light and the light from a light bulb. (C)

5 Suggest how the cornea of a long-sighted patient might be reshaped. (A*)

Key words

optical fibre, endoscope, laser

Did you know...?

Blu-ray disc players use blue lasers (hence the 'Blu' part of their name). Most CD and DVD players use red light. Blue laser light has a smaller wavelength, which means you can put more information on the disc. This is how it is possible to store a high-definition movie on a Blu-ray disc whereas it would need about six or seven normal DVDs.

B Give two uses for a laser.

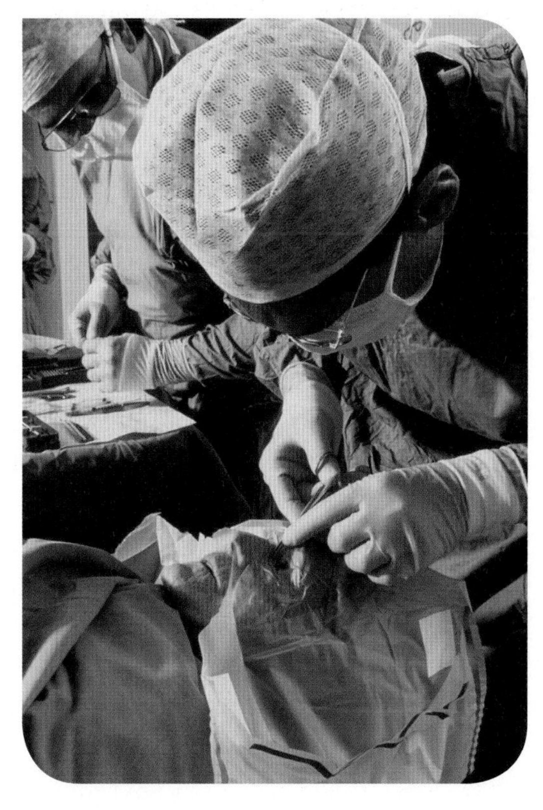

▲ Special lasers can be used to reshape the cornea, eliminating the need for glasses

Course catch-up

Revision checklist

- X-rays are electromagnetic waves with a short wavelength. They are ionising, so potentially dangerous.
- Ultrasound waves are sound waves with a frequency above 20 000 Hz. Ultrasound reflects off boundaries between materials.
- Refraction is the bending of light when it moves from one medium to another. Lenses use refraction to form an image.
- Refractive index is a measure of the speed of light through a material compared with the speed of light in a vacuum.
- The nature of an image is determined by its distance from the lens, its size, its orientation, and whether it is real or virtual.
- The image produced by converging lenses is real and inverted, or virtual, upright, and magnified.
- The image produced by diverging lenses is virtual, upright, and diminished.
- Magnification = image height/object height.
- The human eye has a range of vision from 25 cm to infinity. Features of the eye: retina; cornea; iris; pupil; ciliary muscles; suspensory ligaments.
- Short sight is inability to focus on far away objects, corrected by diverging lenses.
- Long sight is inability to focus on near objects, corrected by converging lenses.
- In a camera a converging lens focuses light onto photographic film or a CCD (charge-coupled device) to produce a real image.
- More powerful lenses have shorter focal lengths. Lens power is measured in dioptres. Power = 1/focal length.
- If the angle of light travelling from a denser medium towards a less dense medium exceeds the critical angle, total internal reflection (TIR) occurs.
- Optical fibres channel light through total internal reflection, and are used for communication and in endoscopes.
- Laser beams remain very narrow even over long distances. Their intensity means they can be used to cut, etch, or cauterise objects (including in laser eye surgery).

safety precautions

treating cancers

created by electrons bombarding metal

ULTRASOUND

sound waves above 20 000Hz

refractive index = 1/sin c

critical angle

optical fibres

endoscopes

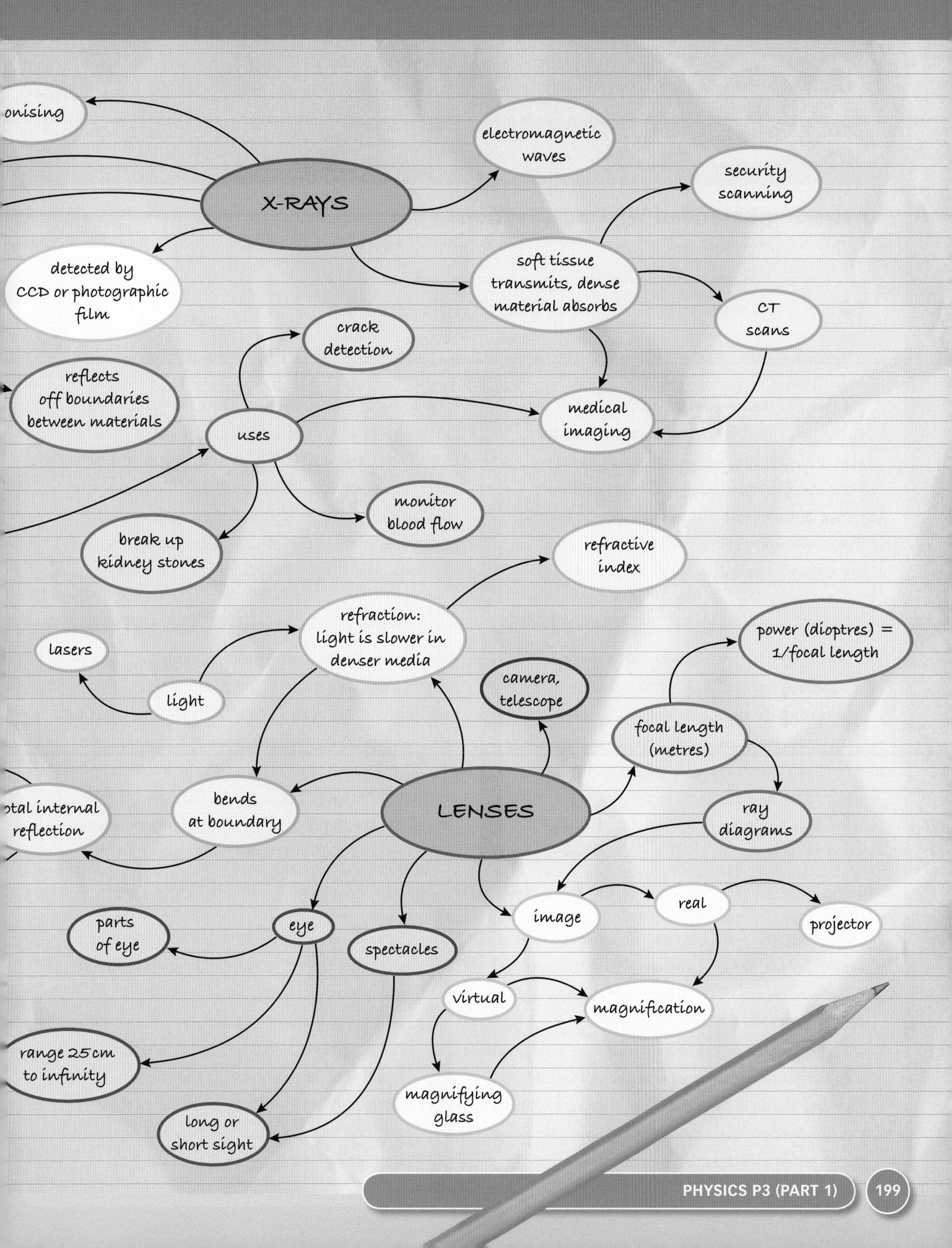
onising
electromagnetic waves
X-RAYS
security scanning
detected by CCD or photographic film
soft tissue transmits, dense material absorbs
CT scans
crack detection
reflects off boundaries between materials
uses
medical imaging
monitor blood flow
break up kidney stones
refractive index
refraction: light is slower in denser media
lasers
power (dioptres) = 1/focal length
camera, telescope
light
focal length (metres)
otal internal reflection
bends at boundary
LENSES
ray diagrams
real
parts of eye
eye
image
projector
spectacles
virtual
magnification
range 25 cm to infinity
magnifying glass
long or short sight

Answering Extended Writing questions

QUESTION

Explain the processes involved when you take a photo using a digital camera.

The quality of written communication will be assessed in your answer to this question.

G–E

The lense makes an image on the pixls at the back, you can zoom in or out to change the size of the image you get It's like your eye the picture is upside down but your brane turns it back again the CCD makes it into a digtal number for the card to store it there are milions of pixls to make the pictur

Examiner: This answer is jumbled with ideas. A few are relevant, but most are not. The answer is not structured well. The starting and ending statements are roughly accurate, though not well linked. The digression into zooming and the eye comparison are unhelpful. Spelling and grammar are poor, and there is a lack of punctuation.

D–C

Light comes into the camera and the lens focuses it onto the back of the camera – it's a real image cos the light is achully there. The CCD takes the light and turns it into electricity. There are millions of CCDs called pixls,, each one sends electricity to the memory which stores it digitally till you want to look at it or send it to your computer.

Examiner: Most of the described physics is correct. The sequence is logical, but the candidate tends to include irrelevant information (such as justifying the realness of the image, or describing the stored information). There is confusion between pixels and CCDs, and vague use of the term 'electricity'. There are occasional errors in spelling, grammar, and punctuation.

B–A*

Light reflected from the field of view enters the lens of the camera. The lens creates a focussed upside-down real image of the field of view on the light sensor at the back of the camera. The sensing CCD (charge-coupled device) has millions of pixels on it – each pixel is one little spot on the image. The CCD converts the light into electrical signals which go to the camera's processor. There the signals are turned into binary numbers (strings of 1s and 0s).

Examiner: This answer is well ordered and accurate. Use of technical terms is detailed and almost perfect (though maybe 'scattered' would be better than 'reflected'). The process inside the camera is well explained, for example the distinction between pixel and CCD, and CCD is spelt out. The term 'digital' is explained. Spelling, punctuation, and grammar are all good.

Exam-style questions

1 A01 Match these parts of a human eye with their function:

Part	Function
lens	Hole that controls the amount of light entering.
cornea	Surface consisting of light-sensitive cells.
pupil	Front surface where first refraction occurs.
retina	Adjusts focal length to achieve focussed image.

2 A01 **a** Explain how a converging lens affects light.

A01 **b** Explain the meaning of focal length for a converging lens.

A01 **c** Explain the meaning of real image.

A02 **d** Using a particular projector, a transparency of width 15 cm produces an image 1.8 m wide on a screen. Calculate the magnification achieved.

A01 **e** What is a virtual image?

3 A02 **a** An object 4 cm tall is 8 cm from a converging lens with focal length 5 cm. Draw a ray diagram at actual size on a sheet of A4 graph paper to find the size and nature of the image. Show all three possible construction rays. (Hint: use the paper in landscape orientation, and put the lens about 10 cm from the left edge of the paper.)

A02 **b** Describe how the image changes if the object is brought gradually closer to the lens.

A02 **c** Calculate the power of this lens.

4 A02 **a** A ray of light enters a glass block at an angle of incidence of 40°. Calculate the angle of refraction of the ray within the block. The refractive index for glass is 1.5.

A02 **b** A ray travelling in the glass strikes the glass/air surface at an angle of 35° to the normal. Calculate the angle at which the ray emerges from the glass.

A02 **c** A ray travelling in the glass strikes the glass/air surface at an angle of 55° to the normal. Explain what happens to this ray.

G–E D–C B–A*

Extended Writing

5 A01 What is ultrasound? Describe some of the uses of ultrasound in medicine.

6 A01 Explain how total internal reflection (TIR) occurs. Why is TIR not possible if light is travelling in air towards a glass surface?

7 A01 What are X-rays? Describe how they are created, some of their medical uses, and why radiographers have to take care when using them.

G–E D–C B–A*

A01 Recall the science
A02 Apply your knowledge
A03 Evaluate and analyse the evidence

P3 Part 2

Making things work

AQA specification match

P3.2	3.17	Centre of mass
P3.2	3.18	Pendulums
P3.2	3.19	Moments
P3.2	3.20	Principle of moments
P3.2	3.21	Levers
P3.2	3.22	Toppling objects
P3.2	3.23	Pressure in liquids
P3.2	3.24	Hydraulics
P3.2	3.25	Circular motion
P3.2	3.26	Size of centripetal force
P3.2	3.27	Electromagnets and the motor effect
P3.2	3.28	Using the motor effect
P3.2	3.29	Transformers
P3.2	3.30	Step-up and step-down transformers
P3.2	3.31	Transformer power equation
P3.2	3.32	Switch mode transformers
Course catch-up		
AQA Upgrade		

Why study this unit?

You can use physics to explain why many simple machines around you work. Many things, such as toys and fairground rides, are based on principles such as the lever. Simple machines are also used in many other things that we see and use in everyday life, from scissors to the cranes on building sites. Engineers need to be able to understand the principles behind how and why these simple machines work, so that they can apply them in designing new ones.

In this unit you will learn about the centre of mass of an object, and why some objects are more stable than others. You will look at the forces acting on levers. You will also be introduced to the principles of hydraulics systems, and how objects move in circles. You will learn how the electric motors that power some of these systems work, and how the electrical energy they need is transferred to them. You will also learn about the different kinds of transformers, and how they are used in many electronic devices.

You should remember

1. How forces can act on objects.
2. The effect of gravity on objects.
3. How electricity is generated and transmitted to users.
4. The effects of a magnetic field on charged particles.
5. That the power used by an electrical device depends on the potential difference across it and the current flowing through it.

The largest trucks in the world are used at mines in the USA, Australia, Chile, and South Africa. This truck is over 7 metres tall and has a mass, when empty, of 203 tonnes. It can carry a load of over 350 tonnes.

Even though this truck is unusually large, it is still designed with the basic principles of physics used to make much smaller vehicles work. It has a hydraulic system to push the dumper part up to empty it. The hydraulic system exerts a force on the dumper, causing it to rotate about the pivot at the back of the vehicle. The system works in the same way as in a much smaller dumper truck – it is just much more powerful!

Learning objectives

After studying this topic, you should be able to:

- explain what the centre of mass of an object is
- explain how to find the centre of mass of a thin object by suspending it
- describe where the centre of mass is in a symmetrical object

Key words

centre of mass

This child's centre of mass is directly below the hand that he is hanging from

Centre of mass

Can you balance a book on the end of your finger? A ruler? A pencil? When the object is balanced, all its mass seems to be acting through a point that is above your fingertip. This point is called the **centre of mass**.

Every object has a mass, and the mass is spread out throughout the object. Every object has a point where all of the mass appears to be concentrated.

A What is the centre of mass of an object?

When you hang an object from something, it may swing for a little while from side to side, but then it will come to rest. The centre of mass will be directly below the point it is hanging from. You can use this to find the centre of mass of an object. The diagram shows how you can find the centre of mass of a thin sheet of material that has an irregular shape.

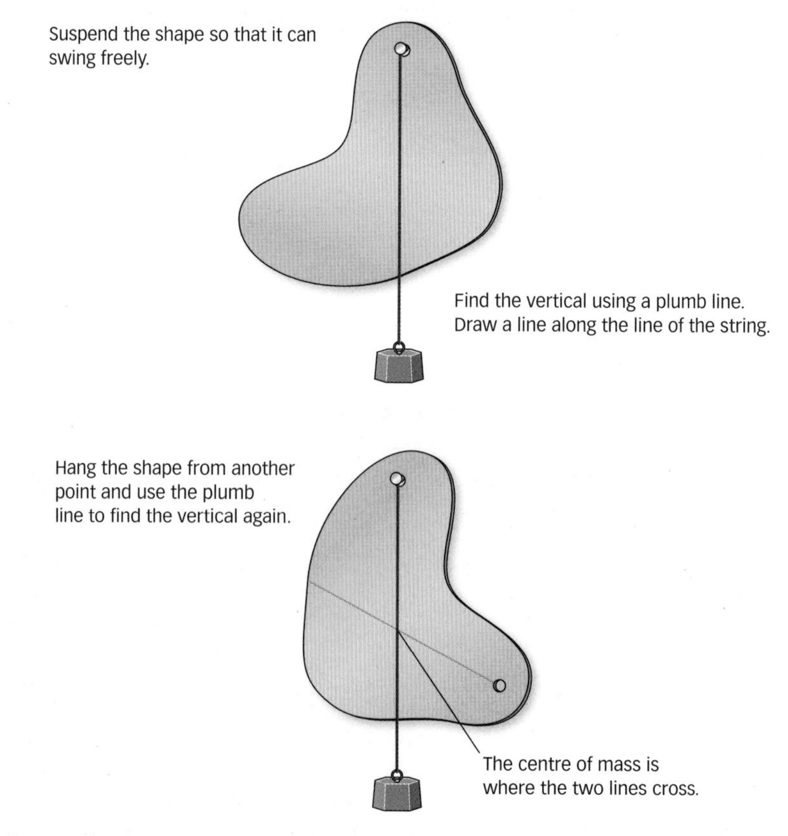

How to find the centre of mass of a thin sheet

Symmetry and centre of mass

When the mass of an object is evenly distributed throughout the object, you can use symmetry to find the centre of mass. For example, a pool ball is a sphere that is symmetrical. Its centre of mass lies where the lines of symmetry intersect.

There are equal amounts of mass all around the centre of mass. In a shape with an axis of symmetry, there are equal amounts of mass on either side of the axis.

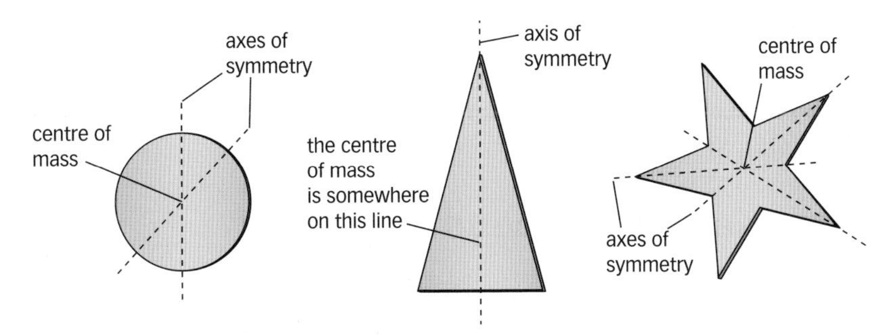

▲ Centres of mass for various regular shapes. We need two lines of symmetry to pinpoint the centre of mass.

B Where would you find the centre of mass of a rectangle?

The centre of mass does not have to be inside the object itself. For example, the centre of mass of a ring is not in the solid part of the ring.

Questions

1 Where is the centre of mass of a football?

2 Explain how you could find the centre of mass of an irregular shape cut out of thin card.

3 Josh hangs a mobile over his baby's cot. Where will the centre of mass of the mobile lie?

4 Why will the centre of mass of the triangle in the diagram above be nearer to the base of the triangle?

5 Explain where the centre of mass of a doughnut is. You could draw a diagram to help your explanation.

E ↓ C ↓ A*

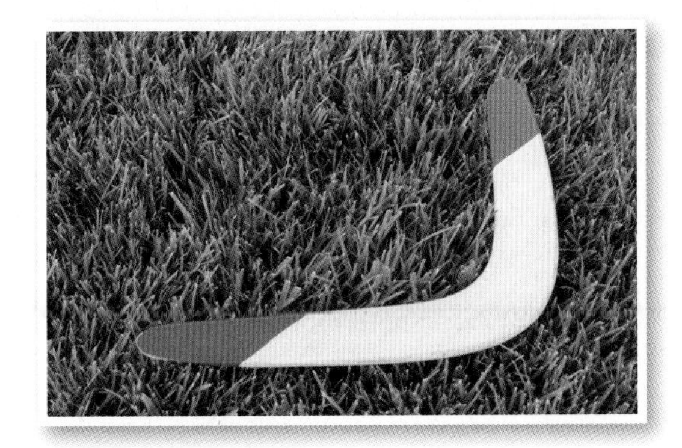

▲ Where might the centre of mass of this boomerang be?

Did you know...?

This toy balances because its centre of mass is below the point that the toy rests on.

▲ This toy balances on the point that the bottom of the horse is resting on

C How could you find the centre of mass of a regular hexagon?

Exam tip

✔ Remember that an object always has a centre of mass, but it might not be inside the solid part of the object.

Learning objectives

After studying this topic, you should be able to:

- describe the motion of a pendulum.
- explain how the time period of a pendulum is related to its length

Key words

pendulum, time period, frequency

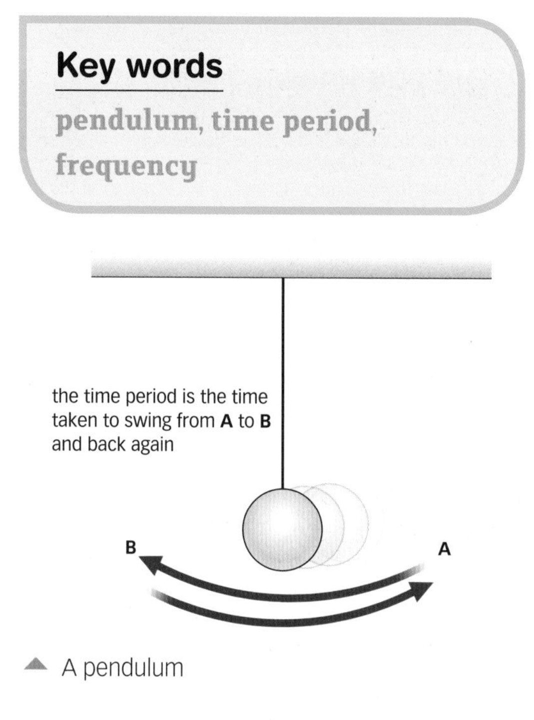

A pendulum

The pendulum in this clock has a period of 1 second

What is a pendulum?

A simple **pendulum** consists of an object attached to one end of a long thin piece of material (this can be a bar or a string). The object can swing freely under gravity. The mass of the object is much larger than the mass of the bar or string. The other end of the bar or string is attached securely to something.

When you pull the pendulum slightly to one side and let go, it will swing from side to side with a regular movement. Pendulums have been used in clocks for hundreds of years.

Time period of a pendulum

The **time period** of a pendulum is the time it takes to move from one side of its swing to the other and all the way back again. This is another way of saying that the period is the time taken for one complete cycle of the pendulum's movement.

The number of cycles that the pendulum completes in one second is called the **frequency**, so:

$$\text{time period (seconds, s)} = \frac{1}{\text{frequency (hertz, Hz)}}$$

If T is the time period in seconds, and f is the frequency in hertz (cycles per second), then:

$$T = \frac{1}{f}$$

Worked example

The frequency of a pendulum is 0.5 Hz. What is its time period?

$$\text{time period} = \frac{1}{\text{frequency}} \quad \text{or} \quad T = \frac{1}{f}$$

$f = 0.5$ Hz

$$T = \frac{1}{0.5} = 2 \text{ seconds}$$

A A pendulum has a frequency of 1 Hz. What is its time period?

The time period of a pendulum depends on its length. A short pendulum has a short time period. A long pendulum has a long time period. The longer the pendulum is, the longer its time period.

Pendulums around us

You can find simple pendulums around you. For example, a swing is a simple pendulum. Some fairground rides are also just large pendulums.

A Foucault pendulum is a simple pendulum with a long wire – usually several tens of metres long. It is free to swing in any direction. As the pendulum swings, the direction of the motion changes slowly during the day. There are many examples of Foucault pendulums around the world.

▲ A swing is a pendulum

▲ The Foucault pendulum in the Science Museum in London

▲ This fairground ride is a large pendulum

Questions

E
1. What is a pendulum?

C
2. Explain how the time period of a simple pendulum is varied.
3. A pendulum has a frequency of 0.1 Hz. What is its time period?
4. A pendulum has a time period of 0.25 seconds. What is its frequency?

A*
5. Sam says that a fairground ride like the one in the photo above will swing faster when there are more people on it. Explain why he is wrong.

B What features of the swinging boat in the photo make it a pendulum?

Did you know...?

The Taipei 101 skyscraper in Taiwan contains a large pendulum that stretches over 10 floors. It is used to help keep the building stable in high winds. It absorbs energy when the building sways in the wind.

19: Moments

Learning objectives

After studying this topic, you should be able to:

- ✔ understand that the turning effect of a force is a moment
- ✔ calculate the size of a moment

Key words

pivot, moment, perpendicular distance

A What is a moment?

B Where is the pivot for a door?

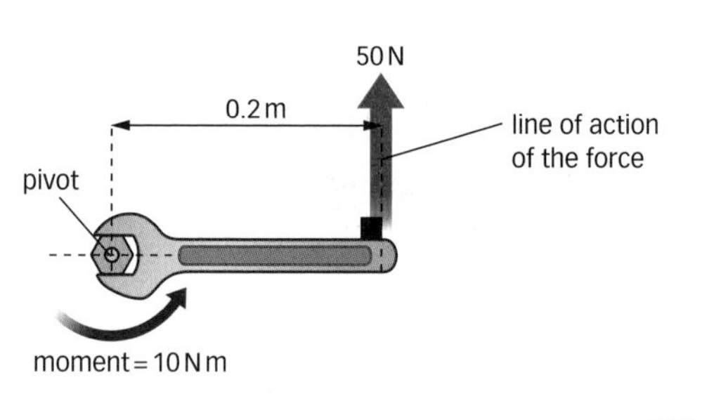

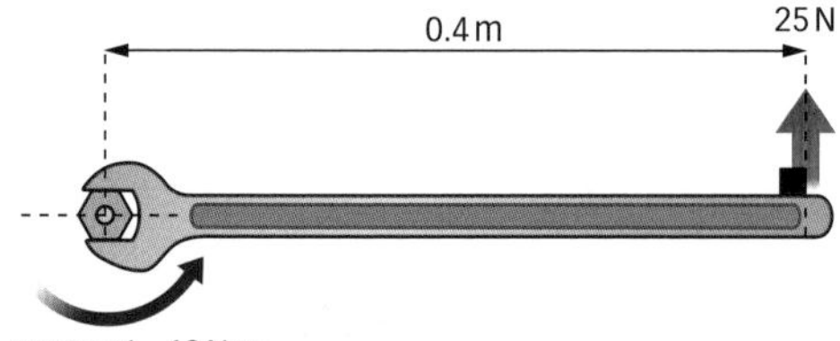

▲ The moment about each nut shown here is the same, but the forces and perpendicular distances from the pivot are different

C Sam exerts a force of 25 N at the end of a 40 cm long spanner when tightening a nut. What is the moment on the nut?

Turning effect of a force: the moment

When you apply a force to the revolving door in the photo, it turns about a fixed point or **pivot**. The turning effect of the force is called a **moment**.

◀ People use a turning effect (moment) to move this revolving door

Every time you use a force to turn something about a pivot, there is a moment. When you open a door in your house, the force you use to turn the door knob has a moment, and then the force you use to push open the door also has a turning effect about the hinges.

The size of the force you need depends on where you push the door. You need to use a larger force if you push the door nearer the hinge.

The size of a moment is given by the equation:

moment of the force (newton-metres, Nm) = force (newtons, N) × **perpendicular distance** from the line of action of the force to the pivot (metres, m)

If M is the moment in newton-metres, F is the force in newtons, and d is the perpendicular distance in metres from the line of action of the force to the pivot, then:

$$M = F \times d$$

Worked example

A force of 10 N acts at a perpendicular distance of 50 cm from the pivot. What is the turning effect (moment) of the force?

$$M = F \times d$$

$F = 10\text{ N}$

$d = 50\text{ cm} = 0.5\text{ m}$

$$M = 10\text{ N} \times 0.5\text{ m}$$
$$= 5\text{ N m}$$

When you measure the distance from the force to the pivot, it must be measured correctly. The perpendicular distance is measured along a line that is at right angles to the direction of the force.

In the revolving door shown in the diagram below, the force is not acting at right angles to the door. This means that the perpendicular distance is less than the actual distance from the point where the force is acting to the pivot.

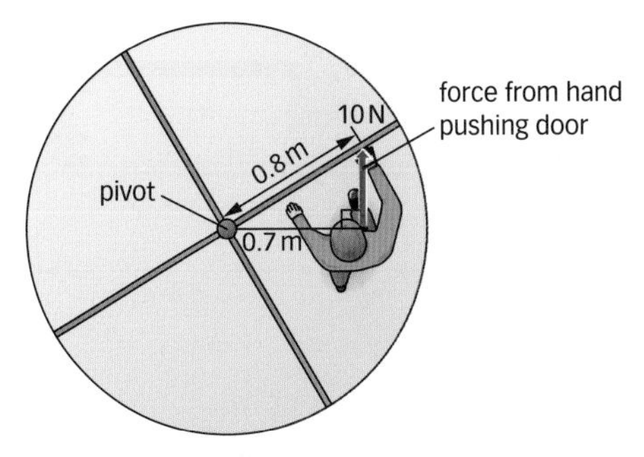

▲ The distance from the line of action of the force to the pivot is measured perpendicularly from the line. Here the perpendicular distance from the line of action of the pushing force to the pivot is 0.7 m.

D What is the perpendicular distance?

Did you know...?

Mechanics sometimes use a torque wrench to apply a specific moment to a nut or a bolt. ('Torque' is another word for 'moment'.) A torque wrench is used to prevent nuts and bolts being overtightened. The moment to be applied is set in the wrench and the nut or bolt is tightened. When the moment needed to turn the spanner reaches the value set, the spanner does not move any more.

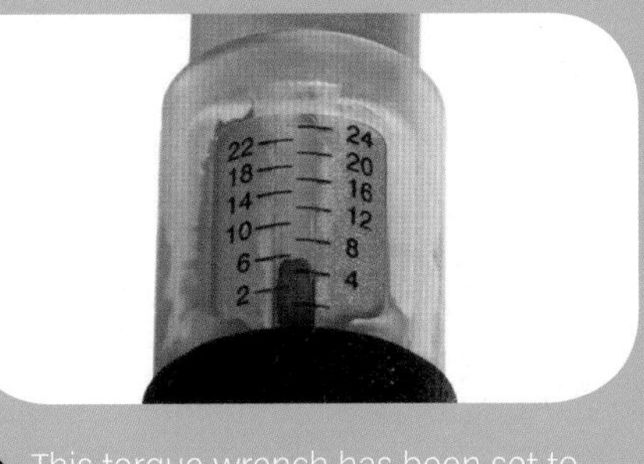

▲ This torque wrench has been set to limit the moment to 6 Nm

Questions

1. What is a pivot?
2. Give three examples of applying a moment.
3. Ben pulls on a door handle with a force of 5 N. The handle is 60 cm from the hinges. What is the moment about the hinges?
4. What is the moment about the pivot of the revolving door in the diagram above?
5. The diagram shows a person lifting the handles of a wheelbarrow. What turning effect about the wheel axis is the person applying?

Exam tip

- ✔ Remember that a moment is a turning effect – many students lose a mark because they do not realise this.
- ✔ Make sure you use the correct units for moments in answers to calculations. The units are newton-metres (Nm).

Learning objectives

After studying this topic, you should be able to:

- explain the principle of moments
- calculate the size of a force or its distance from the pivot for an object that is balanced

This crane uses a counterweight to produce a balanced moment to allow it to lift heavy objects

A If an object is balanced, what can you say about the moments acting on it?

B Are moments balanced when an object is turning? Explain your answer.

Balanced moments

In the seesaw shown in the diagram below, there are two forces that produce moments (turning effects) about the pivot.

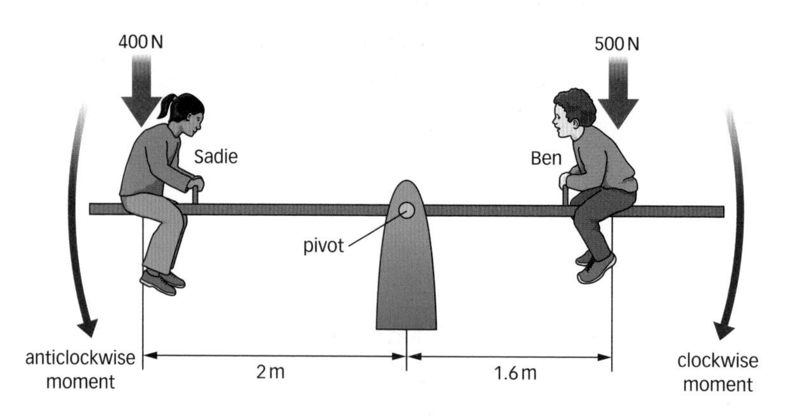

The seesaw is balanced – the clockwise and anticlockwise moments are equal

The weight of the girl has an anticlockwise turning effect on the seesaw, and the weight of the boy has a clockwise turning effect.

$$\text{anticlockwise moment} = F \times d = 400\ \text{N} \times 2\ \text{m} = 800\ \text{N m}$$

$$\text{clockwise moment} = F \times d = 500\ \text{N} \times 1.6\ \text{m} = 800\ \text{N m}$$

The moments are equal, but acting in opposite directions – they are balanced and the seesaw is not moving.

When an object is balanced, the total clockwise moment about the pivot is equal to the total anticlockwise moment.

The clockwise and anticlockwise moments on this seesaw are balanced

Calculations with balanced moments

The clockwise and anticlockwise moments do not have to be on different sides of the pivot.

In the wheelbarrow shown in the diagram, the person holding the wheelbarrow handles is exerting a clockwise moment about the pivot. The 400 N weight of the wheelbarrow is exerting an anticlockwise moment about the pivot.

Worked example

Looking at the diagram, what force does the person need to exert to hold the wheelbarrow steadily with its legs off the ground?

When the person is holding the wheelbarrow steady as shown, the clockwise and anticlockwise moments are balanced.

clockwise moment = anticlockwise moment

clockwise moment = $F \times 1.2$ m

anticlockwise moment = 400 N × 0.3 m = 120 N m

$$F \times 1.2\ \text{m} = 120\ \text{N m}$$

$$F = \frac{120\ \text{N}}{1.2\ \text{N}}$$

$$= 100\ \text{N}$$

The force exerted by the person is 100 N.

C Where is the pivot on the wheelbarrow?

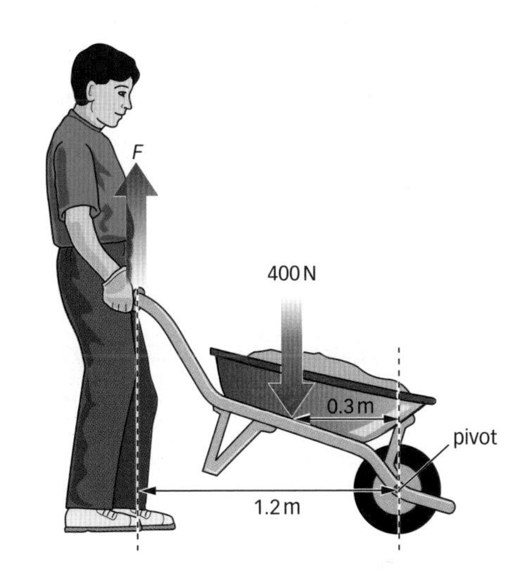

Moments on a wheelbarrow that is being held steady

Questions

1 What is the turning effect of a force called? (E)

2 Why might an object not turn when moments are acting on it?

3 Show that the moments on the seesaw shown here are balanced

30 N
90 N
1.2 m
0.4 m

(C)

4 In the seesaw diagram at the top of the previous page, Sadie moves so that she is 1.5 m from the pivot. How far does Ben need to be from the pivot for the seesaw to be balanced? (A*)

Exam tip AQA

✓ Whan an object is balanced, the clockwise moments must equal the anticlockwise ones.

21: Levers

Learning objectives

After studying this topic, you should be able to:

- give some examples of levers
- explain how a lever works to multiply a force

Key words

lever, force multiplier

A What is a lever?

B Where is the pivot on the crowbar?

Examples of levers

When you use a crowbar to remove a nail from a piece of wood, you are using a simple machine that increases the force you apply. The crowbar is acting as a **lever**.

This crowbar is being used as a lever

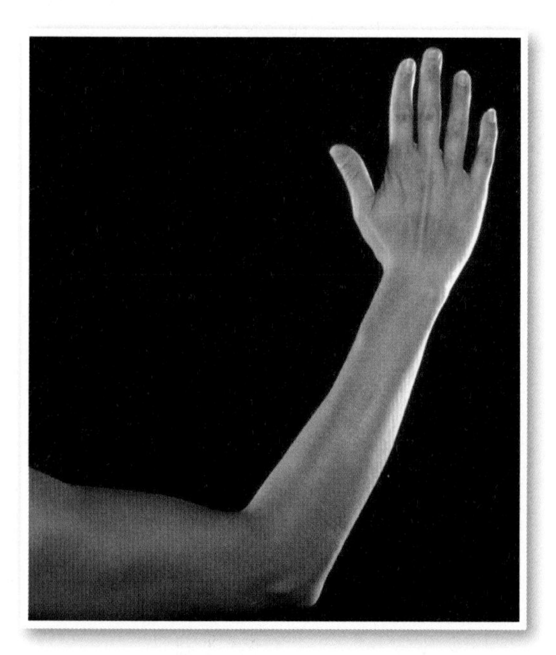

The human body contains a number of simple levers that help us to lift a wide range of objects.

A lever uses a force to turn something around a **pivot**. As you know, the turning effect of a force is called a moment. In the scissors shown in the diagram below, the force that is applied to the handles has a moment about the pivot.

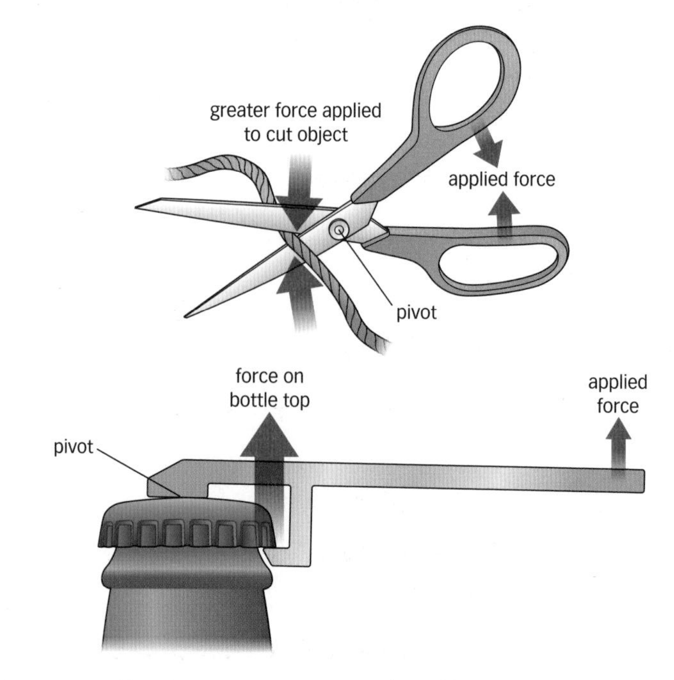

Scissors and bottle openers are examples of levers

C What do you notice about the pivot of the bottle opener?

Increasing the force

The moment about the pivot is given by the force multiplied by the perpendicular distance from the line of action of the force to the pivot.

When you use scissors as shown in the diagram on the previous page, the moment of the force that you apply to the handles is equal to the moment of the force acting on the object.

But the distance of the handles from the pivot is greater than the distance of the material from the pivot. As the moments are the same, this means that the force applied to the object is greater. The scissors are acting as a **force multiplier**.

Levers are used to change smaller forces into larger ones in many everyday devices.

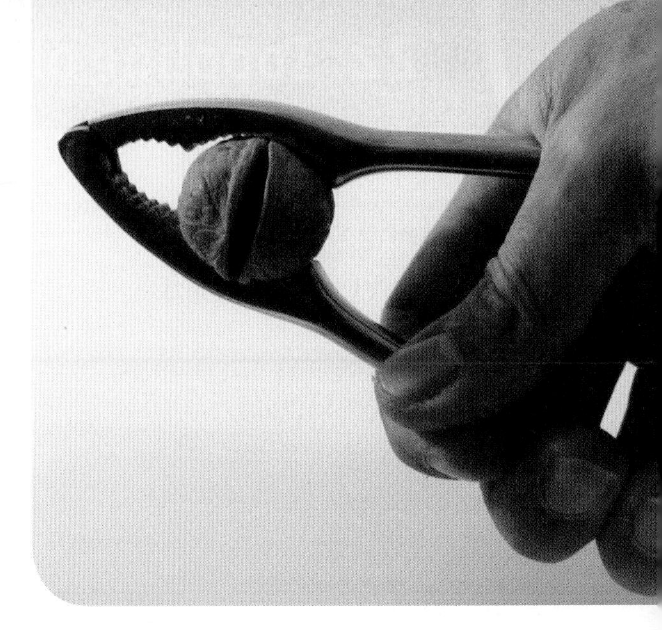

▲ You can't crack a nut easily with your hands. This nutcracker acts as a force multiplier. The force you apply at a distance from the pivot is applied as a greater force on the nut nearer the pivot. The nut cracks open.

D What is a force multiplier?

▲ The screwdriver is being used as a lever

Did you know...?

Archimedes wrote about the principles of levers over 2200 years ago. It is assumed that levers would have been used by the ancient Egyptians to move large lumps of rock with masses of up to 100 tonnes.

Questions

E

1 Give two examples of levers.

↓ C

2 Why is a lever usually a force multiplier?

3 Look at the picture of a screwdriver being used to open a can of paint. Where is the pivot?

4 How can you increase the size of the force applied to the load in a lever?

A*

5 When can a seesaw act as a force multiplier?

Exam tip

✓ When answering an exam question, make sure you have read the question properly and that you are actually answering the question.

P3 22: Toppling objects

Learning objectives

After studying this topic, you should be able to:

- ✔ know the factors that affect the stability of an object
- ✔ explain when an object will or will not topple

Key words

line of action, **stability**

▲ Traffic cones are very stable

▲ This pencil is very unstable. A small push would move the line of action of the weight outside the base, toppling the pencil.

Increasing stability

Stability is a measure of how hard it is to make an object topple over.

Some objects are much more stable than others – they do not topple over easily.

You can design objects to be more stable by:

- making the centre of mass lower
- making the base of the object wider.

For example, traffic cones have a wide, heavy base. Making the base heavier lowers the centre of mass. This makes it much more difficult for them to be toppled over.

A What features make a traffic cone very stable?

Why objects fall over

The diagram below shows a box on a flat surface, The weight of the box always acts in the same direction. The **line of action** of the weight is through the centre of mass and is always downwards.

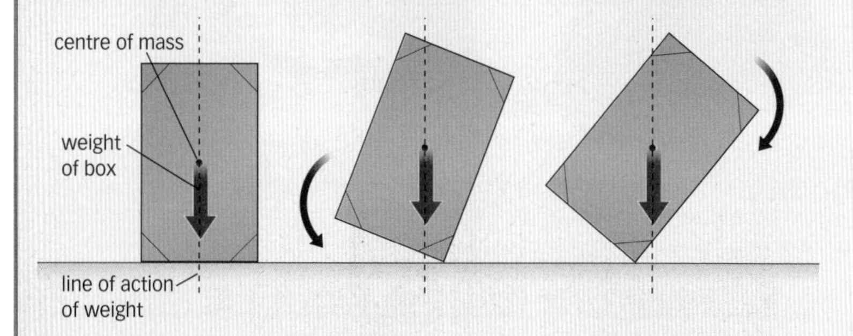

On the left, the line of action of the weight is inside the base of the box. The box is stable – it does not move.

B Where is the line of action of the weight of the boxes?

In the middle diagram, the box has been tilted. The line of action of the weight is still inside the base of the box, so the box will fall back to its original position. We could say that there is a resultant anticlockwise moment about the bottom right-hand edge of the box.

On the right, the box has been tilted further, so that the line of action of the weight is now outside the base of the box. The box will topple over to the right. There is a resultant clockwise moment about the bottom right-hand edge of the box.

C Where is the pivot of the box in the middle diagram?

D How would the stability of the box on the right be affected if it were lying on its side?

Questions

1. Where is the line of action of the weight in a stable object?
2. Why do drinking glasses sometimes have thick heavy bases?
3. Look at the photos of bar stools. Which has the most stable design? Explain your answer.

▲ Two designs for bar stools

4. Look at the diagram of the box on the flat surface on the previous page. Explain in terms of moments what happens when you tilt the box and let it go:
   (a) when the line of action of the weight is inside the base
   (b) when the line of action of the weight is outside the base.

E ↓ C ↓ A*

Did you know...?

Road vehicles are tested for stability by tilting them. Designers see how far they can be tilted before they fall over. They need to be stable so that it is less likely that they will fall over in an accident.

▲ The stability of this bus is being tested at a test centre in Bedfordshire

Exam tip

✔ The base of an object does not have to be solid. For example, a chair has four legs – it is the area between the bases of the legs of the chair that affects how stable the chair is.

23: Pressure in liquids

Learning objectives

After studying this topic, you should be able to:

- ✔ recall that liquids are almost incompressible
- ✔ explain that pressure in liquids is transmitted equally in all directions

Key words

incompressible, hydraulic system

▲ The pressure in a fluid increases with depth. Large dams like the Hoover Dam have to be thicker at the base to cope with the extra pressure.

Incompressible liquids

As you know, the particles in a liquid are very close together, touching one another, although they are moving about all the time. This means that liquids are almost **incompressible**.

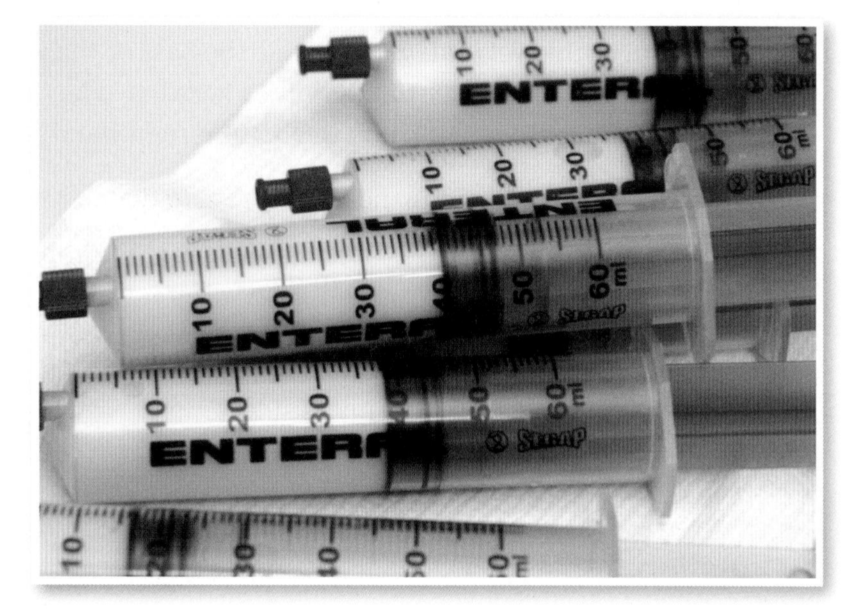

▲ If you put your finger over the open end of a filled syringe and try to push the plunger, it will not move

A Why are liquids almost incompressible?

The pressure at each point in a liquid acts equally in all directions. This means that if you apply a force to a liquid at one point, the pressure from the force will be transmitted equally throughout the liquid.

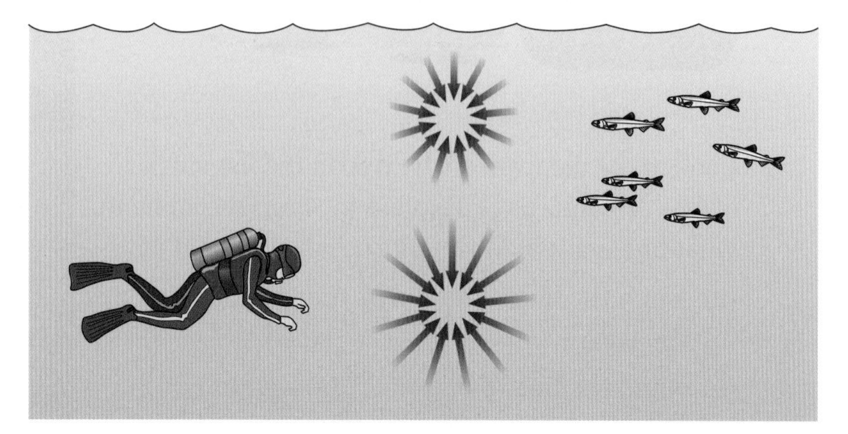

▲ At any point in a liquid, the pressure acts equally in all directions

Hydraulic systems

Hydraulic systems use liquids in pipes to take advantage of the fact that pressure in a liquid will be transmitted equally in all directions. They are used to move devices remotely, controlling the size of the forces applied.

B What is a hydraulic system?

For example, in cars a hydraulic system is used to transfer the action of the driver pushing the brake pedal to the application of the brakes.

The pushing force on the brake pedal applies pressure to the liquid in the braking system. This pressure is transmitted throughout the liquid, and a much larger force is applied to the brake pads at the wheels.

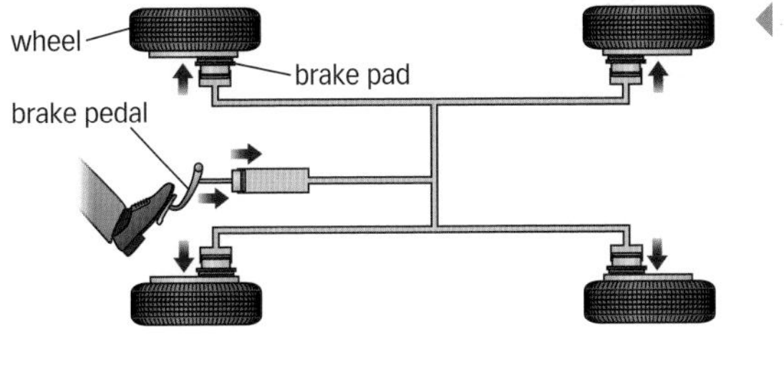

When a car brakes, pressure from the driver's foot is transmitted through the hydraulic system to apply the brakes at the wheels

C What happens when the brake pedal is pushed?

Hydraulic systems are also used in many other machines such as hydraulic car jacks, inspection platforms and excavating machines. They are also used to control the wing flaps and landing gear on aeroplanes.

Did you know...?

In aeroplanes, the rudder (which changes the aeroplane's direction) and elevator (which allows the aeroplane to ascend and descend) used to be connected to the joystick by a hydraulic system. In modern aeroplanes, a fly-by-wire system is used. The joystick sends an electrical signal to a pump that pumps liquid in the hydraulic system.

Hydraulic systems are used to extend the wing flaps on this aircraft and then take them back in again

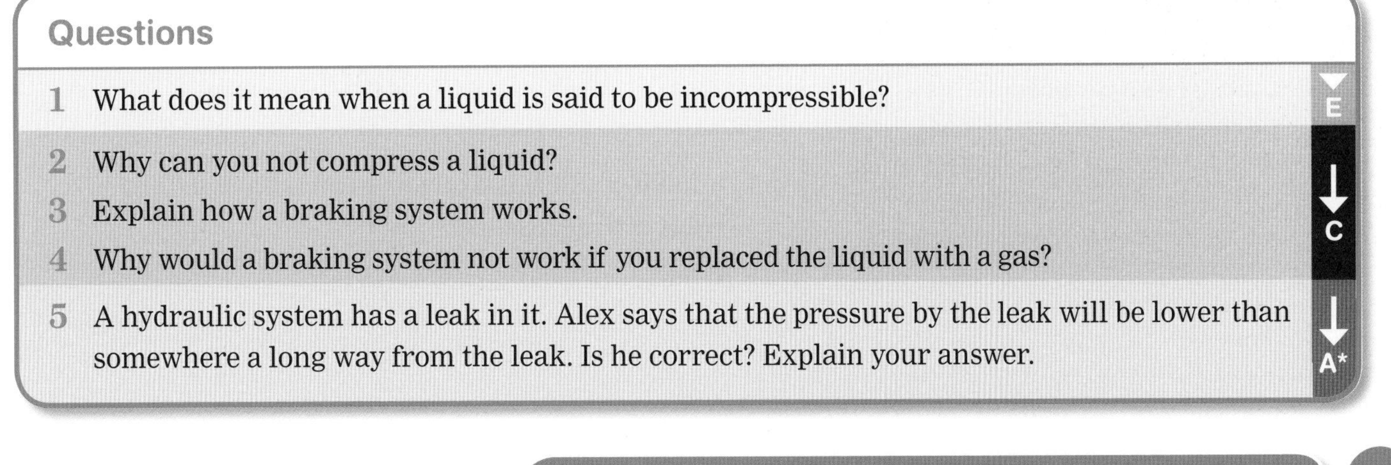

Questions

E
1 What does it mean when a liquid is said to be incompressible?

C
2 Why can you not compress a liquid?
3 Explain how a braking system works.
4 Why would a braking system not work if you replaced the liquid with a gas?

A*
5 A hydraulic system has a leak in it. Alex says that the pressure by the leak will be lower than somewhere a long way from the leak. Is he correct? Explain your answer.

Learning objectives

After studying this topic, you should be able to:

- use the equation linking pressure, force, and area
- explain how using different cross-sectional areas multiplies the force in a hydraulic system

Key words

pascal

Exam tip

- The units of pressure are pascals (Pa), which are newtons per square metre (N/m²). You can use either of these units in your answers.
- If you are not sure what the correct units for your answer are, you can find them by looking at the other units of the quantities in the equation and working out what the combination of units is.

Using a hydraulic system to magnify a force

The picture below shows two pistons connected by a pipe filled with liquid. If you push on piston A, the pressure is transmitted through the liquid and pushes on piston B. You may recall that pressure is force per unit area. The pressure is the same for A and B. The effect of a small force at A, which has a small area, is to produce the same pressure at B. B has a larger area, so the total force is greater.

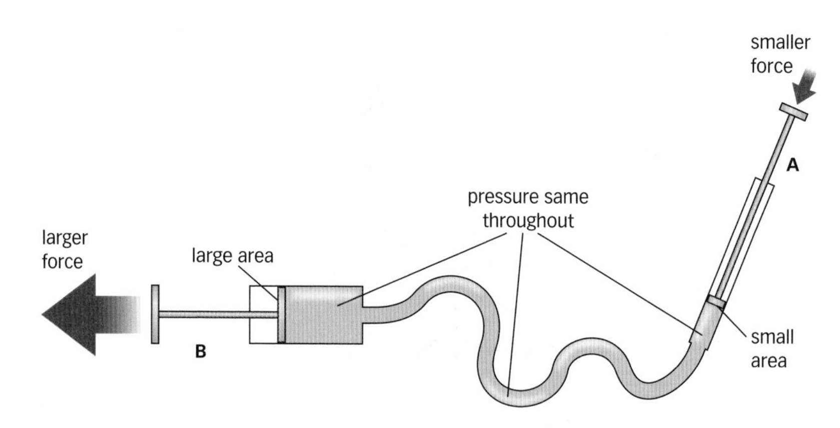

▲ Pressure is transmitted equally throughout an hydraulic system

As pressure is the total force divided by the area over which it acts, it is measured in newtons per square metre (N/m²) or **pascals** (Pa).

The pressure in the different parts of the hydraulic system is given by:

$$\text{pressure (pascals, Pa)} = \frac{\text{force (newtons, N)}}{\text{cross-sectional area (metres squared, m}^2\text{)}}$$

If P is the pressure in pascals, F is the force in newtons, and A is the cross-sectional area in square metres, then:

$$P = \frac{F}{A}$$

A A force of 250 N is exerted on an area of 0.02 m². What is the pressure?

B What is a pascal?

Worked example

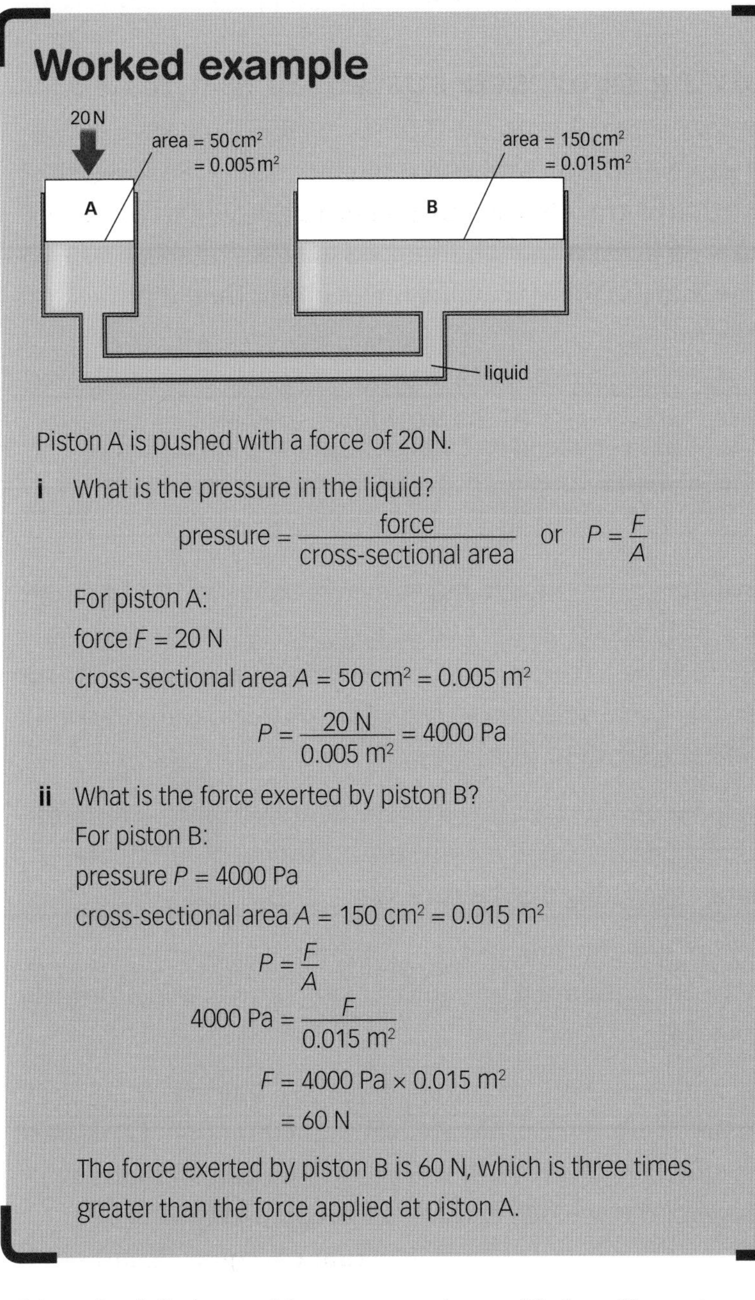

Piston A is pushed with a force of 20 N.

i What is the pressure in the liquid?

$$\text{pressure} = \frac{\text{force}}{\text{cross-sectional area}} \quad \text{or} \quad P = \frac{F}{A}$$

For piston A:

force F = 20 N

cross-sectional area A = 50 cm² = 0.005 m²

$$P = \frac{20\ \text{N}}{0.005\ \text{m}^2} = 4000\ \text{Pa}$$

ii What is the force exerted by piston B?

For piston B:

pressure P = 4000 Pa

cross-sectional area A = 150 cm² = 0.015 m²

$$P = \frac{F}{A}$$

$$4000\ \text{Pa} = \frac{F}{0.015\ \text{m}^2}$$

$$F = 4000\ \text{Pa} \times 0.015\ \text{m}^2$$
$$= 60\ \text{N}$$

The force exerted by piston B is 60 N, which is three times greater than the force applied at piston A.

This principle is used in many systems. Hydraulic systems are often used as force multipliers, that is to increase the effect of the force applied. For example, in a car braking system a relatively small force (effort) is applied to the brake pedal. The hydraulic system multiplies the force so that a much larger force (load) is applied to the brake pads.

Using a hydraulic car jack allows a car to be lifted easily by one person

This excavator uses a hydraulic system to move the excavating bucket

Questions

1. What are the units of pressure? (E)
2. The piston of a hydraulic system has a surface area of 0.03 m². A force of 45 N is applied. What is the pressure?
3. A mechanic uses a force of 50 N to operate a car jack. The surface area of the piston being pumped is 0.001 m². What area of piston is needed to lift a car with weight 12 500 N? (C)
4. Which should have the largest surface area: the cylinder where the effort is applied or the cylinder where the load is?
5. In the worked example, the small piston moves three times as far as the large piston. How could you use the equation for work done to show that energy is conserved in the hydraulic system? (A*)

Learning objectives

After studying this topic, you should be able to:

- explain that an object moving in a circle is accelerating continuously
- describe what a centripetal force is

Key words

centripetal force, tension

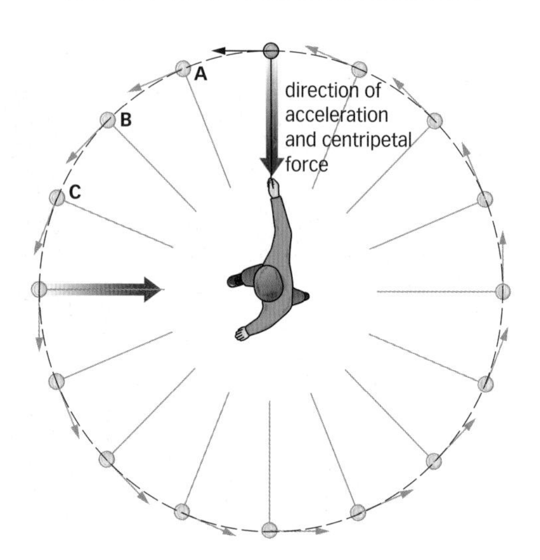

The cork is being whirled around at a constant speed, but its direction, and so its velocity, is changing all the time. It is accelerating.

Moving in a circle

You may remember that a force is needed to change either the speed or the direction of a moving object. So when the direction of travel of an object is changing, a force must be acting on it, even though its speed may be constant. You also know that if a force is acting on an object then it is accelerating.

A What is needed to change either the speed or the direction of something that is moving?

The diagram on the left shows a cork being whirled around at a steady speed on the end of a string. We know that the tension force in the string changes the direction of travel of the cork, because if the string were to break, the cork would fly off in a straight line.

The force on the cork towards the centre, from the **tension** in the string, is called a **centripetal force**. ('Centripetal' means 'searching for the centre'.)

As there is a continuous force on the cork towards the centre, the cork must be continuously accelerating towards the centre.

This makes sense – we can see from the diagram that the velocity of the cork is always changing direction. Looking at the velocity of the cork at A, B, and C we can see that the change in the velocity is always towards the centre.

Any object that is moving in a circle at a steady speed is continuously accelerating towards the centre. This acceleration is a change of direction but not speed.

This racing car is going round a bend. Its direction is changing, so it is accelerating.

B Why must the racing car in the photo be accelerating?

C In which direction does the centripetal force on the racing car act?

Where is the centripetal force coming from?

The centripetal force is provided by, to take some examples, gravity, friction between two surfaces, or tension in a rope or string. It does not exist on its own.

The centripetal force for the racing car in the photo on the previous page is provided by the force of friction between the tyres and the road surface. The force of friction stops the car from carrying on moving in a straight line. The centripetal force is acting towards the centre of the circle. (While it is going round a bend the racing car is travelling in part of a circle.)

A centripetal force could also be provided by your hanging on to something, for example if you hang on to a roundabout with your hands. The centripetal force is then the tension in your arms.

▲ The tension in the child's arms is the centripetal force here

▲ The seats on this fairground ride are attached to the central part of the carousel by a chain

D What provides the centripetal force on the fairground ride in the photo on the right?

Questions

1 What is a centripetal force?

2 An object is moving in a circle at a steady speed. Describe how its velocity is changing.

3 For each of the following, state what is providing the centripetal force:

(a) A rollercoaster where the cars hang from a rail and the cars go round a loop.

(b) A train going round a bend.

4 Explain why the direction of the acceleration of an object moving in a circle and the direction of the centripetal force are the same.

5 Most machines have a maximum speed at which they can operate. What do you think might happen if they rotated much faster than the maximum speed?

Exam tip

✔ Remember that acceleration does not just involve a change in speed. When an object is changing direction it is accelerating, even if its speed is constant.

26: Size of centripetal force

Learning objectives

After studying this topic, you should be able to:

- explain how increasing the mass of an object increases the centripetal force needed for the object to move in a circle
- explain how increasing the speed of an object increases the centripetal force required for circular motion
- explain how decreasing the radius of the circle increases the centripetal force needed for circular motion

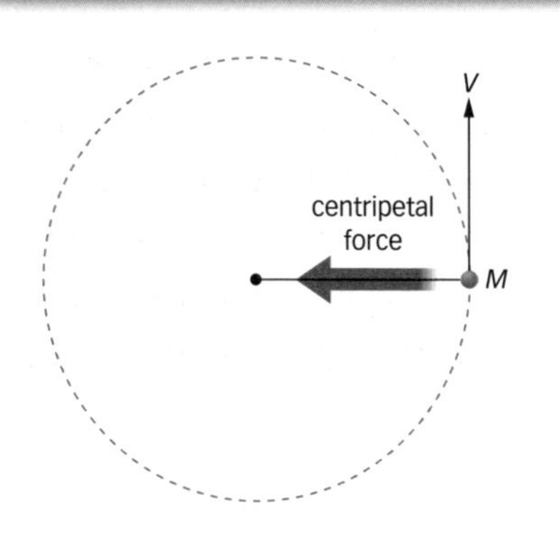

▲ The object *M* is in circular motion. If the mass is increased, a greater centripetal force will be needed for the same motion.

B Two cars with the same mass go round the same bend. One is going faster than the other. Which one needs the higher centripetal force?

You may recall that the greater the mass of an object, the greater the force needed to provide a particular acceleration.

This can help you to understand the changes in the centripetal force needed to keep an object moving in a circle, if the mass, speed, or radius of the circle change.

Increasing mass

A mass M is moving in a circle, as shown in the diagram on the left. If the mass is increased, the centripetal force needed to keep the same acceleration will increase. Doubling the mass will double the centripetal force required.

A What does increasing the mass do to the size of the centripetal force needed for circular motion?

Increasing speed

The diagram below shows two objects with the same mass moving in circles with the same radius. The speed of one object is twice the speed of the other object. The diagram also shows how far each object travels in a given time.

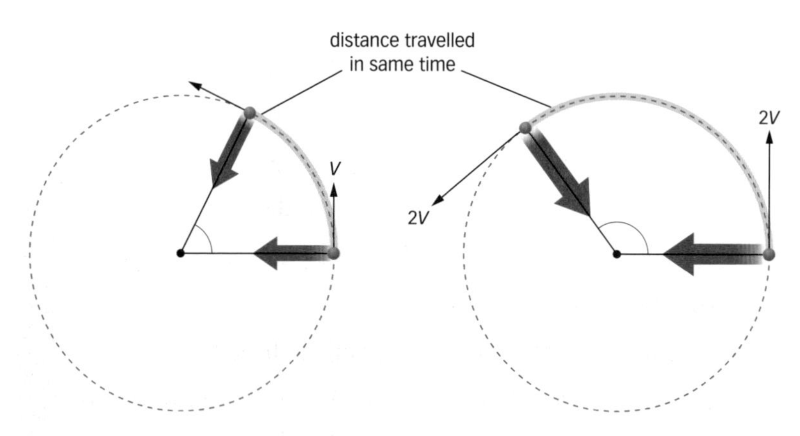

▲ Changing speed in circular motion

The faster object has moved through a much bigger angle than the slower one. This means that the rate at which its direction has changed is higher, so its acceleration is also higher. So the centripetal force needed to keep it moving in a circle will also be higher. As the speed of an object increases, the centripetal force needed to keep it in circular motion also increases.

Decreasing radius

Two objects with the same mass are moving at the same speed in circles with different radii. The object moving around the smaller circle moves through a bigger angle in a given time than the object moving in the larger circle. The change of direction has been greater, so the acceleration is higher. Again, the centripetal force needed for the higher acceleration is also higher.

The centripetal force needed to keep an object in circular motion increases as the radius decreases.

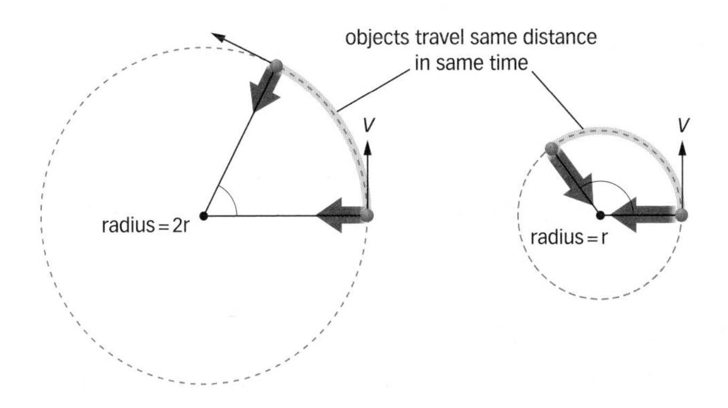

Changing the radius of circular motion

There is a centripetal force on these rollercoaster cars

C Two runners with the same mass are racing at the same speed round a bend in a running track. One is on the inside track. Which runner needs the greater centripetal force?

Questions

1 What affects the size of the centripetal force needed to make an object move in a circle? **E**

2 Look at the photo of the rollercoaster above. How would the centripetal force needed change if:
(a) all the riders were adults rather than children
(b) the cars went faster
(c) the cars went round a smaller loop? **C**

3 A car and a bus travel round the same bend at the same speed. Which one needs the higher centripetal force? Explain your answer.

4 How would you decrease the centripetal force needed for the fairground ride shown on spread P3.25? **A***

Exam tip

✔ Remember that the centripetal force needed for an object to perform circular motion increases as:
- the mass of the object increases
- the spread of the object increases
- the radius of the circle decreases.

Learning objectives

After studying this topic, you should be able to:

- understand that a current through a wire produces a magnetic field
- describe how a force is produced using the motor effect
- describe how to change the size and direction of this force

Key words

magnetic field, motor effect, Fleming's left-hand rule

Electromagnets are used to lift and move large objects in scrapyards

Currents create magnetic fields

When an electric current flows in a wire, it creates a **magnetic field** around the wire. This magnetic field is only there when there is a current. If the current stops, the magnetic field collapses. It is a bit like the wake from a moving boat. When the boat is moving, it creates a wake behind it. When the boat is not moving, there is no wake.

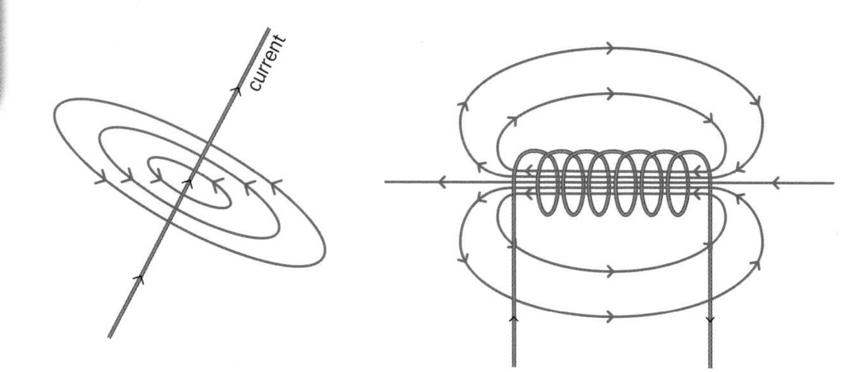

An electric current in a wire has a magnetic field around it. The shape of the field depends on how the wire is arranged.

The shape of the magnetic field around a single wire is concentric circles.

If the wire is looped into a coil (making a solenoid), the magnetic fields for all the turns of the wire combine. The shape of the magnetic field looks like the magnetic field of a bar magnet.

Electromagnets make use of this effect. They are designed to produce a strong magnetic field when there is a current in their wires. When the current is switched off, the electromagnet loses its magnetism.

A Describe the shape of the magnetic field around a coil of wire.

B Give one use of an electromagnet.

The motor effect

If a current-carrying wire is placed inside another magnetic field (for example in between the poles of two other magnets), the two magnetic fields interact. The magnetic field from the current and the magnetic field from the magnets push on each other, creating a force on the wire. This is called the **motor effect**.

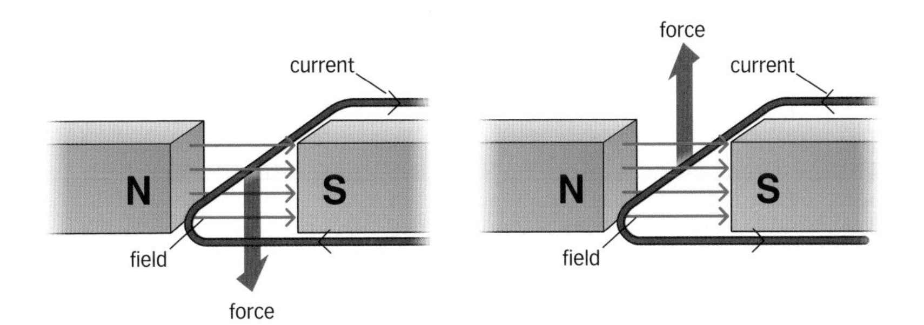

▲ The motor effect. The magnetic field from the current in the wire interacts with the field from the magnets, making a force that pushes the wire, making it move. Reversing the direction of the current flips the direction of the force.

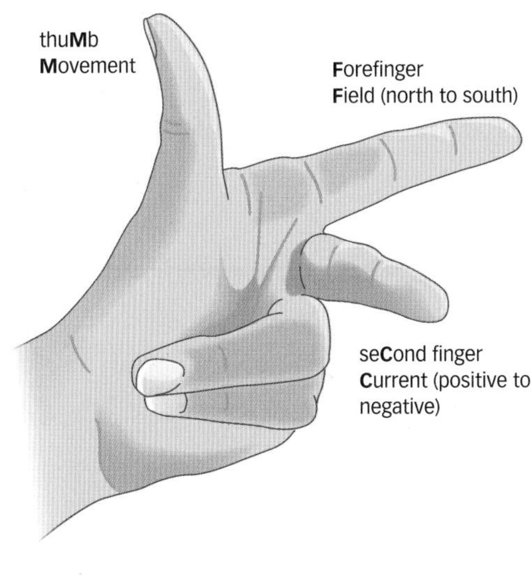

▲ Fleming's left-hand rule. The thumb shows the direction that the wire moves in (and therefore the direction of the force.)

The direction of the force acting on the wire can be found using **Fleming's left-hand rule**.

It is important to note that if the wire is parallel to the magnetic field then there is no force acting on the wire.

C What is the direction of the forces in the motor effect diagram?

D A wire is placed in a magnetic field so that it is parallel to the field. A current is passed through the wire. What happens?

Changing the size or direction of the force

The size of the force on the wire can be increased by:

- increasing the strength of the magnetic field (for example, by using a stronger magnet)
- increasing the size of the current in the wire.

The direction of the force on the wire can be reversed by:

- reversing the direction of the magnetic field (for example, by swapping the poles of the magnets round)
- reversing the direction of the current in the wire.

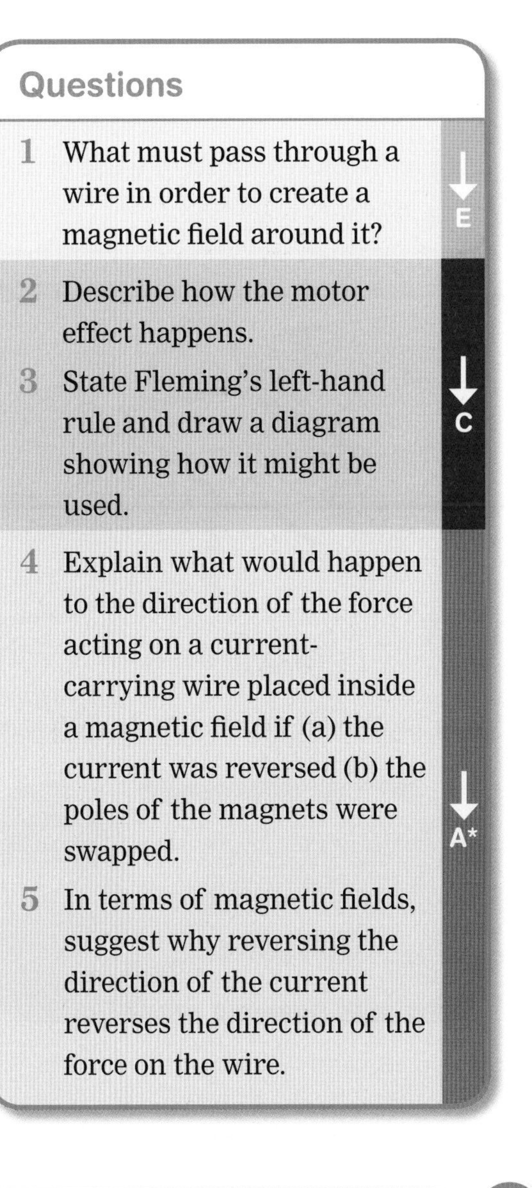

Questions

1. What must pass through a wire in order to create a magnetic field around it?
2. Describe how the motor effect happens.
3. State Fleming's left-hand rule and draw a diagram showing how it might be used.
4. Explain what would happen to the direction of the force acting on a current-carrying wire placed inside a magnetic field if (a) the current was reversed (b) the poles of the magnets were swapped.
5. In terms of magnetic fields, suggest why reversing the direction of the current reverses the direction of the force on the wire.

28: Using the motor effect

Learning objectives

After studying this topic, you should be able to:

- describe some examples of uses of the motor effect
- use the principle of the motor effect to explain its uses in different situations

Key words

electric motor, split-ring commutator

A Give one example of a use of an electric motor.

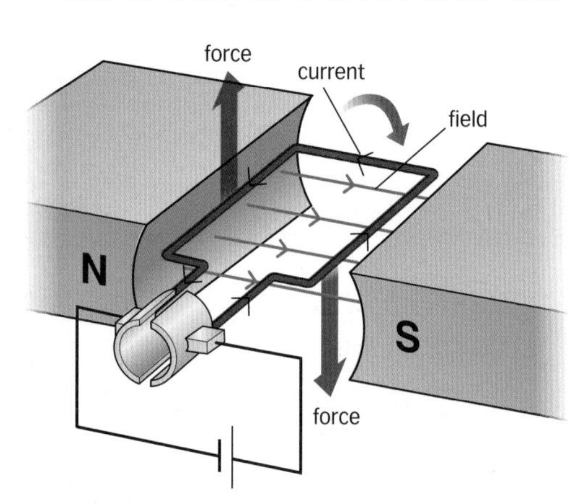

A simple motor spins due to the motor effect

Electric motors are not just found in small devices. They can be used in large machines, including trains and cars.

Electric motors

Electric motors have a wide range of uses, from providing the tiny vibrations in game controllers and some mobile phones to powering hybrid cars and even some high-speed trains.

All electric motors make use of the motor effect. A simple electric motor has a loop of wire inside a magnetic field between two magnets. When there is a current in the wire, the magnetic fields interact. One side of the loop is pushed down and the other side is pushed up. This makes the motor spin.

For the motor to continue to spin in the same direction, the current needs to be reversed every half-turn of the loop. This is done using a device called a **split-ring commutator**. Each time the loop is vertical, the current inside it reverses. The side of the loop that was pushed up is now pushed down. This allows the motor to continue to spin.

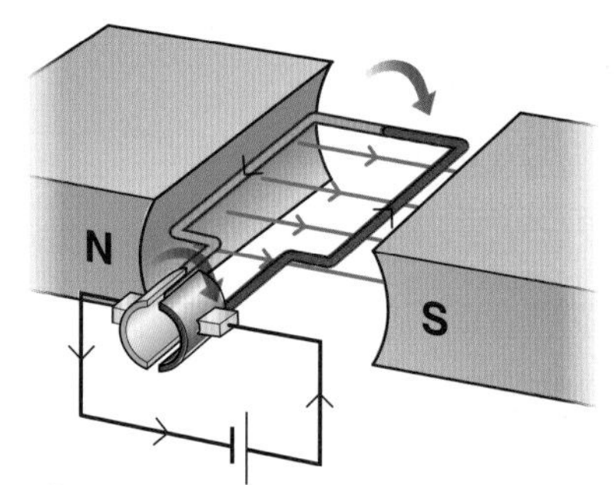

① The blue part of this coil is pushed upwards and the red half downwards. (Check with Fleming's left-hand rule.)

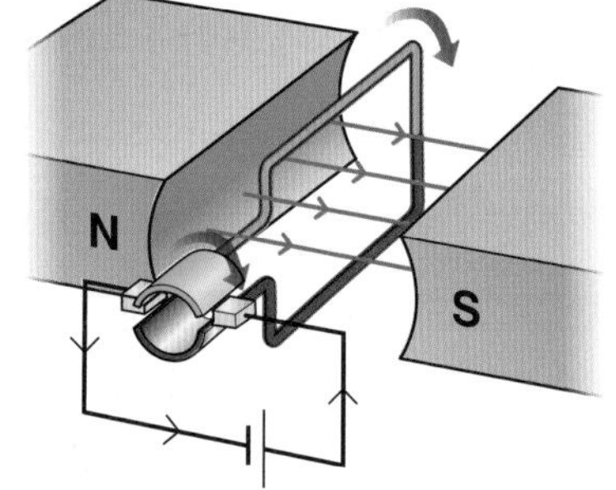

② No current, but the coil continues to turn because of its own momentum.

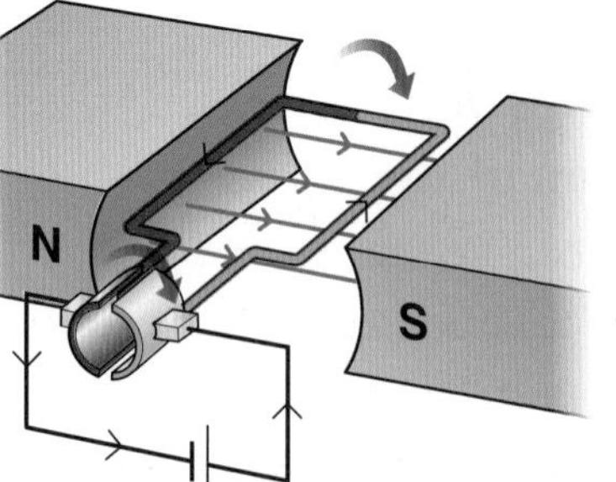

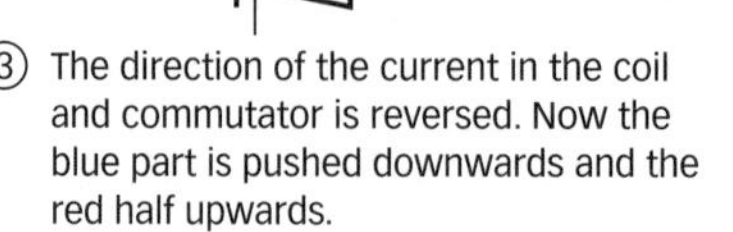

③ The direction of the current in the coil and commutator is reversed. Now the blue part is pushed downwards and the red half upwards.

coil wire

gaps

contact to DC power supply

Reversing the current in the wire loop. Look at the direction of the current in the part of the commutator and coil that is shaded blue. The direction of the current changes with each half-turn of the commutator. The same applies to the part shaded red.

Other uses of the motor effect

A traditional analogue ammeter makes use of the motor effect. It has a coil attached to a small spring in between a pair of small magnets. This spring is in turn attached to a needle that points to a value on a dial showing the size of the current.

When there is a current in the coil it rotates; compressing the spring. The greater the current, the stronger the force and the more the spring is compressed. This has the effect of moving the needle further. The greater the current the further the needle moves, giving a higher reading on the ammeter.

B Explain why a higher current causes the needle in an analogue ammeter to move further.

Loudspeakers also use the motor effect. Inside each speaker there is a small paper or plastic cone (you can sometimes see this cone moving in and out). On the back of this cone there is a coil of wire loosely wrapped around a magnet.

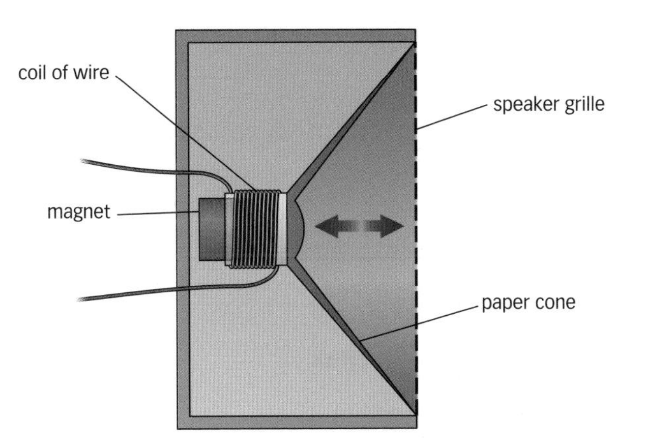

A loudspeaker cone moves in and out due to the motor effect

When there is a current in the coil, the magnetic field it creates interacts with the field from the magnet. This creates a force and the coil is pushed out. This in turn pushes the cone out. If the current then reverses, the cone is pulled back in.

When the loudspeaker is being used, the electrical signal that it receives, which creates the current in the coil, is changing very rapidly in size and direction. So the cone vibrates in and out several thousand times each second, creating the sound waves that we hear.

The motor effect can be used in an analogue ammeter to determine the current

Did you know...?

Your headphones contain tiny magnets and coils of wires. They work exactly the same way as a loudspeaker. It is amazing that the sound you hear from them is produced by moving a tiny paper cone forwards and backwards!

Questions

1 Other than the electric motor, give one use of the motor effect.

E

2 Describe the purpose of the split-ring commutator found in most simple motors.

3 Describe how the motor effect is used inside a simple analogue ammeter.

C

4 Use Fleming's left-hand rule to explain how, inside a simple motor, one side of the loop of wire is pushed down while the other side is pushed up.

5 Suggest how the current in the coil in a speaker changes to make sounds with different pitches and different volumes.

A*

29: Transformers

Learning objectives

After studying this topic, you should be able to:

- describe how a potential difference is induced when a conductor moves in a magnetic field or when a magnet moves in a coil of wire
- describe the structure of a transformer
- explain how a transformer works

Key words

induced, electromagnetic induction, iron core, primary coil, secondary coil

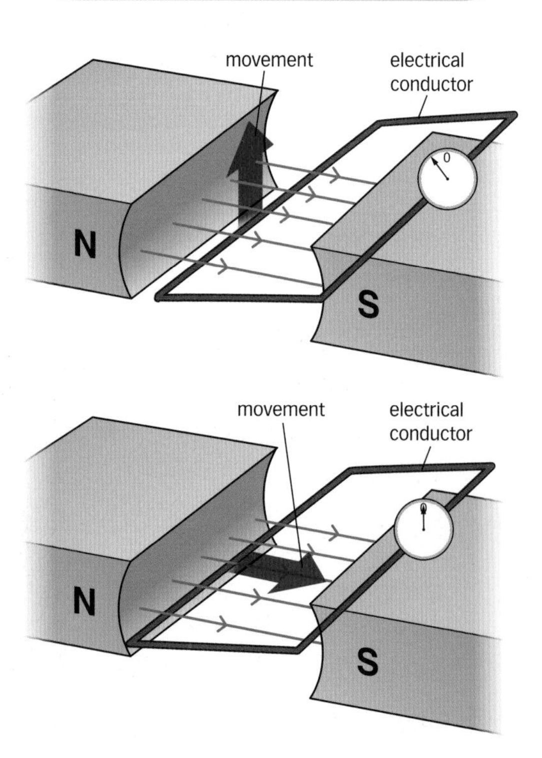

When a conductor is moved so that it cuts across a magnetic field, an induced current will flow if there is a complete circuit. If the conductor is moved parallel to the field, there is no induced current.

Electromagnetic induction

You have already learnt that when an electric current flows through a wire, a magnetic field is produced around it. So, might the reverse effect happen? Might it be possible to get electricity from magnetism?

The answer is yes. When a wire is moved so that it cuts across a magnetic field, a potential difference is **induced** across the ends of the conducting wire. If the circuit is complete, a current flows. This is called **electromagnetic induction**.

> A What is electromagnetic induction?

The wire must be moved so that it cuts across the magnetic field. If it is moved in the same direction as the magnetic field (parallel to the magnetic field), no potential difference is induced.

We can get the same effects if we keep the wire in the same place but move the magnet instead. The wire is still cutting across a magnetic field.

So when a magnet is moved into a coil of wire, a potential difference is induced across the ends of the coil.

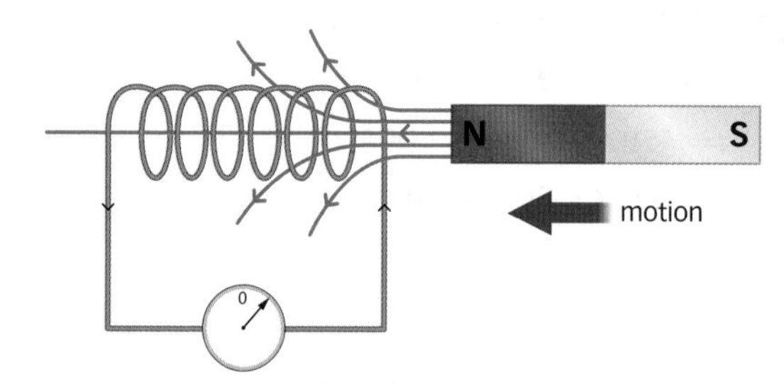

Moving a magnet into or out of a coil induces a potential difference

Transformers

Transformers use both the idea that an electric current can produce a magnetic field and the idea that a magnetic field can be used to create a potential difference.

You already know that a current flowing in a coil of wire will create a magnetic field shaped like the field of a bar magnet. If the coil is wrapped round an **iron core**, the iron becomes magnetised and the strength of the field seems to become much greater.

The coil in the diagram has been wrapped round an iron core shaped like a ring, and the magnetic field caused by the current runs all the way round the core.

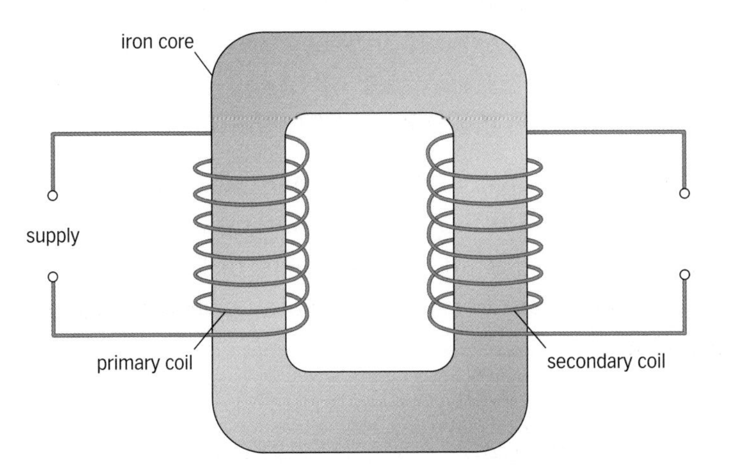

▲ The main parts of a transformer

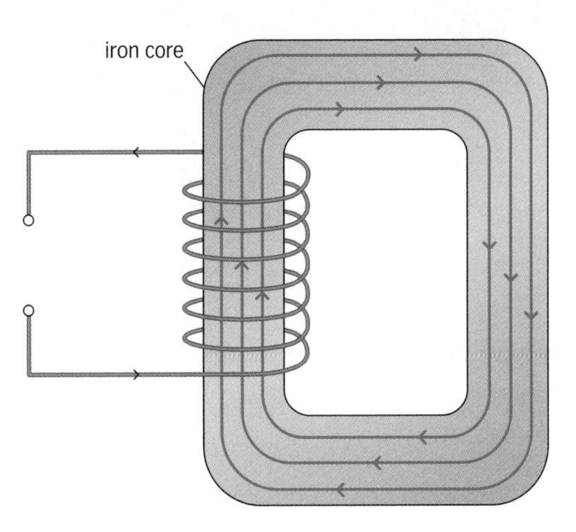

▲ The current in the coil creates a magnetic field. The iron core makes the effect of the field much stronger.

A transformer has an iron core with two coils of wire wound around it. The **primary coil** is connected to a power supply. The current flowing through the primary coil sets up a magnetic field – it works like an electromagnet.

If the current through the primary coil is steady, the magnetic field in the core does not change. There is no effect on the **secondary coil**.

If the primary coil is connected to an alternating (a.c.) supply, the current changes continuously, so the magnetic field also changes continuously. This has the same effect as moving a magnet in and out of the secondary coil. An alternating potential difference is set up between the ends of the secondary coil, and when the secondary coil is connected to a circuit, a current flows.

The two coils of wire are separate – they are not connected so current cannot flow from the primary coil to the secondary coil. They are only connected by the magnetic field.

B What happens when there is an alternating current in the primary coil?

C Why does an electric current not flow from the primary coil to the secondary coil?

A transformer only works with an alternating current – it does not work with direct current.

Exam tip

- ✔ Remember all the parts of a transformer.
- ✔ No current flows between the primary and secondary coils – they are not connected electrically.
- ✔ Transformers do not change alternating current to direct current.

Questions

1 What happens when a wire is moved so that it cuts across a magnetic field?

2 What are the main parts of a transformer?

↓ E

3 Explain how a transformer works.

4 Why will a transformer not work with direct current?

↓ C

5 Explain why no potential difference is induced across the ends of a wire when the wire is moved in a direction parallel to the magnetic field.

↓ A*

30: Step-up and step-down transformers

Learning objectives

After studying this topic, you should be able to:

- explain what is meant by a step-up and a step-down transformer
- use the equation linking potential difference across the primary and secondary coils and the number of turns in the coils

Key words

number of turns

A In a step-up transformer, which coil has the greater potential difference?

B In a step-down transformer, which coil has the greater number of turns?

▲ The inside of a transformer

Types of transformer

In a step-up transformer, the potential difference across the secondary coil is greater than the potential difference across the primary coil. The **number of turns** on the secondary coil is greater than the number of turns on the primary coil.

In a step-down transformer, the potential difference across the secondary coil is less than the potential difference across the primary coil. The number of turns on the secondary coil is less than the number of turns on the primary coil.

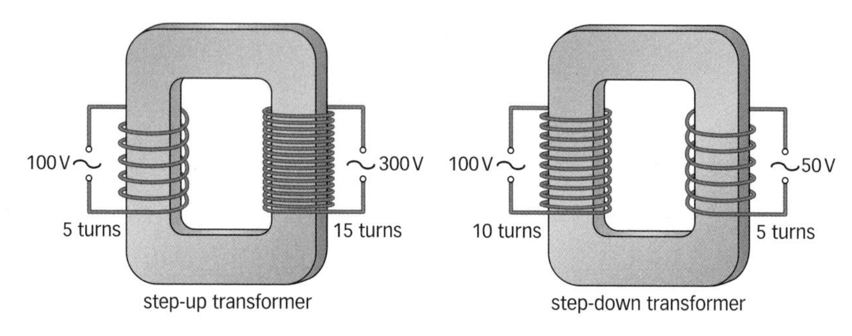

▲ A step-up transformer and a step-down transformer

Transformer equation

The potential difference across the primary and secondary coils of a transformer are related by the equation:

$$\frac{\text{potential difference across primary coil}}{\text{potential difference across secondary coil}} = \frac{\text{number of turns on primary coil}}{\text{number of turns on secondary coil}}$$

If:

- V_p is the potential difference across the primary coil in volts, V
- V_s is the potential difference across the secondary coil in volts, V
- n_p is the number of turns on the primary coil
- n_s is the number of turns on the secondary coil

then:

$$\frac{V_p}{V_s} = \frac{n_p}{n_s}$$

Worked example

A computer runs off the mains supply, but only needs a potential difference of 11.5 V. There are 1000 turns on the secondary coil. How many turns are there on the primary coil?

$$\frac{V_p}{V_s} = \frac{n_p}{n_s}$$

V_p = 230 V (mains voltage)
V_s = 11.5 V
n_s = 1000

Substituting the values into the equation, we get:

$$\frac{230\text{ V}}{11.5\text{ V}} = \frac{n_p}{1000}$$

$$n_p = 1000 \times \frac{230\text{ V}}{11.5\text{ V}}$$

$$= 1000 \times 20$$

$$= 20\,000 \text{ turns.}$$

There are 20 000 turns on the primary coil.

▲ A transformer in an electronic circuit

C The potential difference across the secondary coil of a transformer is 6.1 V, and the number of turns is 10. The number of turns on the primary coil is 380. What is the potential difference across the primary coil?

Questions

1 What are the two types of transformer?

2 A transformer has 200 turns on the primary coil and 2400 turns on the secondary coil. The potential difference across the primary coil is 11 kV. What is the potential difference across the secondary coil?

3 A transformer has 1000 turns on the primary coil and 20 turns on the secondary coil. The potential difference across the primary coil is 230 V. What is the potential difference across the secondary coil?

4 The primary coil of a transformer has 100 turns, and the mains supply is 230 V. What is a suitable number of turns on the secondary coil for:
(a) a laptop needing a supply of 19 V
(b) a mobile phone charger needing 6.5 V
(c) a battery charger needing 4.2 V.

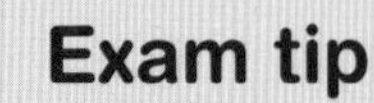

✔ Remember that the correct term is the number of turns on a coil.

31: Transformer power equation

Learning objectives

After studying this topic, you should be able to:

- ✔ use the transformer power equation
- ✔ compare the types of transformer used for different applications

An industrial transformer that steps the potential difference down from 11 kV to 433 V

Worked example

The potential difference across the primary coil of a transformer is 230 V and the current through the primary coil is 2 A. The potential difference across the secondary coil is 12 V. What is the current through the secondary coil?

$$V_p \times I_p = V_s \times I_s$$

$V_p = 230\,\text{V}$, $I_p = 2\,\text{A}$, $V_s = 110\,\text{V}$

Substituting the values into the equation, we get:

$$230\,\text{V} \times 2\,\text{A} = 110\,\text{V} \times I_s$$

$$I_s = \frac{230\,\text{V}}{110\,\text{V}} \times 2\,\text{A}$$

$$= 4.2\,\text{A}$$

The current in the secondary coil is 4.2 A.

Transformers and power

You have learnt earlier that power, potential difference (or voltage), and current are related by the equation:

power (watts, W) = potential difference (volts, V) × current (amperes, A)

If power is P, potential difference is V and current is I, then:

$$P = VI$$

This equation can also be used for transformers.

Transformers are very efficient. If a transformer is assumed to have an efficiency of 100%, the electrical power input applied to the primary coil is equal to the power output from the secondary coil:

power in primary coil, P_p = power in secondary coil, P_s

$$\text{or } P_p = P_s$$

- V_p is the potential difference across the primary coil in volts, V
- I_p is the current in the primary coil in amperes, A
- V_s is the potential difference across the secondary coil in volts, V
- I_s is the current in the secondary coil in amperes, A
- n_p is the number of turns on the primary coil
- n_s is the number of turns on the secondary coil.

From the equations linking power, potential difference and current:

$$V_p \times I_p = V_s \times I_s$$

A The potential difference across the secondary coil of a transformer is half of the potential difference across the primary coil. If the transformer is 100% efficient, what can we say about the currents in the secondary coil and primary coil?

Applications of transformers

Some electrical devices use the 230 V mains electricity supply directly. These include cookers, washing machines, microwave ovens, and electric heaters. Other devices need a much lower potential difference. Most electronic devices use a much lower potential difference. Step-down transformers are used in their power units to reduce the potential difference to what is needed.

▲ The power supplies for all of these devices include a transformer. The laptop needs 19 V, the phone charger 6.5 V, and the battery charger 4.2 V.

B Why do some devices need transformers?

Questions

1. What assumption is made about transformers in the power equation? ↓E
2. Calculate the potential difference across the primary coil of a transformer when the current through the primary coil is 0.25 A, the potential difference across the secondary coil is 230 V, and the current through the secondary coil is 13 A. ↓C
3. Calculate the current in the secondary coil of a transformer when the potential difference across the secondary coil is 132 kV, the potential difference across the primary coil is 11 kV, and the current in the primary coil is 100 A.
4. Why is the assumption made when using the transformer power equation likely to be incorrect? ↓A*

Did you know...?

A transformer forms only part of the battery charger for a mobile phone. There is also a rectifier that converts the alternating current to direct current. However, these battery chargers are not 100% efficient – they heat up when they have been left on for a while. Some of the electrical energy is transferred to thermal energy.

▲ A mobile phone charger warms up when it is switched on because its efficiency is less than 100%

Exam tip

✔ Always show your working in any calculations. Even if you get the answer wrong, you will get a mark if you have shown that you have used the correct method.

32: Switch mode transformers

Learning objectives

After studying this topic, you should be able to:

- explain that switch mode transformers operate at high frequencies
- describe the advantages of switch mode transformers over traditional transformers

Key words

switch mode transformer, **load**

Transformers for electronic devices

Electronic devices such as mobile phones, digital cameras, TVs and computers need a supply that is much lower than the 230 V of the mains electricity supply, often a potential difference of somewhere between 3 V and 20 V.

They also use direct current rather than alternating current. The power units they use to operate, or to recharge their batteries must reduce the potential difference of the a.c. mains supply and convert it to direct current. They may use a conventional transformer or a **switch mode transformer**.

These devices also transform potential differences, but use very complex circuits to do this. A standard transformer operates at the frequency of the alternating mains supply, which is 50 Hz. Switch mode transformers usually operate at frequencies between 50 kHz and 1000 kHz. They are used extensively within electronic devices.

▲ This power unit for a PC uses a switch mode transformer

A What is a switch mode transformer?

Advantages of switch mode transformers

Switch mode transformers are usually much smaller and lighter than a conventional transformer.

▲ This power supply for a mobile phone contains a switch mode transformer and is the same size as a normal plug

The efficiency in a typical a.c. to d.c. power supply for an electronic device that uses a conventional transformer is between 30% and 40%. The efficiency of a switch mode transformer is between 60% and 70%. A well-designed switch mode transformer can have an efficiency of 95%.

B What is the efficiency of a switch mode transformer?

Switch mode transformers use very little power when they are switched on but no **load** is being applied. For example, a switch mode transformer for a battery charger will use very little power when it is switched on but no batteries are actually being charged.

The potential difference and frequency of the mains supply varies around the world. In the UK it is 230 V, 50 Hz. In North America it is 110 V, 60 Hz. Switch mode transformers can be used with all the different mains supplies around the world and produce the same output. So manufacturers do not have to make different models for each mains electricity supply.

However, switch mode transformers have much more complex circuits than conventional transformers.

Did you know...?

When a transformer is working, you can sometimes hear 'mains hum'. The transformer may vibrate at the frequency of the mains supply. If you stand outside an electricity substation, you can sometimes hear the transformers.

Questions

1 What frequency range do switch mode transformers use?

E

2 What are the advantages of a switch mode transformer?

3 What are the differences between a conventional transformer and a switch mode transformer?

4 What happens to the energy that is wasted in a transformer?

C

5 The upper limit of hearing is less than 20 kHz for most humans.
What are the advantages of using a frequency between 50 kHz and 200 kHz for mains electricity supplies?

A*

Course catch-up

Revision checklist

- An object's centre of mass is the point at which all the object's mass seems concentrated.
- The time period of a pendulum is the time taken to complete a full swing.
- Moment is the turning effect of a force.
- When an object is balanced, the total clockwise moment about the pivot is equal to the total anticlockwise moment. This is the principle of moments. Levers multiply forces using moments.
- An object will topple if the line of action of its weight falls outside the base of the object. Stability is increased by making the base wider and the centre of mass lower.
- Liquids are almost incompressible. Pressure in liquids is transmitted equally in all directions.
- Hydraulic systems multiply force.
- An object moving in a circle is constantly changing direction, so is constantly accelerating towards the centre of the circle.
- The force between an object in circular motion and the centre of the circle is a centripetal force.
- Flow of electric current through a wire produces a magnetic field around the wire.
- If a current-carrying wire is placed within another magnetic field, the magnetic fields interact, creating the motor effect.
- When a wire cuts across a magnetic field, a potential difference is induced across the ends of the wire.
- A transformer has an iron core with two coils of wire around it. If the primary coil is connected to an a.c. supply an alternating potential difference is induced in the secondary coil.
- In a step-up transformer, potential difference is higher across the secondary coil than the primary coil.
- In a step-down transformer, potential difference is lower across the secondary coil than the primary coil.
- Potential differences across the primary and secondary coils and the number of turns in the coils are related.
- In a transformer with 100% efficiency, the power input to the primary coil is equal to the output from the secondary coil.
- Most electronic devices use a lower potential difference than the 230 V mains supply. Step-down transformers are used.

clocks

time period = $\frac{1}{\text{frequency}}$

pendulum

moment = force × perpendicular distance

acceleration

CIRCULAR MOTION

satellites

centripetal force

depends on mass, speed, radius

$V_p I_p = V_s I_s$

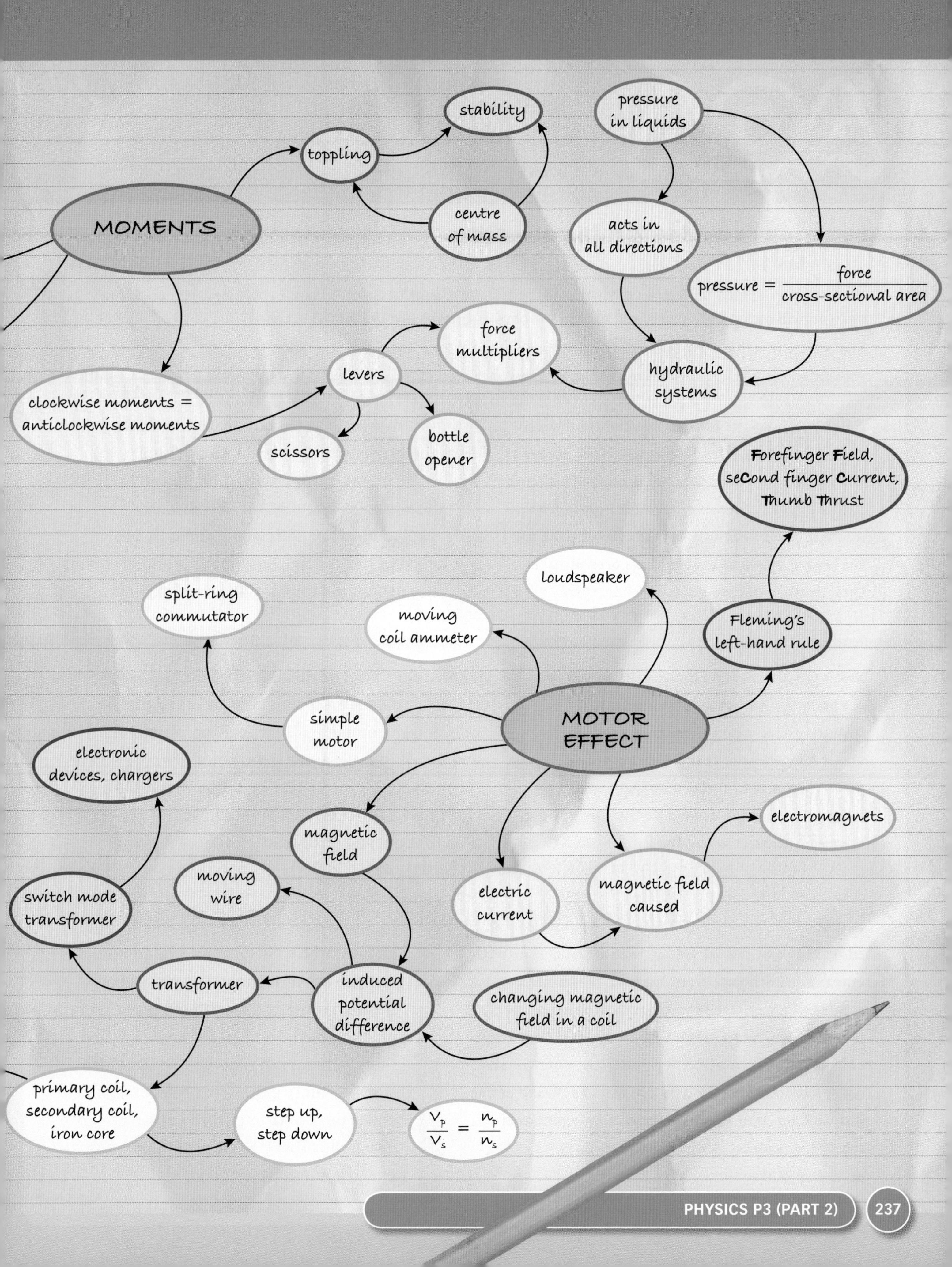
stability
pressure in liquids
toppling
MOMENTS
centre of mass
acts in all directions
pressure = force / cross-sectional area
force multipliers
levers
hydraulic systems
clockwise moments = anticlockwise moments
scissors
bottle opener
Forefinger Field, seCond finger Current, Thumb Thrust
loudspeaker
split-ring commutator
moving coil ammeter
Fleming's left-hand rule
simple motor
MOTOR EFFECT
electronic devices, chargers
electromagnets
magnetic field
moving wire
switch mode transformer
electric current
magnetic field caused
transformer
induced potential difference
changing magnetic field in a coil
primary coil, secondary coil, iron core
step up, step down
$\frac{V_p}{V_s} = \frac{n_p}{n_s}$

Answering Extended Writing questions

QUESTION

You are given a piece of Perspex cut into the shape of the islands of Great Britain. Explain how you could find the centre of mass of the piece, and explain the physics on which your method is based.

The quality of written communication will be assessed in your answer to this question.

G–E

Stick a pin in the Perspex somewhere it dusnt mater where and let it hang down the centre of mass is underneeth so draw it on that's becose gravity pulls it down then do it again from another place an another if you have time an the centre of mass is wher thay cros maybe they wont exacly so its somewhere there

Examiner: The candidate clearly remembers the experiment. However, the written description is unplanned and ungrammatical. The practical details and the physics explanation are both vague and lack the correct technical terms. Spelling and grammar are poor, and punctuation is non-existent.

D–C

Put a pin through a place on the corner of the perspex and hang it in a clamp, gravity pulls it so the centr of mass is strait down under the pin. Draw a line down (use a waight on a string). Hang it from another place and do the same, the centre of mass is where the lines cross.

Examiner: Most of the described physics is correct. However, the use of technical terms is vague: no reference to 'moments', 'vertical', or 'plumb-line', and use of 'place' rather than 'point'. The physics of the method is glossed over and other details are left out. There are occasional errors in spelling, punctuation, and grammar.

B–A*

Suspend the piece from a point somewhere on the edge, so it can swivel freely, with a pin through it. Hang a plumb-line from the pin. The Perspex will rotate till its centre of mass is vertically below the pin – this happens because the weight of the Perspex causes a moment about the pivot (the pin) which makes it rotate. Draw a line along the thread of the plumb-line. Use another suspension point and repeat the process. Where the two lines cross is the centre of mass.

Examiner: This answer is well ordered and accurate. The physics explanation and the use of technical terms is good throughout. This is a difficult question to answer briefly, and all the key ideas are there – though the candidate might also suggest safety glasses to protect against sharp pins, and/or a third suspension to check the others. Spelling, punctuation, and grammar are all good.

Exam-style questions

1 Match the words on the left with the correct description on the right:

A01

Term	Description
centripetal force	Point where all an object's mass appears to act
centre of mass	Force multiplier that uses the principle of moments
pressure	Turning effect of a force about a pivot
moment	System to exert a larger force, starting with a smaller one
force multiplier	Force needed to keep an object in a circular path
lever	Exerts force on a surface within a liquid

2 The diagram illustrates a hydraulic system using two pistons.

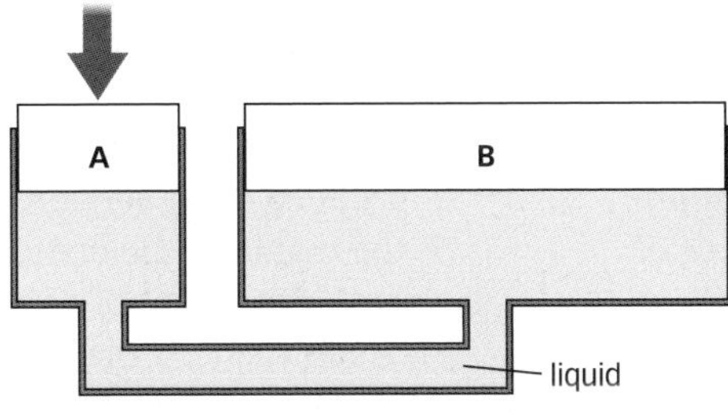

Piston A cross-section = 20 cm^2.
Piston B cross-section = 80 cm^2.
Piston A is pushed with a force of 300 N.

A02 **a** Calculate the pressure in the liquid behind piston A:
i in N/cm^2 **ii** in Pa.

A01 **b** What is the pressure in the liquid behind piston B in N/cm^2? Explain this answer.

A02 **c** Calculate the force on piston B.

A02 **d** Piston A moves a distance of 60 cm. How far does piston B move back?

A02 **e** Explain how this system can be called a force multiplier.

G–E D–C

3 The diagram below shows the main parts of a simple electric motor.

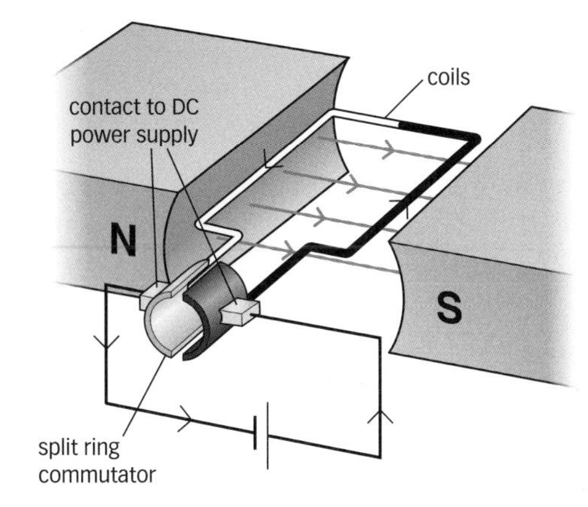

A01 **a** Explain why there is a force on the white part of the coil.

A02 **b** Use Fleming's left-hand rule to work out the direction of the force.

A02 **c** What is the direction of the force on the black part of the coil?

A02 **d** What is the combined effect of the forces on the two parts of the coil?

A01 **e** Explain how the split ring commutator allows the motor to work as required.

B–A*

Extended Writing

4 What is meant by the principle of moments? Describe two examples where the principle applies.
A01 G–E

5 A GPS satellite is in a circular orbit around the Earth. Explain why it keeps moving in its circular path.
A01 D–C

6 Many household appliances include a simple step-down transformer. Explain how a step-down transformer works, and why one might be used.
A01 B–A*

A01 Recall the science
A02 Apply your knowledge
A03 Evaluate and analyse the evidence

Glossary

accelerate To speed up, slow down, or change direction.
acceleration Speeding up, slowing down, or changing direction. Change in velocity per second, measured in m/s^2.
activity Number of radioactive decays per second.
air resistance Resistive force that slows down objects moving through air.
alpha particle Very ionising, but not very penetrating form of ionising radiation. Made up of 2 protons and 2 neutrons (a helium nucleus).
alternating current, a.c. Current that continually reverses its direction.
amplitude The maximum displacement of a wave or oscillation.
atomic number The number of protons in a nucleus of an atom.
attract Pull towards.
background radiation Radiation around us all of the time from a variety of natural and man-made sources.
barrage Dam built across a river estuary.
battery Scientific name for two or more cells connected together.
beta particle Ionising form of radiation. It is a fast electron from the nucleus.
Big Bang The supposed origin of the Universe, when all matter and energy emerged from one place.
biomass Plant material that can be used as an energy resource.
black hole A very small, dense object with gravity so strong not even light can escape. The result of the supernova of a massive star.
braking distance Distance moved by a vehicle when it slows down and stops after the brakes have been applied.
braking force Used to stop or slow down a moving object such as a car.
camera Device for capturing images.
carbon capture Storage of carbon dioxide gas from combustion so that it doesn't enter the atmosphere.
CCD Electronic sensor of visible light that can be used to record an image.
cell (electrical) A device that transfers chemical energy into electrical energy.
centre of mass Point at which all of the mass of an object appears to be acting.
centripetal force Force that keeps an object moving in a circle.
chain reaction One reaction going on to create another, which creates another, and so on, such as a nuclear fission chain reaction inside a nuclear reactor.
charge Property of some objects (like electrons and protons). There are two types of charge: negative and positive.
chemical energy Energy stored in things like batteries, food, and fuels.
ciliary muscles Muscles that pull and relax to change the shape of the lens in an eye.
compression Pushing together of particles caused by the passing of a sound wave.
concave Describes a lens whose thickness increases with increasing distance from its centre.
condensation When particles in a gas lose energy and turn into a liquid.
conduction One way of transferring energy by vibrations being passed on from one particle to another.
convection How thermal energy is transferred by the movement of particles in fluids.
convection current The movement of particles by convection when heat energy is transferred.
conventional current Model used to represent charges moving through a circuit. Conventional current goes from positive to negative (opposite direction to the electron flow).
converging Describes a lens that changes the direction of parallel rays of light so that they meet at a focus once they have passed through the lens.
convex Describes a lens whose thickness decreases with increasing distance from its centre.
cornea Transparent curved surface where light enters the eye.
corrective lenses Lenses placed in front of the eyes to improve their performance.
cosmic microwave background radiation Microwaves received from all directions in space.
cost-effective Describes when the cost of making a change is low in relation to the resultant savings in energy bills.
coulomb Unit of electric charge.
critical angle Light will only totally internally reflect if its angle of incidence is greater than the critical angle.
CT scan Uses a series of X-ray scans to produce cross-sectional images of the inside of the body.
current Movement of charged particles (usually electrons) through a material.
current-potential difference graph Graph with current on the y-axis and potential difference on the x-axis. It shows how the current varies for different potential differences.
deceleration Negative acceleration, that is acceleration in the opposite direction to the direction of motion, so slowing down.
decommissioning Dismantling of a power station at the end of its useful life.
diffraction The spreading out of waves when they pass through a gap.
diminished Made smaller.
diode Circuit component that conducts electricity in one

direction only. It has a very high resistance in the reverse direction.

dioptres Unit of lens power.

direct current, d.c. Current that goes in only one direction.

directly proportional Describes the relationship between two variables. If one increases by a factor, the other increases by the same factor (i.e. if one doubles, the other doubles).

dissipate To spread out.

distance-time graph Graph showing distance on the *y*-axis and time on the *x*-axis. It shows the distance travelled from a certain point at a particular moment.

diverging Describes a lens that changes the direction of parallel rays of light so that they appear to come from a focus once they have passed through the lens.

Doppler effect The change of wavelength of a wave when it is produced or reflected from a moving object.

double-insulated Describes an appliance in which all the live components are sealed away from the case, so the case cannot become live.

earth Pin/wire that carries energy safely away from the device to the ground if there is a fault.

elastic potential energy Energy stored in stretched or compressed objects.

electric motor Device with a rotating shaft powered by electric current interacting with a magnet.

electromagnetic induction Creation of a potential difference across the ends of a wire when it moves through a magnetic field, or when the magnetic field is changing.

electromagnetic spectrum All the different types of electromagnetic wave: radio waves, microwaves, infrared, visible, ultraviolet, X-rays, and gamma rays.

electromagnetic wave Wave that has oscillating electric and magnetic fields at right angles to its direction of motion.

electron Tiny sub-atomic particle with a negative charge found orbiting the nucleus.

electrostatic charge Charge from electrons that have been moved to or from an insulator.

emit To give out.

endoscope Instrument for imaging inside the body using optical fibres.

energy A measure of how much work something can do.

equal and opposite Describes the pair of forces produced when objects interact: equal in size and acting in opposite directions.

evaluate To examine something and work out its properties or value.

evaporation When particles escape from the surface of a liquid to become a gas.

evidence Scientific observations that can be used to disprove a scientific hypothesis.

extension Change in length of an object when a force is applied.

filament lamp Lamp containing a coil of wire that glows when an electric current flows through it.

Fleming's left-hand rule Uses thumb and first two fingers of left hand to show directions of force, current, and magnetic field in a motor.

fluid A liquid or a gas.

focal length Distance from the centre of a lens to the principal focus.

focus Point in space where rays originating from a point on an object come together after passing through a lens.

force Push or pull that changes the way an object is moving, or its shape. Measured in newtons.

force multiplier Something that increases the size of a force.

free electrons Electrons that are free to move through a substance.

frequency (waves) The number of oscillations per second for a vibration.

frequency (electrical) Number of cycles of potential difference per second for an alternating current.

frequency (pendulum) The number of cycles that a pendulum completes in one second.

friction Force that acts to stop or slow down two objects that are sliding against each other.

fuse Thin piece of wire that melts (and breaks the circuit) if too much current flows through it.

gamma knife A special machine found in hospitals that uses gamma rays to kill cancerous tumours by focusing them on the tumour and minimising the exposure to the surrounding tissues.

gamma ray Very penetrating but not very ionising form of ionising radiation. It is a high frequency electromagnetic wave.

gas State of matter that has no fixed size or shape. Its particles move in straight lines, bouncing off each other when they make contact.

generator Machine that rotates to generate electricity.

geothermal power Energy from hot rocks in the Earth.

gradient Slope of a graph.

gravitational field strength Strength of the force of gravity on a planet. On Earth it is 10 N/kg.

gravitational potential energy Energy stored in an object that is in a higher position than its surroundings.

half-life The time taken for half of the radioactive nuclei in a substance to decay, or the time taken for the activity from a substance to halve.

hydraulic system System that uses a liquid to transfer forces.

hydroelectric Generating electricity from the kinetic energy in moving water.

incompressible Describes something that cannot be squashed.

induced One thing having caused something else to happen in a system.

infrared radiation Form of electromagnetic radiation that you can feel as heat. It has a shorter wavelength than microwaves but a longer wavelength than visible light.

insulating material Does not allow electrons to move easily through it.

intensity Measure of the energy received by an area each second.

inverted Describes an image that is upside down compared with the object.

ion Atom with electrons added or removed, giving the atom an overall charge.

ionising Describes radiation that removes electrons from atoms to create ions.

ionising radiation Alpha, beta, and gamma radiation that removes electrons from the atoms of the material it passes through.

iris Coloured disc in front of lens that controls amount of light reaching the retina by altering the size of the hole at its centre.

iron core Central part of a transformer around which the primary and secondary coils are wrapped.

isotope Atoms with the same number of protons but different numbers of neutrons.

joule Unit of energy or work done.

kidney stone Lump of material that forms in the kidney, blocking ducts and causing pain.

kilowatt 1000 watts.

kilowatt-hour The amount of energy transferred by a device using 1 kW in 1 hour.

kinetic energy Energy due to the movement of an object.

kinetic theory Theory that explains the properties of solids, liquids, and gases by the movement of the particles they are made of.

laser Device that emits light of a single wavelength as a narrow intense beam.

laterally inverted Description of an image in a mirror that has been flipped horizontally.

law of conservation of energy Law stating that energy cannot be created or destroyed.

law of conservation of momentum Law stating that momentum is conserved in collisions. Total momentum of a system before a collision or explosion is the same as the total momentum of the system after a collision or explosion.

law of reflection The angle between the incident ray and the normal is the same as the angle between the reflected ray and the normal.

lens power Equal to 1 divided by its focal length in metres.

lever Simple machine with a pivoted rod that can increase the size of a force.

light-dependent resistor Special type of resistor whose resistance decreases when the intensity of light falling on it increases.

light-emitting diode Diode that emits light when an electric current flows through it.

limit of proportionality Point beyond which the extension of an object is no longer directly proportional to the force applied.

line of action Line passing through an object in the direction of the force on that object.

liquid State of matter that has a fixed size, but can have any shape. Its particles can move around, but must stay close to each other.

live Pin/wire that transfers energy to an appliance.

load Device attached to a power source that draws electrical energy from it.

longitudinal wave Wave whose vibrations and energy flow are in the same direction.

long-sightedness Inability to see close objects clearly.

magnetic field Region of space around a magnet where magnetic forces act on objects.

magnification Height of an image of an object divided by the height of the object itself.

magnified Made larger.

mains supply The electricity supply available in most homes.

mass Amount of matter in an object. Measured in kilograms.

mass number The number of protons and neutrons in a nucleus of an atom.

medium Any substance that a wave passes through, eg air, water, metal.

moment Turning force, calculated as force x distance of its line of action from the pivot.

momentum Mass of an object x velocity.

motor effect The push on an electric current placed in a magnetic field.

National Grid UK electricity distribution system that transfers electrical energy from power stations to consumers.

nebula Huge cloud of gas and dust in space (mainly hydrogen).

neutral Pin/wire that completes the circuit.

neutron Small sub-atomic particle with a neutral charge found in the nucleus.

neutron star A small, dense object that remains after a supernova.

newton Unit of force.

non-renewable Describes an energy resource that can only be used once.

normal A construction line at right angles to a surface, drawn at the point where a ray meets the surface.

nuclear energy Energy stored inside the nucleus in atoms.

nuclear fission Splitting a nucleus into two smaller nuclei, releasing energy.

nuclear fusion Fusing two atoms to form a single, larger nucleus, releasing energy.

nuclear reactor Part of a power station fuelled by uranium that transfers nuclear energy to heat energy.

nucleus Small centre of an atom containing the protons and neutrons.

number of turns Number of times a wire is wrapped round in a transformer to create a coil.

optical fibre Long thin cylinder made from very transparent glass down which light can be sent over long distances.
oscillation Back-and-forth motion that repeats over and over again.
oscilloscope Device that can be used to determine the frequency and potential difference of an a.c. supply.
pair of forces Two equal and opposite forces produced when objects interact.
parallel circuit Circuit with components connected side by side so that there is more than one path to take around the circuit.
pascal Unit of pressure, newtons per square metre.
payback time Time taken to recover the amount of money paid out in making a change from the resultant savings in energy bills.
pendulum Heavy object attached to one end of a long, thin piece of material.
perpendicular distance Distance that is at right angles to the line of action of the force to a pivot.
pickup Sudden large increase in the demand for electricity.
pitch The position of a sound on a musical scale. It is closely linked to frequency.
pivot Point around which an object can rotate.
plutonium 239 Isotope of plutonium that can be used to generate heat in nuclear power stations.
potential difference Difference in voltage between two points. Measured in volts. Also a measure of the difference in energy carried by electrons between two points.
power Work done (or energy transferred) in a given time. Measured in watts.
power rating Indicates the amount of energy a device uses each second. Can be found by multiplying the voltage and current.
primary coil Coil in a transformer where the input is applied.
principal axis Line through the centre of a lens, at right angles to its plane.
principal focus Point to which parallel rays of light are focussed by a converging lens.
proton Small sub-atomic particle with a positive charge found in the nucleus.
protostar Ball of hot, dense gas, on its way to becoming a star.
pumped storage A hydroelectric power station that can pump water to a higher reservoir when there is low demand for electricity.
pupil Hole in the iris that lets light into the eye.
radioactive decay Breakdown of an unstable nucleus, giving out alpha particles, beta particles, or gamma rays.
radioactive waste By-product of nuclear reactors that contains a high concentration of radioactive atoms.
radiographer Technician who uses X-rays or gamma rays for medical procedures.
radiotherapy Use of radioactive materials to cure people with cancer.
random Having no set pattern in which atoms will decay.
range of vision Difference between the furthest and nearest distance over which a person can see clearly.
rarefaction Pulling apart of particles caused by the passing of a sound wave.
ray diagram Diagram showing the passage of rays of light through a lens.
reaction time Time taken from seeing a hazard to starting to press the brake pedal.
real Type of image that can be displayed on a screen.
red giant The result when a smaller star begins to die. It expands and cools, forming a red giant.
red super giant When a large star begins to die it expands and cools, forming a gigantic red super giant.
red-shift The increase in wavelength of light emitted from an object that is moving away from the observer.
reflection The change of direction of a wave when it bounces off a surface.
refraction The change of speed and direction of a wave when it goes from one medium to another.
refractive index Sine of the angle of incidence of light divided by the sine of the angle of refraction; a measure of the speed of light in a transparent material.
regenerative braking A type of vehicle braking that converts some of the kinetic energy into a useful form (eg chemical energy in batteries).
renewable Describes an energy resource that does not run out. It usually relies on energy from the Sun to replace what you take out.
repel Push away.
residual current circuit breaker Circuit breaker that detects a difference in current between the live and neutral wire.
resistance Measure of how difficult it is for electrons to pass through a component. Measured in ohms (Ω).
resistive force Force in the opposite direction to the direction of motion of an object.
resistor Circuit component that reduces the current flowing in a circuit.
resultant force Single force that would have the same overall effect of all the forces combined.
retina Light-sensitive surface at the back of the eyeball.
secondary coil Coil in a transformer from where the output is taken.
series Circuit with components that are connected end to end so there is only one way around the circuit.
short-sightedness Inability to see distant objects clearly.
solar cell Device that transfers light energy into electricity.

solar panel Black panel that absorbs infrared radiation from the Sun and transfers heat energy to water in pipes in the panel.

solar thermal tower Tower where a group of mirrors reflects light energy from the Sun to heat water to turn it into steam.

solid State of matter that has a fixed shape and size. Its particles are held in place by each other.

sound waves Waves formed by vibrations in solids, liquids, or gases. They result in pressure oscillations in the same direction as their flow of energy.

specific heat capacity The amount of energy needed to increase the temperature of 1 kg of a material by 1 °C.

speed Distance travelled in a certain time. Usually measured in m/s.

speed camera Camera used to find the speed of a vehicle.

split-ring commutator Rotary switch in an electric motor that reverses current in the coil every half revolution.

stability The greater an object's stability, the more likely it is that an object will return to its original position when displaced.

start up time Time delay between starting up a power station and it generating electricity.

states of matter There are three states of matter: solid, liquid, and gas.

static electricity Build up of charge from electrons that have been moved to or from an insulator.

stationary Not moving, at rest.

step-down transformer Device that reduces the voltage of an electricity supply.

step-up transformer Device that increases the voltage of an electricity supply.

sterilising Using ionising radiation (usually gamma rays) to kill the microorganisms on medical equipment.

stopping distance Sum of the thinking distance and braking distance.

substation Place that contains one or more transformers.

supernova What happens when a massive star explodes at the end of its life.

suspensory ligaments Tissue holding the lens in place within the eye.

switch Circuit component that can switch the circuit on or off by closing or making a gap in the circuit.

switch mode transformer Transformer that operates at a high frequency.

tension Force that is trying to pull something apart.

terminal velocity Maximum velocity of an object when the forces on it are balanced. Usually applied to objects falling under gravity.

thermal conductor Material that conducts heat well.

thermal insulator Material that does not conduct heat well.

thermal power station Power station that uses a fuel to heat water to turn it into steam.

thermistor Special type of resistor whose resistance decreases as the temperature increases.

thermogram Image taken by an infrared camera that shows how much infrared radiation is being emitted by an object.

thinking distance Distance travelled by a vehicle in the reaction time.

tidal stream turbine Turbine that transforms kinetic energy in tidal streams into electrical energy.

time base Time interval represented by each horizontal square on an oscilloscope.

time period Time for one complete cycle of potential difference for an alternating current.

time period Time taken for a pendulum to swing from one extreme to the other and back again.

total internal reflection (TIR) Where light inside a transparent medium reflects off its surface without any refraction out of the medium.

transfer The act of changing energy from one form to another.

transmitted Passed through.

transverse wave Wave whose vibrations are at right angles to the direction of energy flow.

turbine Machine that is turned by something moving through it, eg air or water.

ultrasound Sound wave that has too high a frequency (above 20 kHz) for it to be heard by humans.

ultrasound scan Detection of ultrasound echoes at the skin's surface to construct an image of the inside of the body.

unit (of electricity) Unit for measuring the amount of energy transferred. It is the same as a kilowatt-hour.

upright Describes an image that is the same way up as the object.

uranium 235 Isotope of uranium that can be used to generate heat in nuclear power stations.

useful energy Energy transfer from a system that is used to perform a task.

U-value A measure of how much energy something transfers.

vacuum Region that contains no atoms or other particles. Not possible to create in practice, but outer space gets close.

variable resistor Resistor whose resistance can be changed.

velocity How fast something is moving in a certain direction.

velocity-time graph Graph with velocity on the *y*-axis and time on the *x*-axis. It shows the velocity of an object at a particular moment.

vibration Back-and-forth motion of an object that is caused by a wave or oscillation.

virtual Type of image that cannot be displayed on a screen.

virtual focus Point from which parallel rays of light appear to come after they have passed through a diverging lens.

virtual image Image formed by a lens that cannot be projected onto a screen.

voltage A measure of the energy given to charge by a power supply.

voltage *See* potential difference.

volume The loudness of a sound, represented by the amplitude of the sound wave.

wasted energy Energy transfer from a system that doesn't perform a task. It is usually in the form of heat or sound.

watt The power of a device that transfers one joule of energy per second.

watt Unit of power.

wave equation Wave speed = frequency × wavelength.

wave source Source of a wave's energy and frequency.

wavelength Distance from one peak of a wave to the next peak.

weight Force on an object due to the force of gravity on a planet.

white dwarf White-hot core left after the outer layers of a red giant break away.

wind turbine Turbine that generates electricity from the kinetic energy in the wind.

work done Energy transferred, measured in joules. Work done is equal to force applied × distance moved in the direction of the force.

X-rays Electromagnetic waves of frequency between that of gamma rays and ultraviolet.

Index

Reference material

Equations	
$a = \frac{F}{m}$ or $F = m \times a$	F is the resultant force in newtons, N m is the mass in kilograms, kg a is the acceleration in metres per second squared, m/s^2
$a = \frac{v - u}{t}$	a is the acceleration in metres per second squared, m/s^2 v is the final velocity in metres per second, m/s u is the initial velocity in metres per second, m/s t is the time taken in seconds, s
$W = m \times g$	W is the weight in newtons, N m is the mass in kilograms, kg g is the gravitational field strength in newtons per kilogram, N/kg
$F = k \times e$	F is the force in newtons, N k is the spring constant in newtons per metre, N/m e is the extension in metres, m
$W = F \times d$	W is the work done in joules, J F is the force applied in newtons, N d is the distance moved in the direction of the force in metres, m
$P = \frac{E}{t}$	P is the power in watts, W E is the energy transferred in joules, J t is the time taken in seconds, s
$E_p = m \times g \times h$	E_p is the change in gravitational potential energy in joules, J m is the mass in kilograms, kg g is the gravitational field strength in newtons per kilogram, N/kg h is the change in height in metres, m
$E_K = \frac{1}{2} \times m \times v^2$	E_K is the kinetic energy in joules, J m is the mass in kilograms, kg v is the speed in metres per second, m/s
$p = m \times v$	p is the momentum in kilograms metres per second, kg m/s m is the mass in kilograms, kg v is the velocity in metres per second, m/s
$I = \frac{Q}{t}$	I is the current in amperes (amps), A Q is the charge in coulombs, C t is the time in seconds, s
$V = \frac{W}{Q}$	V is the potential difference in volts, V W is the work done in joules, J Q is the charge in coulombs, C
$V = I \times R$	V is the potential difference in volts, V I is the current in amperes (amps), A R is the resistance in ohms, Ω
$P = \frac{E}{t}$	P is power in watts, W E is the energy in joules, J t is the time in seconds, s
$P = I \times V$	P is power in watts, W I is the current in amperes (amps), A V is the potential difference in volts, V
$E = V \times Q$	E is the energy in joules, J V is the potential difference in volts, V Q is the charge in coulombs, C

Fundamental physical quantities	
Physical quantity	**Unit(s)**
length	metre (m) kilometre (km) centimetre (cm) millimetre (mm)
mass	kilogram (kg) gram (g) milligram (mg)
time	second (s) millisecond (ms)
temperature	degree Celsius (°C) kelvin (K)
current	ampere (A) milliampere (mA)
voltage	volt (V) millivolt (mV)

Derived quantities and units	
Physical quantity	**Unit(s)**
area	cm^2; m^2
volume	cm^3; dm^3; m^3; litre (l); millilitre (ml)
density	kg/m^3; g/cm^3
force	newton (N)
speed	m/s; km/h
energy	joule (J); kilojoule (kJ); megajoule (MJ)
power	watt (W); kilowatt (kW); megawatt (MW)
frequency	hertz (Hz); kilohertz (kHz)
gravitational field strength	N/kg
radioactivity	becquerel (Bq)
acceleration	m/s^2; km/h^2
specific heat capacity	J/kg°C
specific latent heat	J/kg

Electrical symbols			
junction of conductors	ammeter	diode	capacitor
switch	voltmeter	electrolytic capacitor	relay (NO, COM, NC)
primary or secondary cell	indicator or light source	LDR	LED
battery of cells	or	thermistor	NOT gate
power supply	motor	AND gate	OR gate
fuse	generator	NOR gate	NAND gate
fixed resistor	variable resistor		

Acknowledgements

The publisher and authors would like to thank the following for their permission to reproduce photographs and other copyright material:

p8T Martyn F. Chillmaid/SPL; **p8B** Chris Pearsall/Alamy; **p13** James Brittain/VIEW/Corbis; **p14T** T-Service/SPL; **p14B** Victor De Schwanberg/SPL; **p15L** Lillisphotography/Istockphoto; **p15R** Alex Segre/Alamy; **p16** Jonathan Wilson/Istockphoto; **p17L** Martin McCarthy/Istockphoto; **p17R** Martin McCarthy/Istockphoto; **p19** Ashley Cooper, Visuals Unlimited/SPL; **p20** John Pitcher/Istockphoto; **p21** Ange/Alamy; **p22** Stratesigns, Inc./Istockphoto; **p23** Paul Rapson/SPL; **p24R** Jacob Wackerhausen/Istockphoto; **p24L** Reinhold Tscherwitschke/Istockphoto; **p25** Robert Hardholt/Istockphoto; **p26M** Nico Smit/Istockphoto; **p26R** John Beatty/SPL; **p26L** Tony Craddock/SPL; **p27** Gabor Izso/Istockphoto; **p28** Tony Mcconnell/SPL; **p30** Rob Hill/Istockphoto; **p32** Dave King/Dorling Kindersley/Getty Images; **p33** Trevor Clifford Photography/SPL; **p34L** Peter Mukherjee/Istockphoto; **p34R** © 2008 Dimplex North America Limited; **p35** Perets/Istockphoto; **p36ML** Torsten Lorenz/Shutterstock; **p36MR** J. Helgason/Shutterstock; **p36R** Mariola Kraczowska/Shutterstock; **p36L** Denise Bush/Istockphoto; **p37** Janda75/Istockphoto; **p38R** Bonita Hein/Istockphoto; **p38L** Jiri Hera/Shutterstock; **p39** Sheila Terry/SPL; **p40R** Vovan/Shutterstock; **p40TL** Murat Giray Kaya/Istockphoto; **p40BL** Emmeline Watkins/SPL; **p41** Sheila Terry/SPL; **p42** MBPhoto, Inc/Istockphoto; **p43T** Manuel velasco/Istockphoto; **p43M** Realimage/Alamy; **p43B** Emmeline Watkins/SPL; **p49** Lidacheng/Dreamstime; **p50R** Alexandru Romanciuc/Istockphoto; **p50L** Macmaniac/Istockphoto; **p51** gprentice/Istockphoto; **p52L** dgmata/Shutterstock; **p52R** Zhuda/Shutterstock; **p54** Ria Novosti/SPL; **p55** Steve Allen/SPL; **p56** Martin Bond/SPL; **p57** Picture Contact/Alamy; **p58R** Nikada/Istockphoto; **p58TL** Julio Etchart/Photolibrary; **p58BL** kshishtof/Istockphoto; **p59** Irina Belousa/Istockphoto; **p60R** Martin Bond/SPL; **p60TL** Tides Stream; **p60BL** Martin Bond/SPL; **p61** Alberto Pomares/Istockphoto; **p62** SPL; **p63** Max Blain/Shutterstock; **p65** Martin Bond/SPL; **p66B** Amra Pasic/Shutterstock; **p66T** Sally Woods; **p68** Iain Frazer/Dreamstime; **p69** Sciencephotos/Alamy; **p71R** Andrew Lambert Photography/SPL; **p71L** UK Aerial Photography; **p73L** Pascal Goetgheluck/SPL; **p73R** SPL; **p74L** Richard Hutchings/SPL; **p74R** Matt Henry Gunther/Taxi/Getty Images; **p75** YellowRobin/Bigstock; **p76** Joss/Fotolia; **p77T** Jeremy Horner/Corbis; **p77B** Bob Crook/Photographers Direct; **p78** American Honda Motor Co., Inc.; **p79** Doug Baines/Shutterstock; **p81R** NASA/JPL-Caltech/ESA/Harvard-Smithsonian CfA; **p81L** NASA; **p82** Vakhrushev Pavel/Shutterstock; **p83** Goddard Space Flight Center/NASA; **p89** Nürburgring Automotive GmbH/Fotoagentur Urner; **p90** Jim Grossman/NASA; **p91** Heathcliff O'Malley/Rex Features; **p92** Akihiro Sugimoto/Photolibrary; **p93** Ian Cuming/SPL; **p94R** Andrew Wong/Getty Images Sport/Getty Images; **p94L** Cordelia Molloy/SPL; **p95** Edward Shaw/Istockphoto; **p98B** Kim Kirby/Photolibrary; **p98T** Alan & Sandy Carey/SPL; **p99** nicmac.ca/Fotolia; **p103** Sankei/Getty Images News/Getty Images; **p104** Barry Phillips/Evening Standard/Rex Features; **p105L** Ken McKay/Rex Features; **p105R** James D. Morgan/Rex Features; **p109** Patrick Eden/Alamy; **p110** Martyn F. Chillmaid/SPL; **p112** Orange Line Media/Shutterstock; **p113B** NASA/SPL; **p113T** Paul Bernhardt/Alamy; **p114TR** Stephen Hird/Reuters; **p114BR** Kondrashov MIkhail Evgenevich/Shutterstock; **p114L** Andrew Lambert Photography/SPL; **p117** Yuriko Nakao/Reuters; **p118** Joshua Hodge Photography/Istockphoto; **p125** Keith Kent/SPL; **p127R** Leslie Banks/Istockphoto; **p127L** Charles D. Winters/SPL; **p130T1** Andrew Lambert Photography/SPL; **p130T2** Paul Reid/Shutterstock; **p130T3** Chris Hutchison/Istockphoto; **p130T4** Andrew Lambert Photography/SPL; **p130T5** Andrew Lambert Photography/SPL; **p130T6** Trevor Clifford Photography/SPL; **p130T7** Doug Martin/SPL; **p130T8** Webking/Istockphoto; **p130T9** Алексей Брагин/Istockphoto; **p130T10** Andrew Lambert Photography/SPL; **p130T11** Martyn F. Chillmaid/SPL; **p130T12** Martyn F. Chillmaid/SPL; **p131** Trevor Clifford Photography/SPL; **p132** Trevor Clifford Photography/SPL; **p134** Doug Martin/SPL; **p136** Doug Martin/SPL; **p137** Trevor Clifford Photography/SPL; **p138** Pali Rao/Istockphoto; **p139** Martyn F. Chillmaid/SPL; **p140** Martyn F. Chillmaid/SPL; **p141T** Martyn F. Chillmaid/SPL; **p141B** Eyewave/Istockphoto; **p142** Studiomode/Alamy; **p143** Andrew Lambert Photography/SPL; **p144TL** Krasowit/Shutterstock; **p144TR** Jorge Farres Sanchez/Dreamstime; **p144B** oksana2010/Shutterstock; **p145** David J. Green/Alamy; **p146** Andrew Lambert Photography/SPL; **p151L** Philippe Psaila/SPL; **p151R** SPL; **p153** Patrick Landmann/SPL; **p155** Prof. J. Leveille/SPL/SPL; **p157** EFDA-JET/SPL; **p158** NASA; **p159** NASA; **p165** Mark Kostich/Istockphoto; **p166** ©Diamond Light Source 2011 Ltd.; **p167T** Don Bayley/Istockphoto; **p167B** Mauro Fermariello/SPL; **p168L** Zephyr/SPL; **p168R** George Bernard/SPL; **p169B** SPL; **p169T** CNRI/SPL; **p170** Doncaster and Bassetlaw Hospials/SPL; **p172** Gustoimages/SPL; **p173R** AJ Photo/Hop Americain/SPL; **p173L** Astier/SPL; **p174** Erich Schrempp/SPL; **p176** Jeffrey L. Rotman/Photolibrary; **p177** Phil Degginger/Alamy; **p178** Imagemore Co., Ltd./Alamy; **p180** Peter Burnett/Istockphoto; **p182** Suzanne Grala/SPL; **p183** Aaron Amat/Shutterstock; **p185** Diego Cervo/Shutterstock; **p186** Constant/Shutterstock; **p187** Mauro Fermariello/SPL; **p188** Jaume Gual/Photolibrary; **p189** Serg64/Shutterstock; **p190** Miroslav Georgijevic/Istockphoto; **p191** Giphotostock/SPL; **p193** Micro10x/Shutterstock; **p194** Giphotostock/SPL; **p196R** Tek Image/SPL; **p196TR** Dr P. Marazzi/SPL; **p196BR** David M. Martin, Md/SPL; **p197L** Pascal Goetgheluck/SPL; **p197R** Geoff Tompkinson/SPL; **p203** Paulo Fridman/Corbis; **p204** Kristen Olenick/Photolibrary; **p205** Kkant1937/Istockphoto; **p206** Garret Bautista/Istockphoto; **p207BR** Aaron Haupt/SPL; **p207TR** Yuriy Brykaylo/Istockphoto; **p207L** Javier Larrea/Photolibrary; **p208** David H. Lewis/Istockphoto; **p209** Michael Maher/Istockphoto; **p210R** Pavel Losevsky/Istockphoto; **p210L** Esemelwe/Istockphoto; **p212R** Roman Milert/Fotolia; **p212L** Southern Illinois University/SPL; **p213L** Tetra Images/Corbis; **p213R** Ulisse/Dreamstime; **p214** Mustafa deliormanli/Istockphoto; **p215L** NatUlrich/Shutterstock; **p215M** Paul Matthew Photography/Shutterstock; **p215R** Millbrook Proving Ground Ltd. and Wrightbus Limited.; **p216R** AJ Photo/SPL; **p216L** Ron Niebrugge/Alamy **p217** Brett Gonzales/Istockphoto; **p219R** Elena Elisseeva/Istockphoto; **p219L** Steve Nagy/Photolibrary; **p220** Inhaus Creative/Istockphoto; **p221B** Matt Jeacock/Istockphoto; **p221T** Spark62/Alamy; **p223** Lisa Denise Hillström/Istockphoto; **p224** David R. Frazier Photolibrary,Inc./Alamy; **p226** Zcw/Shutterstock; **p227** Andrew Lambert Photography/SPL; **p230** Jeronimo Create/Istockphoto; **p231** Vitaly Shabalyn/Istockphoto; **p232** Sheila Terry/SPL; **p233L** Edhar/Shutterstock; **p233LM** Mark Goble/Alamy; **p233RM** Judith Collins/Alamy; **p233R** Mark Sykes Energy and Power/Alamy; **p234T** Timur Arbaev/Istockphoto; **p234B** Parrus/Istockphoto; **p235** Paul Rapson/Alamy.

Cover image courtesy of GUSTO IMAGES/SCIENCE PHOTO LIBRARY

Illustrations by Wearset Ltd, HL Studios, Peter Bull Art Studio.

Although we have made every effort to trace and contact all copyright holders before publication this has not been possible in all cases. If notified, the publisher will rectify any errors or omissions at the earliest opportunity.

OXFORD
UNIVERSITY PRESS

Great Clarendon Street, Oxford OX2 6DP

Oxford University Press is a department of the University of Oxford.
It furthers the University's objective of excellence in research,
scholarship, and education by publishing worldwide in

Oxford New York

Auckland Cape Town Dar es Salaam Hong Kong Karachi
Kuala Lumpur Madrid Melbourne Mexico City Nairobi
New Delhi Shanghai Taipei Toronto

With offices in
Argentina Austria Brazil Chile Czech Republic France Greece
Guatemala Hungary Italy Japan Poland Portugal Singapore
South Korea Switzerland Thailand Turkey Ukraine Vietnam

First published 2011

British Library Cataloguing in Publication Data

Data available

ISBN 978-0-19-913608-7

10 9 8 7 6 5 4 3

Printed in Great Britain by Bell and Bain, Glasgow

Paper used in the production of this book is a natural, recyclable product
made from wood grown in sustainable forests. The manufacturing process
conforms to the environmental regulations of the country of origin.